Advances in

ORGANOMETALLIC CHEMISTRY

VOLUME 8

Advances in
ORGANOMETALLIC
CHEMISTRY

EDITED BY

F. G. A. STONE
DEPARTMENT OF INORGANIC CHEMISTRY
SCHOOL OF CHEMISTRY
THE UNIVERSITY
BRISTOL, ENGLAND

ROBERT WEST
DEPARTMENT OF CHEMISTRY
UNIVERSITY OF WISCONSIN
MADISON, WISCONSIN

VOLUME 8

1970

ACADEMIC PRESS　　　　New York • London

ACADEMIC PRESS, INC.
111 Fifth Avenue, New York, New York 10003

United Kingdom Edition published by
ACADEMIC PRESS, INC. (LONDON) LTD.
Berkeley Square House, London W1X 6BA

LIBRARY OF CONGRESS CATALOG CARD NUMBER: 64–16030

PRINTED IN THE UNITED STATES OF AMERICA

List of Contributors

Numbers in parentheses indicate the pages on which the authors' contributions begin.

E. W. Abel (117), *Department of Inorganic Chemistry, The University, Bristol, England*

M. Cais (211), *Department of Chemistry, Technion-Israel Institute of Technology, Haifa, Israel*

P. Heimbach (29), *Max-Planck-Institut für Kohlenforschung, Mülheim-Ruhr, West Germany*

Walter Hieber (1), *Anorganisch-Chemisches Laboratorium der Technischen Hochschule, München, Germany*

P. W. Jolly (29), *Max-Planck-Institut für Kohlenforschung, Mülheim-Ruhr, West Germany*

M. S. Lupin (211), *Department of Chemistry, Technion-Israel Institute of Technology, Haifa, Israel*

John P. Oliver (167), *Department of Chemistry, Wayne State University, Detroit, Michigan*

Lee J. Todd (87), *Department of Chemistry, Indiana University, Bloomington, Indiana*

S. P. Tyfield (117), *Department of Inorganic Chemistry, The University, Bristol, England*

G. Wilke (29), *Max-Planck-Institut für Kohlenforschung, Mülheim-Ruhr, West Germany*

v

Contents

Metal Carbonyls, Forty Years of Research

WALTER HIEBER

π-Allylnickel Intermediates in Organic Synthesis

P. HEIMBACH, P. W. JOLLY, and G. WILKE

Transition Metal–Carborane Complexes

LEE J. TODD

Metal Carbonyl Cations

E. W. ABEL and S. P. TYFIELD

Fast Exchange Reactions of Group I, II, and III
Organometallic Compounds

JOHN P. OLIVER

Mass Spectra of Metallocenes and Related Compounds

M. CAIS and M. S. LUPIN

Metal Carbonyls, Forty Years of Research[1]

WALTER HIEBER

*Anorganisch-Chemisches Laboratorium
der Technischen Hochschule,
München, Germany*

I

INTRODUCTION

A most appreciated invitation from the Editors of *Advances in Organometallic Chemistry* has provided the motive for this chapter, namely to provide readers with the thoughts and impressions which have stimulated me in the foundation and evolution of this branch of inorganic chemistry, which has now become so well developed. I should state that I have never

[1] Translated from the German by M. McGlinchey, University of Bristol, Bristol, England.

1

been an enthusiastic author, avoiding numerous suggestions for the authorship of a monograph about our work, the intensive experimental work with my co-workers keeping me constantly in the laboratory, at least in spirit. Only original discussions describing new findings and developments were routinely published.

Metal carbonyls arous d my special interest early in my career. Even as a student at the University of Tübingen I was excited by the story of this strange class of compounds. During my Assistantship, and before my Habilitation at the Chemical Institute of the University of Würzburg, my unforgettable teacher R. F. Weinland encouraged me to set up a demonstration of a synthesis of nickel carbonyl by passing carbon monoxide over activated metal in 1921. Since that time I have always been interested in experimental studies on metal carbonyls. In the early years little was known of the chemical behavior of these compounds in spite of the important development of the Mond-Langer process for the processing of Canadian magnetic pyrites ore.

As the years passed, I became largely concerned with research on metal complexes formed from various organic ligands, and it was only in the autumn of 1927 at the Institute of Chemistry of the University of Heidelberg that I took up research experiments with iron pentacarbonyl, which was kindly provided by Dr. A. Mittasch of B.A.S.F. in neighboring Ludwigshafen-am-Rhein. On the basis of his own experience with nickel carbonyl he warned me emphatically of the danger inherent in the use of these highly toxic substances, coupling his warning with the comment that in this field one could only expect a great deal of trouble and results of little scientific value! However, my first experiments, described below, soon proved to be profitable. Shortly afterwards great interest was aroused by A. Mittasch's report (I)[2] on the 31st May 1928 to the General Assembly of the Verein Deutscher Chemiker in Dresden, in which he described his impressive "rediscovery" of iron carbonyl and the development, under his leadership, of the eventual large-scale production of this compound. It became my firm resolve to extend this field which I had just taken up.

The metal carbonyls were at that time far from being firmly categorized as a class of compounds, and were considered by many as curiosities.

[2] In this chapter, Roman numerals refer to those references which are comprehensive and general reviews. Arabic numbers quoted immediately after a Roman numeral refer to references cited in the particular review article to which reference is made. Arabic numbers *only* refer to papers describing experimental work and reviews on specific topics and these references are also included in the list at the end of this chapter.

P. Pfeiffer (*II*) regarded them "as coordination compounds of the elements." R. Weinland (*III*) considered metal carbonyls a "special class of mononuclear compounds which existed because of secondary valencies of carbon." In contrast, the hypothesis was held by H. Reihlen (*1*), and others, that the metal carbonyls were metal salts of organic *pseudo*-acids, the carbon atoms being in a chain with neutralization of the metal by oxygen atoms. H. Freundlich and co-workers (*2*) provided valuable experimental beginnings in their papers, but without systematically following up and interpreting their findings. The same is true for a series of papers discussed elsewhere (*IV, V*).

From the beginning I intended to open a field of "pure coordination compounds," the complex chemistry of metal carbonyls. During this era of classical coordination chemistry, only saltlike or decidedly polar compounds were under investigation, notably those with ligands giving rise to metal–oxygen or metal–nitrogen bonds. Apart from certain cyano complexes, transition metal compounds with direct metal–carbon bonds had scarcely been examined systematically. In contrast to many well-known scientists of that period, I believed metal carbonyls to be organometallics, and thereby anticipated the now known special position of the transition metals in forming bonds to organic groups. Plainly from the standpoint of chemistry as a whole this field was destined to grow in importance.

In collaboration with outstanding co-workers, who have independently also contributed to the development of metal carbonyl chemistry, it was granted to me to obtain novel results for some decades before any serious competition arose from other researchers! I can even remember the occasional allusions to the strange nature of the work which was directed upon a field where success was hardly to be expected. Yet, on the occasion of his 80th birthday in January 1950, Mittasch told me "an extensive chemistry of metal carbonyls has arisen, and you are the initiator of this highly interesting field."

Because of the properties of the metal carbonyls, the experimental work required new techniques, such as the use of carbon monoxide under pressure in the laboratory, and the design of special apparatus. The results obtained always kept us in a state of excitement, thereby stimulating us to make further studies. Starting from modest beginnings it has been shown that the metal carbonyls in no way represent an isolated area of chemistry. Herein major results of our work have been summarized with no pretension to making a complete review.

II

SUBSTITUTION AND DISPROPORTIONATION REACTIONS OF THE METAL CARBONYLS

A. Reactions with Nitrogen and Oxygen Bases

In the summer of 1927, at the Heidelberg Institute, I took up the study of "Reaktionen und Derivate des Eisencarbonyls" (3), especially experiments to substitute the carbon monoxide groups in iron pentacarbonyl with ethylenediamine, in order to demonstrate that the metal carbonyls were "element complexes." The known stability of cyclic complex compounds (chelate effect), which at that time made possible the preparation of stable thallium(III) ethylenediamine, and silver(II) o-phenanthroline complexes (4), suggested formation of an iron carbonyl derivative of this type. I still clearly remember formation of the characteristic red solutions of iron penta- or tetracarbonyl with ethylenediamine or pyridine, from which we were able to isolate the corresponding amine-containing iron carbonyl compounds. In the course of further researches at the Inorganic Chemistry Institute of the Technische Hochschule in Munich we recognized that the very first reactions of the iron carbonyls which we investigated were very complex in character. Later the reactions of amines and ammonia with dicobalt octacarbonyl were studied (VII, 10, 13, 29) and in this manner valence disproportionation reactions were recognized and established, e.g.,

$$3 \, Co_2(CO)_8 + 12 \, NH_3 \; \rightarrow \; 2 \, [Co(NH_3)_6][Co(CO)_4]_2 + 8 \, CO$$

$$3 \, Co^0 \; \rightarrow \; Co^{2+} + 2 \, Co^-$$

My co-worker J. Sedlmeier then held the view that the amine-containing iron carbonyl complexes were also ionic compounds (VII, 14, 21). Hence the compound $Fe_2(CO)_4(en)_3$ (en = 1,2-ethylenediamine) was formulated as $[Fe(en)_3]^{2+}[Fe(CO)_4]^{2-}$. Systematic investigations revealed that reactions of the iron carbonyls with other nitrogen and oxygen donors likewise involved valency disproportionation of the metal with concomitant formation of mono- and polynuclear carbonylferrates, viz., $[Fe(CO)_4]^{2-}$, $[Fe_2(CO)_8]^{2-}$, $[Fe_3(CO)_{11}]^{2-}$. R. Werner (VII, 15, 17, 19, 20) even discovered and characterized compounds containing the tetranuclear anion $[Fe_4(CO)_{13}]^{2-}$, the first being that from pyridine and iron carbonyl, viz.,

$$[py_6Fe][Fe_4(CO)_{13}]$$

According to our work with nitrogen bases, which was done in Heidelberg and in Stuttgart (1933/34), it was shown that pyridine, o-phenanthroline, and 1,2-ethylenediamine react in a different manner with the hexacarbonyls of chromium, molybdenum, and tungsten with, in each case, substitution of the carbon monoxide ligands (*VII, 27*).

$$M(CO)_6 + n\,py \rightarrow M(CO)_{6-n}py_n + n\,CO$$
$$M = Cr, Mo; W; n = 1, 2, 3$$

Especially characteristic were the tricarbonyl derivatives $M(CO)_3L_3$ (L = amine, NH_3, etc.), but tetra- and pentacarbonyl compounds were also discovered. Recently, even dicarbonyls $M(CO)_2(L—L)_2$ and fully substituted derivatives $M(L—L)_3$ and $M(L—L—L)_2$ (L—L = dipyridyl, o-phenanthroline; L—L—L = tripyridyl) have been synthesized from the hexacarbonyls by my former co-worker H. Behrens (5). Monosubstituted derivatives of iron pentacarbonyl $Fe(CO)_4L$ (L = NH_3, amines, etc.) have been identified relatively recently (6).

B. Reactions with Isonitriles and Phosphorus(III) Compounds

Since 1948–50, by using as reactants isonitriles, phosphorus trihalides, and tertiary phosphines, we have gained important insight into the dependence of reactions of metal carbonyls with bases upon the nature of the ligand. Organophosphines were introduced into carbonyl chemistry even prior to 1948 by Reppe and Schweckendiek (7). In general, these ligands react only by substitution of CO, and do not cause disproportionation. Thus nickel carbonyl frequently reacts with complete displacement of carbon monoxide, as we were first able to demonstrate in the reaction with phenyl isonitrile (8).

$$Ni(CO)_4 + 4\,C_6H_5NC \rightarrow Ni(CNC_6H_5)_4 + 4\,CO$$

Wilkinson (9) isolated the tetrakis(trihalogenophosphine)nickel compounds $Ni(PX_3)_4$ (X = F, Cl, Br), and Behrens (10) isolated the triphenylphosphine complex $Ni[P(C_6H_5)_3]_4$ via $[Ni(CN)_4]^{4-}$. With iron pentacarbonyl, isonitriles and phosphines yield (11) mono- and disubstituted derivatives, $Fe(CO)_4L$ and $Fe(CO)_3L_2$, respectively, the latter being the well-known cyclization catalyst of Reppe (7). With the same ligands, carbonyls of the chromium group afforded pentacarbonyl derivatives $M(CO)_5L$. However, a series of other compounds of the type $M(CO)_m L_n$ ($m+n=4$ to 6) has been prepared by L. Malatesta (12), who systematically extended the chemistry of isonitrile complexes.

C. General Conclusions

In sum, we have shown that the tendency of the metal carbonyls to react with Lewis bases with valency disproportionation of the metal atom increases along the series $Cr(CO)_6 \ll Ni(CO)_4 < Fe(CO)_5 < Mn_2(CO)_{10} < Co_2(CO)_8 \sim V(CO)_6$ (VIII). As with iron carbonyls, we found that treatment of nickel carbonyl with amines gives rise to a series of polynuclear carbonyl nickelates $[Ni_2(CO)_6]^{2-}$, $[Ni_3(CO)_8]^{2-}$, $[Ni_4(CO)_9]^{2-}$, and $[Ni_5(CO)_9]^{2-}$ (13). With the carbonyls of elements of odd atomic number (V, Mn, Co) mononuclear, singly charged carbonylmetallate anions $[V(CO)_6]^-$, $[Mn(CO)_5]^-$, and $[Co(CO)_4]^-$ are always formed (VII). Substitution occurs preponderantly with π-acceptor ligands of which trifluorophosphine (14) and diphenyl telluride (15) may be mentioned; carbon monoxide is also replaced by sulfur compounds (16), by aromatic and olefin systems (17), and by other ligands. Dicobalt octacarbonyl and vanadium hexacarbonyl, which are remarkable for their tendency to undergo valency disproportionation, even with phosphines, under certain reaction conditions give saltlike compounds $[Co(CO)_3(PR_3)_2][Co(CO)_4]$ (VII, 13) or $[V(L—L)_3][V(CO)_6]_2$ ($R=C_6H_5$; $L—L=R_2PCH_2CH_2PR_2$), respectively (18), in addition to substitution products. The choice of solvent and of special reaction conditions, such as thermolysis or ultraviolet irradiation, can be decisive in determining the course of reaction (19).

III

METAL CARBONYL HYDRIDES

Already in the early experiments involving treatment of iron tetracarbonyl with ethylenediamine hydrate, we observed that upon acidification of the solutions an extraordinarily intense repugnant smell was released (3). I can still clearly remember the day when I, together with my co-workers Leutert and Vetter, at the Heidelberg Institute, was able to freeze out a volatile water-clear liquid from the decomposition of the ethylenediamine-containing iron carbonyls, identifying it as $H_2Fe(CO)_4$ *(20)*. To our considerable advantage, we found that we could prepare this absolutely novel compound—the first complex metal hydride having a formally negative oxidation number on a metal atom—via the reaction of iron pentacarbonyl with alkali ("Basenreaktion") with subsequent acidification of the solution *(V, 14)*.

$$Fe(CO)_5 + 3\ OH^- \longrightarrow [HFe(CO)_4]^- + CO_3^{2-} + H_2O$$

$$\downarrow H^+$$

$$H_2Fe(CO)_4$$

Moreover, oxidation of the solutions, e.g., with MnO_2, provided a rational synthesis of iron tetracarbonyl $[Fe(CO)_4]_3$ *(21)*. Already at that time I suggested *(22)* that in these hydrides, such as $H_2Fe(CO)_4$, which are extremely sensitive to oxygen and temperature, the hydrogen atoms are directly attached to the metal. Recent studies, in particular those involving infrared and NMR measurements, have confirmed this unequivocally. Since the early days I and my co-workers have actively studied the "Basen-reaktionen" of metal carbonyls and the chemistry of the carbonyl hydrides. Soon after the discovery of $H_2Fe(CO)_4$, while at the Inorganic Chemistry Institute of the Technische Hochschule in Stuttgart in 1934, we were able to isolate the likewise unstable cobalt hydride $HCo(CO)_4$ *(23)*. First evidence for the formation of this hydride was obtained in Heidelberg in 1932 from acidification of amine-containing cobalt carbonyls *(24)*, just like the synthesis of $H_2Fe(CO)_4$. The reaction between dicobalt octacarbonyl and alkali was only understood later when formation of the anion in a homogeneous alcohol–water system was recognized (VII, *10*).

$$3\ Co_2(CO)_8 \rightarrow 2\ Co^{2+} + 4\ [Co(CO)_4]^- + 8\ CO$$

$$8\ Co_2(CO)_8 + 32\ OH^- + 8\ CO \rightarrow 16\ [Co(CO)_4]^- + 8\ CO_3^{2-} + 16\ H_2O$$

We recognized that the isoelectronic hydrides $H_2Fe(CO)_4$ and $HCo(CO)_4$ behaved as *pseudo*-nickel tetracarbonyls (hydride displacement principle). Moreover, manganese pentacarbonyl hydride (*VII, 11*), on which we later worked in Munich, was similar in many physical properties to iron pentacarbonyl, leading us to postulate a "Drawing-in of the hydrogen atom into the electron shell of the metal ($MnH \sim \psi Fe$)." Today, we know, from spectroscopic and X-ray structure determinations, that the hydrogen atom in the metal carbonyl hydrides occupies a definite coordination position in the coordination polyhedron (25); for example, the Mn atom in $HMn(CO)_5$ is in an octahedral environment, the hydrogen atom occupying one of the octahedral positions. The hydrides $HMn(CO)_5$ and $HRe(CO)_5$ are different from the cobalt and iron hydrides in having considerable thermal stability (26). Deuterium derivatives have also been studied, especially by infrared spectroscopy (26, 27). Cobalt tetracarbonyl hydride is remarkable for the ease with which it is formed, being similar to $Ni(CO)_4$ in this respect. Thus it may be synthesized from cobalt metal, CO, and H_2 under pressure, or from cobalt compounds (CoS, CoI_2) and carbon monoxide in the presence of moisture or hydrogen-containing substances (V, 31). Later we also observed the thermally labile carbonyl hydrides $HRh(CO)_4$ and $HIr(CO)_4$ (28) of the homologous rare metals Rh and Ir. The existence of $H_2Os(CO)_4$, which was recently characterized by Calderazzo et al. (29) and proved to be surprisingly stable, was also observed in our laboratory when we developed the synthesis of osmium carbonyl from OsO_4 (30). Recently, Stone et al. (31) obtained $H_2Ru(CO)_4$ as a very unstable colorless liquid.

After our discovery of the metal carbonyl hydrides, other authors (32) pointed out their acidic character in aqueous solution. Potentiometric titrations by Reppe and later by us, showed that in water $HCo(CO)_4$ possesses an acidity ($pK_a \sim 1$) comparable to that of nitric acid. The first ionization stage for $H_2Fe(CO)_4$ corresponds approximately to that of acetic acid (33), whereas the pentacarbonyl hydrides $HM(CO)_5$ (M = Mn or Re) (VII, 11, 26) are hardly acidic at all. The redox potentials of the cobalt and iron carbonyl hydrides were also measured (33).

We also prepared the hydrides $H_2Fe_2(CO)_8$ and $H_2Fe_3(CO)_{11}$ which corresponded to the polynuclear carbonyl ferrates (VII, 14, 15, 21). From crystalline deep brown pyrophoric $H_2Fe_4(CO)_{13}$, which is soluble in organic solvents, we prepared ammonium and pyridinium salts. Presently much work is being done on polynuclear hydrides of ruthenium (31), of manganese and rhenium (34, 37), and on compounds containing different metal atoms such as $HFeCo_3(CO)_{12}$ (35), $HRe_2Mn(CO)_{14}$ (36), and others (37).

We later prepared the more stable phosphine-substituted mononuclear carbonyl hydrides such as $HCo(CO)_{4-n}L_n$ and $HMn(CO)_{5-n}L_n$ [n = 1 and 2;

$L = P(C_6H_5)_3$, $P(OC_6H_5)_3$ etc.] (38, 39). Interestingly, although $HV(CO)_6$ is not known, we were able to isolate the hydride $HV(CO)_5P(C_6H_5)_3$ (40). In contrast to the unsubstituted carbonyl hydrides, we found that the acidity of the aqueous solutions of these compounds was considerably weakened. Recently phosphine-substituted ruthenium carbonyl hydrides, e.g., $H_2Ru(CO)_2(PR_3)_2$ have been reported (31).

IV

ORGANOMETAL CARBONYLS

Following the discovery of the carbonyl hydrides, we succeeded in obtaining carbonyl complexes containing a $3d$ metal-to-carbon σ-bond. Such compounds I had suspected as intermediates in the synthesis of Cr, Mo, and W hexacarbonyls from their chlorides, carbon monoxide, and Grignard reagents (41), and had called them "organometallic carbonyls." We made the first examples of these new compounds from manganese carbonyl (VII, 84). The relatively stable alkyl and acyl pentacarbonyl-manganese compounds $RMn(CO)_5$ and $RCOMn(CO)_5$ were also prepared independently at about the same time by Coffield et al. (42). In addition, after many fruitless attempts, we succeeded in making the thermally unstable $CH_3Co(CO)_4$ from sodium tetracarbonyl cobaltate($-I$) and methyl iodide (VII, 44). In contrast to cobalt, but similarly to manganese, alkyl and acyl rhenium pentacarbonyls are very stable (VII, 84). In the acylpentacarbonylmanganese and -rhenium complexes the CO stretching bands of the keto groups in the infrared lie at remarkably low wavenumbers. This can be attributed to a contribution by the structure (VII, 84)

$$\overset{\delta+}{M} = \overset{\delta-}{C(O)R}$$

Today, through the researches of many groups of workers, a large number of organometal carbonyls are known. To these must be added the carboalkoxymetal carbonyls mentioned below (Section V,C). The considerably more stable perfluoro derivatives have been thoroughly investigated by Stone and his co-workers (43). Our own contribution to this particular field involved the preparation and characterization of phosphine-containing fluoroorganometal carbonyls of cobalt and manganese (38, 39).

V

METAL DERIVATIVES OF CARBONYL HYDRIDES

A. Nonpolar Compounds

The mercury compounds $HgFe(CO)_4$ and $Fe(CO)_4(HgX)_2$ ($X = Cl$, Br, I), which were the first representatives of non-ionic metal derivatives of iron carbonyl hydrides, were discovered by Hock and Stuhlmann (*V, 36*). During investigations into the preparation of cobalt carbonyls from cobalt halides under CO pressure, in the presence of another metal as a halogen acceptor, we discovered the "mixed" metal carbonyls $M[Co(CO)_4]_2$ ($M = Zn$, Cd, Hg, Sn) and $M[Co(CO)_4]_3$ ($M = In$, Tl) (*44*), e.g.,

$$2\ CoBr_2 + 3\ Zn + 8\ CO \xrightarrow[\text{350 atm}]{200°C} Zn[Co(CO)_4]_2 + 2\ ZnBr_2$$

The especially stable "mercury-cobalt carbonyl," which is readily soluble in organic solvents, forms quantitatively according to the equation

$$HgX_2\ (X = Br\ or\ I) + 3\ Co + 8\ CO \xrightarrow[\text{200 atm}]{150°C} Hg[Co(CO)_4]_2 + CoX_2$$

For this compound, which is closely related to the isoelectronic $Hg[Fe(CO)_3NO]_2$ (*45*) (Section IX) and to $Hg[Mn(CO)_5]_2$ (*VII, 79*), a convenient method of preparation in aqueous solution was discovered in conjunction with E. O. Fischer (*46*). Moreover, we could make $Hg[Co(CO)_4]_2$ react, like $Co_2(CO)_8$, with ligands either with substitution of carbon monoxide and formation of $Hg[Co(CO)_3L]_2$ ($L = PR_3$, AsR_3, SbR_3), or by valence disproportionation of cobalt (*VII, 80*) using nitrogen bases or isonitriles, e.g.,

$$3\ Hg[Co(CO)_4]_2 + 12\ L \rightarrow 3\ Hg + 2\ [L_6Co][Co(CO)_4]_2 + 8\ CO$$

Similar chemistry was observed with mercury-iron carbonyl compounds (*47*).

"Organometallic-metal carbonyls" were first isolated by Hein (*VII, 82*) and involved iron, e.g., $[R_2PbFe(CO)_4]_2$. Subsequently cobalt compounds of this type were made by us (*VII, 83*), e.g.,

$$R_3SnCl + NaCo(CO)_4 \rightarrow R_3SnCo(CO)_4 + NaCl$$

This type of metal carbonyl derivative has lately been the object of numerous investigations by many research groups with particular reference to metal–metal bond synthesis.[3] In our laboratory the nitrosyl carbonyliron compounds $R_3MFe(NO)(CO)_3$ (M = Si, Ge, Sn, Pb; R = C_6H_5, C_4H_9) have been prepared, and shown by infrared studies to have a trigonal bipyramidal structure with C_s symmetry (*48*).

We have also been able to obtain nonpolar iron carbonyl metal complexes involving elements of the third to the fifth periodic groups; for example, with arsenic, antimony, and bismuth (*VII*, *81*), the compounds $As_2Fe_3(CO)_{11}$, $SbFe_2(CO)_8$, $SbFe(CO)_4$, and $Bi_2Fe_5(CO)_{20}$ have been prepared, and with four-valent tin and lead, and with thallium compounds, the complexes $Sn_2Fe_5(CO)_{20}$, $PbFe_3(CO)_{12}$, and $Tl_2Fe_3(CO)_{12}$, respectively, have been obtained. As yet the principles which lead to the formation of these polynuclear carbonyl ferrates are not well understood.

B. Anionic Complexes

The ideas which led us to understand the formation of carbonylmetallates in the reactions of metal carbonyls with nitrogen and oxygen Lewis bases have been discussed above, and in addition I have given elsewhere (*VII*) an exhaustive summary of anionic carbon monoxide complexes.

More recently (*49*) the reactions of iron carbonyls with alkali hydroxides and with nitrogen and oxygen bases have been once again thoroughly examined, so that the absorption spectra in the visible region of $[Fe(CO)_4]^{2-}$, $[Fe_2(CO)_8]^{2-}$, $[Fe_3(CO)_{11}]^{2-}$, $[Fe_4(CO)_{13}]^{2-}$, and of the corresponding hydrogen anions have been studied.

Special success has followed the use of alkali metal amalgams in tetrahydrofuran or other ethers (*VII*, *11*, *44*) as reducing agents in the syntheses of mono- and polynuclear metal carbonyls, e.g.,

$$Re_2(CO)_{10} \xrightarrow[Me_2O]{Na/Hg} 2\ NaRe(CO)_5$$

Using the technique of Behrens (*VII*, *43*), liquid ammonia was also found to be a suitable solvent. Our further experiments showed that, depending on the redox potential, it was possible to isolate polynuclear anion complexes of various types. Thus the alkali and alkaline earth metal reduction of

---

[3] For a review, see article by F. G. A. Stone, *in* "New Pathways in Inorganic Chemistry," H. J. Eméleus "Festschrift." Cambridge Univ. Press, London and New York.

$Ni(CO)_4$ in liquid ammonia affords $[Ni_2(CO)_6]^{2-}$, while the amalgams of Li, Na, K, or Mg give rise to Ni_3 or Ni_4 carbonylnickelates (50). Carbonyl-metallates were also obtained from halogenometal carbonyls (see following section) and alkali metals. Even the phosphine-substituted carbonyl halides like $Co(CO)_2(PR_3)_2X$ (38) or $Mn(CO)_3(PR_3)_2X$ (39) were easily reduced with sodium amalgam in tetrahydrofuran to the corresponding "mixed" anions such as $[Co(CO)_2(PR_3)_2]^-$. However, treatment of $Fe(CO)_2(PR_3)_2Br_2$ with sodium amalgam surprisingly leads not to the phosphine-substituted carbonyl ferrate but, with excess phosphine, to the neutral tris(triphenyl-phosphine)dicarbonyliron complex $Fe(CO)_2[P(C_6H_5)_3]_3$, which is not accessible by the substitution reaction of $Fe(CO)_5$ with $P(C_6H_5)_3$ (51). In a similar way, we obtained by the amalgam procedure $Co(NO)[P(C_6H_5)_3]_3$ from $[Co(NO)_2Cl]_2$ with excess triphenylphosphine (52). This is remarkable insofar as a total substitution of the CO groups in $Co(CO)_3NO$ (Section IX) does not occur when using phosphine itself, even under drastic conditions.

C. Cationic Carbon Monoxide Complexes

In contrast to the cationic complexes formed by the more polar isonitriles, e.g., $[Co(CNR)_5]^+$, $[Mn(CNR)_6]^+$ (12), carbonyl metal cations were for many years considered as incapable of existence. In 1961, however, E. O. Fischer et al. (53) characterized the hexacarbonylmanganese(I) cation $[Mn(CO)_6]^+$ from the reaction

$$Mn(CO)_5Cl + CO + AlCl_3 \rightarrow [Mn(CO)_6][AlCl_4]$$

This new type of reaction, involving a carbonyl metal halide, a halogen acceptor, and CO under pressure, is of general utility. Consequently, we were able to prepare a large number of cationic carbon monoxide complexes. With the essential collaboration of T. Kruck, numerous, especially phosphine-containing, compounds were prepared (54), e.g., $[Mn(CO)_4(PR_3)_2]^+$, $[Co(CO)_4PR_3]^+$, and $[Co(CO)_3(PR_3)_2]^+$ ($R = C_6H_5$ or OC_6H_5). These cationic complexes were considerably more stable than the un-substituted derivatives such as $[Mn(CO)_6]^+$.

Recently we have been interested in phosphine-containing carbonyl cations of the noble metals (55) such as $[M(CO)_3(PR_3)_2X]^+$ ($M = Fe$, Ru, Os; $X =$ halogen) and $[M(CO)_2(PR_3)_2]^+$ ($M = Rh$, Ir) which we prepared from the corresponding metal carbonyl halides under atmospheric CO pressure.

The reactions of carbonyl metal cations with alkali metal alkoxides, whereby carboalkoxycarbonyls are formed (*54*), warrant special mention, e.g.,

$$\{Re(CO)_4[P(C_6H_5)_3]_2\}^+ + CH_3O^- \xrightleftharpoons[HX]{KOH} Re(CO)_3[P(C_6H_5)_3]_2COOCH_3$$

The course of these reactions provides an explanation for the nature of the reactions between neutral metal carbonyls and hydroxyl anions. The latter attack the C atom of a CO group (*54, 56*), viz.,

$$(OC)_4Fe{=}C{=}O \xrightarrow{+OH^-} (OC)_4Fe{\overset{\ominus}{=\!\!=}}C\overset{\overset{O}{\diagup}}{\diagdown_{OH}} \xrightarrow[-CO_2]{-H^+} [Fe(CO)_4]^{2-}$$

VI

HALOGENOMETAL CARBONYLS

A. Carbonyl Iron Halides

Almost simultaneously with our first investigations "About Reactions and Derivatives of Iron Carbonyl" (Section II,A) we concerned ourselves at Heidelberg with the effect of halogens on iron pentacarbonyl, initially in the expectation of obtaining a pure surface-active iron(II) halide. However, to our surprise at the time, reaction occurred according to the equation

$$Fe(CO)_5 + X_2 (X = Cl, Br, I) \rightarrow Fe(CO)_4X_2 + CO$$

In this manner, the field of metal carbonyl halides was initiated (*57*). Using low temperatures, we were able to make the addition compounds $Fe(CO)_5X_2$, which subsequent work revealed were intermediary in the exchange of ^{14}CO with the carbonyl groups of $Fe(CO)_4X_2$ (*58*). Meanwhile, $Fe_2(CO)_8I_2$ has also been described (*59*).

By substitution reactions with amines, especially pyridine and *o*-phen-anthroline (*57*), and later with isonitriles (*11*), and phosphines and similar compounds (*60*), we obtained the complete series of iron(II) halide (especially iodide) compounds with one to four CO groups per molecule.

The apparently anomalous behavior of the iron tetracarbonyl halides in terms of the inverse stability gradation iodide > bromide > chloride—the latter is labile above $0°C$—which is the reverse sequence to that found with the nitrogen-base complexes L_4FeX_2 ($L = NH_3$, pyridine, etc.), caused us to study the heats of formation of the compounds FeX_2 and $Fe(CO)_4X_2$ (61). Considering the lattice energies my co-worker E. Levy was able to conclude from thermochemistry measurements that the halogen atoms in the compounds $Fe(CO)_4X_2$ are strongly polarized, in contrast to the halogen atoms in the addition complexes of iron with nitrogen bases L_4FeX_2, for which polarization of the Fe—X bond is reduced.

We were able to prepare (63) iron carbonyl halides of lower carbon monoxide content, $Fe(CO)_2X_2$, which are formally related to the long-known platinum carbonyl compounds (62), e.g.,

$$Fe(CO)_4X_2(X=Br, I)+2\ SOCl_2\ \rightarrow\ Fe(CO)_2Cl_2+X_2+2\ CO+SO_2+SCl_2$$

My co-worker H. Lagally showed that the thermal decomposition of $Fe(CO)_4I_2$ afforded successively $Fe(CO)_2I_2$, $Fe(CO)_2I$, and CO-free FeI. For these decomposition products of $Fe(CO)_4I_2$ a polymeric structure with iodine bridges is postulated. A study of the reaction $FeI_2+4\ CO\rightleftharpoons Fe(CO)_4I_2$, also in collaboration with Lagally, further established the metastable or labile nature of iron carbonyl halides (63).

B. Phosphine-Substituted Cobalt Carbonyl Halides

The monohalogenocobalt carbonyls $Co(CO)_4X$ would correspond to the dihalogenoiron tetracarbonyls. However, such cobalt compounds do not exist under normal conditions. Their existence at low temperature was demonstrated by Bigorgne (64) by infrared spectroscopy, in one of his many valuable contributions to the vibrational spectroscopy of metal carbonyls. Nevertheless by mild halogenation techniques, we obtained the room-temperature-stable phosphine-substituted carbonyl halides $Co(CO)_3(PR_3)X$ and $Co(CO)_2(PR_3)_2X$ ($X=Cl$, Br, I; $R=C_6H_5$ and OC_6H_5) whose modes of formation and decomposition deserve attention (65). Already much earlier we were able to account for the catalytic action of iodine in the high-pressure synthesis of cobalt carbonyl by the initial formation of $CoI_2(CO)$ which is then converted, presumably via labile $Co(CO)_4I$, into $Co_2(CO)_8$ and CoI_2 (V, 22).

C. Carbonyl Halides of Rhenium and of the Platinum Metals

I have always regarded the work of my co-worker H. Schulten (66) on the synthesis of rhenium carbonyl complexes as a landmark in our work.

The synthesis of pentacarbonyl rhenium(I) halides, $Re(CO)_5X$, succeeded from simple and complex rhenium halides below 200 atm of CO at 200° C. The compounds are extraordinarily stable and form easily, often quantitatively, from carbon monoxide and rhenium metal in the presence of other heavy metal halides or halogen sources such as CCl_4. Later we prepared the corresponding carbonyl halides of manganese (67) and technetium (68) from their respective carbonyls. It was found that the corresponding binuclear tetracarbonyl halides $[M(CO)_4X]_2$ (M = Mn, Re) could be made by heating the mononuclear $M(CO)_5X$ complexes (15, 69), as well as by other methods.

These early successes with carbonyl complexes of rhenium encouraged me to undertake systematic research on the carbon monoxide chemistry of the heavy transition metals at our Munich Institute during the period 1939–45, oriented towards purely scientific objectives. The ideas of W. Manchot, whereby in general only dicarbonyl halides of divalent platinum metals should exist, were soon proved inadequate. In addition to the compounds $[Ru(CO)_2X_2]_n$ (70), we were able to prepare, especially from osmium, numerous di- and monohalide complexes with two to four molecules of CO per metal atom (29). From rhodium and iridium (28) we obtained the very stable rhodium(I) complexes $[Rh(CO)_2X]_2$, as well as the series $Ir(CO)_2X_2$, $Ir(CO)_3X$, $[Ir(CO)_3]_x$ (see Section VII,A). With this work the characterization of carbonyl halides of most of the transition metals, including those of the copper group, was completed.

Carbonyl halides of chromium, $Cr(CO)_5I$ and others (71), as well as $Mo(CO)_4Cl_2$ (72), have only recently been added to the list of known carbonyl halides. Moreover, the substitution reactions of the carbonyl halides of manganese and rhenium (73) as well as those of the noble metals (74) have been thoroughly studied. A comprehensive review on this topic has recently been published.[4]

VII

SYNTHESIS OF METAL CARBONYLS

A. Reactions in Which Solvents Are Not Used

Considering the significant discovery of metal carbonyls and their formation in technical processes, it became my intention after our first

[4] F. Calderazzo, in "Halogen Chemistry" (V. Gutmann, ed.), Vol. 3, p. 383. Academic Press, New York, 1967.

successful experiments to develop satisfactory laboratory techniques for reactions of carbon monoxide under pressure. This was made possible after I became Director of the Inorganic Chemical Laboratories of the Technische Hochschule in Munich in 1935 where there was a workshop with trained personnel. It was necessary to take precautions with regard to the toxicity of carbon monoxide and the volatile metal compounds, and to bear in mind that carbon monoxide under severe conditions attacks iron and its alloys. In the course of a few years it became possible to carry out a series of syntheses on a laboratory scale (*V*, *22*, *31*). In contrast to the classical syntheses of metal carbonyls from the metals and carbon monoxide, whereby the metal had to be made in an "active form," we used the halides of metals of the iron subgroup, as well as of molybdenum and tungsten, with the addition of a halogen acceptor (Cu, Ag, Zn, etc.). The reduction of the halides generally occurred at 200 atm of CO and 200°–250° C, e.g.,

$$FeX_2(X=Cl, Br, I) + 2\,Cu + 7\,CO \;\rightarrow\; Fe(CO)_5 + 2\,CuX \cdot CO$$

The thermochemical aspects of these reactions have been discussed in terms of heats of formation of the halides of elements of the iron group, and of the "acceptor" metal (*75*). The yield of carbonyls was especially favored with the iodides and also with sulfides or sulfur-containing materials (*76*). With iron and cobalt iodides the reaction is facilitated by formation of the carbonyl iodide as an intermediate.

With rhenium, however, these researches led not to the pure carbonyl but, on account of their great stability, to the carbonyl halides $XRe(CO)_5$. To our surprise, however, we were able to obtain rhenium carbonyl by reduction of heptoxide (*77*).

$$Re_2O_7 + 17\,CO \;\rightarrow\; Re_2(CO)_{10} + 7\,CO_2$$

Predictably, osmium behaved like rhenium (*29*).

$$OsO_4 + 9\,CO \;\rightarrow\; Os(CO)_5 + 4\,CO_2$$

Studies on the carbonyl halides of the noble metals led us directly to the discovery of the pure carbonyls of these elements. Especially impressive was the formation of tetranuclear iridium tricarbonyl, $[Ir(CO)_3]_4$, via the tricarbonyl chloride $Ir(CO)_3Cl$, as demonstrated by the simultaneous formation of both of these substances from $IrCl_3 \cdot H_2O$ under normal CO pressure at $150°$ C [pictures of the crystals appeared in the original paper (28)]. Rhodium carbonyls were formed from the halides with or without the presence of a halide-acceptor metal, as well as from the finely divided metal at 200–300 atm of carbon monoxide at $280°$ C (28). For the synthesis of ruthenium carbonyl the use of the iodide or sulfide as starting material is favorable (70, 78). We observed immediately that the carbonyls of the noble metals with highest carbon monoxide content were extraordinarily labile and, in general, hard to make. The liquid pentacarbonyls of ruthenium and osmium spontaneously change over to the crystalline tetracarbonyls $[M(CO)_4]_3$ (29, 70). The yellow rhodium and the greenish yellow iridium tetracarbonyls $[M(CO)_4]_2$ are observable only at low temperatures under excess CO pressure; under normal conditions the tetranuclear tricarbonyls $[M(CO)_3]_4$ are formed (28). A further polynuclear rhodium carbonyl, discovered by us, proved after an X-ray crystallographic study to be $Rh_6(CO)_{16}$. It is gratifying that the carbonyls of the noble metals have provided such interesting and valuable knowledge about structural principles and bonding, and have stimulated much further work in the area of metal–metal bonds. For elucidation of structures by X-ray crystallography we thank above all L. F. Dahl, who has investigated, for example, $Os_3(CO)_{12}$ and $[Rh(CO)_2Cl]_2$, as well as hexanuclear $Rh_6(CO)_{16}$ (79). The structure of the trinuclear tetracarbonyls of Ru and Òs differs from that of $Fe_3(CO)_{12}$, the structure of which was also established by Dahl and co-workers (80), after elucidating that of the crystalline hydridocarbonyl anion $[HFe_3(CO)_{11}]^-$. As is well known, the structure of $Fe_3(CO)_{12}$ has been disputed for a long time.

B. Syntheses in the Liquid Phase

Manchot and Gall (81) were the first to observe the formation of $Ni(CO)_4$ upon passing carbon monoxide through an aqueous alkaline suspension of nickel sulfide. As we later established, a similar reaction occurs also with cobalt sulfide to form the anion $Co(CO)_4^-$, the course of the reaction being explained by Behrens and Eisenmann in the Munich Institute (VII, 52).

We further investigated the formation of $Ni(CO)_4$ and cobalt carbonyls by the action of carbon monoxide on a suspension of the appropriate metal hydroxide [Blanchard and Gilmont (82)] by using potassium cyanide (83) (see Section VIII).

With dithionite, alkaline solutions of nickel(II) and cobalt(II) salts give with carbon monoxide the corresponding carbonyl compounds practically quantitatively. This work was carried out with E. O. Fischer (46).

$$Ni^{2+} + S_2O_4{}^{2-} + 4\ OH^- + 4\ CO \rightarrow Ni(CO)_4 + 2\ SO_3{}^{2-} + 2\ H_2O$$

$$Co^{2+} + 1\tfrac{1}{2}\ S_2O_4{}^{2-} + 6\ OH^- + 4\ CO \rightarrow [Co(CO)_4]^- + 3\ SO_3{}^{2-} + 3\ H_2O$$

Eventually we formed carbonyls in the liquid phase by redox disproportionation of nickel and cobalt derivatives of organic thioacids. In the reaction between nickel(II) dithiobenzoate and carbon monoxide in the presence of HS^- ion we assumed the formation of a sulfur-bridged nickel(IV) complex (VII, 32). More recent investigations (84), however, have shown that half the nickel appears as a monomeric nickel(II) complex of the same empirical formulation, formed by insertion of a sulfur atom in the dithio ligand, the other half of the nickel being reduced to nickel(0) by the sulfide.

Other authors, especially Reppe and co-workers (85), developed further methods of synthesizing metal carbonyls in the liquid phase, such as the technically important formation of the carbonyls of the iron group from the aqueous ammoniacal solution of the appropriate metal(II) salts; in this system carbon monoxide itself functions as the reducing agent. For other metal carbonyls, for example, chromium, manganese, and vanadium, which up to that time had only been obtained in trace amounts by complicated reactions, very efficient methods of preparation have since been developed and are described in the general literature.

VIII

OTHER DEVELOPMENTS ORIGINATING FROM THE STUDY OF METAL CARBONYLS

A. Cyanometal Carbonyls

Formation of $Ni(CO)_4$ or $Co(CO)_4{}^-$ by the cyanide method depends upon the stepwise substitution of the anion of the cyano complex by the isoelectronic carbon monoxide molecule. By treating $Co_2(CO)_8$ with potassium cyanide we obtained cyanocarbonyls of cobalt of low oxidation number (83). In reactions of the nitrosyl carbonyls of iron and cobalt, Behrens (86) substituted all the CO groups with CN to give $K_3[Co(NO)(CN)_3]$ or

$K_2[Fe(NO)_2(CN)_2]$. My former co-worker R. Nast (*83*) showed that the reaction of carbon monoxide with $K_4[Ni_2^I(CN)_6]$ and $K_4[Ni^0(CN)_4]$ in liquid ammonia gave nickel cyanocarbonyl complexes with monovalent and zero-valent metal atoms. The isoelectronic hexacyanoiron(III) or tetra-cyanonickel(II) complexes correspond to the cyanocarbonyls $[Fe^{II}(CN)_5 CO]^{3-}$, $[Ni^I(CN)_3CO]^{2-}$, or $[Ni^0(CN)_2(CO)_2]^{2-}$. Cobalt is analogous to nickel in forming the complex $[Co(CN)_3CO]^{2-}$. According to our earlier work, $[Fe^{II}(CN)_5CO]^{3-}$ and $[Fe^{III}(CN)_6]^{3-}$ are isosteric (*87*). Other structural investigations were concerned with tetracyano and tetracarbonyl complexes (*88*).

B. Chalcogen Metal Carbonyls

The known, technically significant, favorable effect of sulfur on the formation of iron pentacarbonyl from the metal and carbon monoxide under pressure (*I*), caused us to undertake a systematic study of the influence of the chalcogens on the formation of iron carbonyl. In this way we were led to discover the polynuclear iron carbonyl complexes $Fe_3X_2(CO)_9$ (X = S, Se, Te), which we later, very advantageously, prepared in an aqueous system by treating $[Fe(CO)_4]^{2-}$ with sulfurous, selenous, or tellurous acids (*89*). The binuclear iron compounds $[Fe(CO)_3X]_2$ (X = S, Se, SR, SeR, TeR) which we also synthesized, possess, as was shown by my former co-worker W. Beck (*90*) by infrared and dipole moment measurements, an interesting structure with a nonplanar arrangement of the $Fe(X)_2Fe$ group. This was subsequently confirmed by an X-ray structural analysis of L. F. Dahl (*91*). It has also been possible to obtain the Mn, Tc, (*68*), and Re (*VII, 57*) compounds $[M(CO)_4SC_6H_5]_2$ with sulfur bridges. In this area of chemistry, Bor and Markó (*92*) have described a series of very remarkable sulfur-containing cobalt carbonyl complexes, e.g., $Co_4(CO)_7(SC_2H_5)_3$.

Following our early studies (*89, 90*), we have recently investigated derivatives of manganese carbonyl obtained from reactions between $Mn(CO)_5X$ (X usually Br) and organic sulfur ligands (*93*). Manganese carbonyl chelate complexes are formed in which sulfur is covalently bonded to manganese, or forms a coordinate bond as an electron donor.

Also, very recent research has shown that the carbonyl halides of rhenium and technetium undergo general substitution reactions with organochalcogenides (*16, 94*). In this way we prepared disubstituted mononuclear or monosubstituted binuclear rhenium compounds, $Re(CO)_3L_2X$ and $[Re(CO)_3LX]_2$ ($L = SR_2$, SeR_2, TeR_2; $R = C_6H_5$, C_2H_5, $n\text{-}C_4H_9$; $X = Cl$, Br, I).

C. Some Remarks Concerning Bridged Metal Carbonyl Compounds

Formation of polynuclear compounds with chalcogen atoms as bridging ligands between metal atoms is a characteristic feature of chalcogen metal carbonyl complexes. Such polynuclear complexes, in which often an additional metal–metal bond is assumed so as to account for the diamagnetism, have become known in increasing numbers because of the work of various research groups, and it is neither possible nor is it within the scope of this article to review this subject here. Only a few of our own results will be mentioned herein.

In contrast to the chalcogen-bridged complexes, no similar oxygen-bridged compounds of iron, cobalt, or nickel exist. However, we obtained such oxo or μ-ol-carbonyl complexes of chromium and its homologs, as well as of rhenium. The compounds are the products of the reactions of the respective metal carbonyls with bases (*VII*).

My co-worker H. Beutner (*95*) was able to isolate from a nitrite-containing carbonylferrate solution, in trace amounts only, a binuclear nitrogen-atom-bridged iron carbonyl compound. This complex is now obtained in good yield by UV irradiation of the reaction solution and was identified mass spectrometrically as di-μ-amino-bis(tricarbonyl)iron, $(OC)_3Fe(NH_2)_2Fe(CO)_3$. The group

$$Fe\begin{array}{c} \diagup N \diagdown \\ \diagdown N \diagup \end{array}Fe$$

possesses a nonplanar structure just like the μ-disulfide compounds (*95*).

The field of phosphorus- or arsenic-bridged compounds is very extensive, especially because of the work of R. G. Hayter and J. Chatt (*96*). We were able to isolate many μ-phosphido and μ-arsenido metal carbonyls by reacting diphosphines R_2PPR_2 or the corresponding diarsines ($R = $ alkyl, phenyl) with vanadium carbonyl, cobalt nitrosyl carbonyl, and manganese pentacarbonyl halides, and also by reacting potassium phenylphosphide with carbonyl halides of rhodium and iridium. Both cis and trans isomers of $[Cl(OC)RhPR_2]_2$ were also prepared (97).

IX

METAL CARBONYL NITROSYLS

I remember with great satisfaction my collaboration with J. S. Anderson in the Heidelberg Institute. In 1932 he made the volatile, previously unrecognized as such, dinitrosyldicarbonyliron by the action of pure nitric oxide on a solution of $Fe_3(CO)_{12}$ in iron pentacarbonyl (98). The complex $Fe(CO)_2(NO)_2$ was a deep red liquid at room temperature. With this compound the isoelectronic series $Ni(CO)_4$, $Co(CO)_3NO$, $Fe(CO)_2(NO)_2$ arose, and in this manner the field of carbonyl nitrosyls was opened up. The next member of this isoelectronic series, $Mn(CO)(NO)_3$, predicted by us in 1932, was discovered recently (99). A study of the chemical behavior of the carbonyl nitrosyls, namely the ready substitution of the CO but not of the NO groups, was essentially established by Anderson (100), with the isolation of the derivatives $Fe(NO)_2py_2$, $Fe(NO)_2(o\text{-phen})$, $Co(NO)(CO)(o\text{-phen})$, and $Co(NO)(CO)(PR_3)_2$, etc.

A rational method of preparation for the nitrosyl carbonyls of iron and cobalt was discovered by my former co-worker F. Seel (101) by acidic decomposition of the appropriate carbonylmetallate solution in the presence of nitrite.

$$[HFe(CO)_4]^- + 2\,NO_2^- + 3\,H^+ \rightarrow Fe(CO)_2(NO)_2 + 2\,CO + 2\,H_2O$$

These results have been discussed by Seel in "Structure and Valence Theory of Inorganic Nitric Oxide Complexes" (102). In such complexes, nitric oxide is covalently bonded to the metal atom as the positive ion NO^+, it being assumed that an electron is transferred to the metal. This allows the isoelectronic groups NO^+, CO, CN^- to be considered together and permits understanding of such known series as $[Fe^{II}(CN)_5NO]^{2-}$, $[Fe^{II}(CN)_5CO]^{3-}$, and $[Fe^{II}(CN)_5CN]^{4-}$.

Together with H. Beutner we succeeded at last in discovering a quantitative synthesis of the anion $[Fe(CO)_3NO]^-$ from the system iron pentacarbonyl/nitrite/methylate (45), and thereby established the isoelectronic series $Fe(CO)_2(NO)_2$, $[Fe(CO)_3NO]^-$, $[Co(CO)_4]^-$. The central position of the nitrosyl tricarbonyl ferrate anion arises on the one hand from its intermediary formation in the nitrite reaction with $HFe(CO)_4^-$, mentioned above in the synthesis of $Fe(CO)_2(NO)_2$ (101), and on the other hand from its behavior, which is analogous to the anion $Co(CO)_4^-$. Thus the nonpolar mercury compound $Hg[Fe(CO)_3NO]_2$ and other heavy metal derivatives can be formed, and are typified by their reactive behavior (47); the corresponding carbonyl nitrosyl hydride is unstable (dec $-45°$ C) (45).

$$2\ HFe(CO)_3NO \rightarrow Fe(CO)_2(NO)_2 + H_2Fe(CO)_4$$

We have also studied the relatively complicated behavior of the nitrosyl carbonyls of iron and cobalt in alkali (103).

The first nitrosyl metal halide was discovered with J. S. Anderson (100).

$$Fe(CO)_2(NO)_2 + \tfrac{1}{2}\ I_2 \rightarrow Fe(NO)_2I + 2\ CO$$

In this manner, the field of monohalide–metal nitric oxide complexes was developed, as well as the so-called "nitroprussiates" extensively studied by Nast and Pröschel (104). Since these complexes do not contain CO groups as ligands a discussion of the noteworthy results falls outside the limits of this survey, although the field is closely related to the metal carbonyl nitrosyls.

Further researches on metal carbonyl nitrosyls have been carried out not only by us, but also by others. Especially extensive is the chemistry of manganese complexes, some of which were obtained by us from phosphine-substituted manganese carbonyls by "nitrosation" with nitric oxide, amyl nitrite, and other reagents (105). Thus the compounds $Mn(NO)_3L$, $Mn(NO)(CO)_3L$, and $Mn(NO)(CO)_2L_2$ originate from the reactions of NO with $Mn(CO)_4L$ ($L = PR_3$). This work led to completion of the isoelectronic series $Mn(NO)_3[P(C_6H_5)_3]$, $Fe(NO)_2[P(C_6H_5)_3]_2$, $Co(NO)[P(C_6H_5)_3]_3$, and $Ni[P(C_6H_5)_3]_4$.

X

PHYSICOCHEMICAL STUDIES

In our preponderantly preparative work it was indispensable to use physical methods for characterizing the compounds and elucidating their constitution. The magnetic studies undertaken with J. G. Floss (106) were successful in demonstrating the polar constitution of the products of reactions between iron and manganese carbonyls with nitrogen and oxygen bases. The mono- and polynuclear carbonyl metallate anions are diamagnetic, while their compounds with paramagnetic complex cations—e.g., [Fe(en)$_3$][Fe$_2$(CO)$_8$], [Fepy$_6$][Fe$_4$(CO)$_{13}$]—show the same known magnetic moment as the cation, and thereby the formula type and the ionic structure is established. In some cases, monomeric or polynuclear structures can be differentiated on the basis of magnetic measurements. Thus, for example, the monomeric dinitrosyl iron halides Fe(NO)$_2$(L)X (L = R$_3$P, R$_3$As, amine) were the first representatives of paramagnetic nitrosyl complexes (107).

Dipole moment measurements, first carried out in our Munich Institute by E. Weiss (108), provided evidence of the very slight polar character of the metal–CO and metal–NO bonds. Similar measurements supplied information about stereochemistry: for example, of cis-I$_2$Fe(CO)$_4$ (108), (Ph$_3$P)$_3$Fe(CO)$_2$ (51), and [Fe(CO)$_3$SEt]$_2$ (90), as well as of the phosphine-containing nitrosyl carbonyl manganese compounds (105).

Of fundamental importance were the infrared spectroscopic studies initiated by Vohler and Jahn (109), and subsequently developed, especially by my former student W. Beck. The special position of the CO or NO stretching vibrations enabled the often unequivocal characterization of newly prepared compounds, for example, a decision as to whether cationic or anionic carbon monoxide, or nitrosyl groups are present (VII, 40). The CO and NO force constants are extraordinarily sensitive to changes in the CO or NO bond environment, such as occur on substitution of a CO group by another ligand. From the number and relative intensity of the carbonyl and nitrosyl stretching bands it has in many cases been possible to obtain evidence as to the stereochemistry of complexes. As examples one may mention RCo(CO)$_3$PR$_3$ (38), cis- and trans-[Co(NO)(L)SR]$_2$ (110), and [Cl(CO)RhPR$_2$]$_2$ (97). The absolute intensities of the carbonyl stretching bands of metal carbonyl complexes were first investigated by Beck (111), and today intensities are the subject of increasing attention.

XI

CONCLUSION

The above account of metal carbonyl chemistry is far from complete. Indeed this is a field which has been developed further not only by my own circle of co-workers, but also by researchers rich with ideas from outside

Germany. Here I must above all mention the work of J. Chatt, F. A. Cotton, L. F. Dahl, J. Lewis, R. S. Nyholm, F. G. A. Stone, and G. Wilkinson, and their respective students. It gives me great satisfaction that this field of research has opened up, and its positively explosive expansion is fascinating to me since it stems from the early studies described in this article. Also I am especially pleased that so many of my students now occupy Chairs of Chemistry: F. Seel (Saarbrücken), R. Nast (Hamburg), H. Behrens (Erlangen), E. O. Fischer (my successor at the Technische Hochschule, München), E. Weiss (Hamburg), W. Beck (Universität München), and T. Kruck (Cologne). It is a great joy for me as a university teacher and research worker to see so many new thoughts budding and to have helped stimulate them, for they will shape the future of this and related fields.

REFERENCES

A. *Comprehensive Reviews and General Literature Papers*

I. Mittasch, A., *Angew. Chem.* **41**, 827 (1928).
II. Pfeiffer, P., "A. Werner's Neuere Anschauungen auf dem Gebiete der Anorganischen Chemie," 5th ed., p. 267. Vieweg, Braunschweig, 1923.
III. Weinland, R., "Einführung in die Chemie der Komplexverbindungen," 2nd ed., p. 341. Enke, Stuttgart, 1924.
IV. Hieber, W., *Z. Elektrochem.* **43**, 390 (1937).
V. Hieber, W., *Angew. Chem.* **55**, 1 (1942).
VI. Hieber, W., Nast, R., and Sedlmeier, J., *Angew. Chem.* **64**, 465 (1952).
VII. Hieber, W., Beck, W., and Braun, G., *Angew. Chem.* **72**, 795 (1960); *Angew. Chem. Intern. Ed. Engl.* p. 65 (1961).
VIII. Hieber, W., Beck, W., and Zeitler, G., *Angew. Chem.* **73**, 364 (1961).

B. *Experimental Work and Reviews on Single Topics*

1. Reihlen, H., *Ann. Chem.* **465**, 72 and 83ff. (1928); **472**, 268 and 275ff. (1929); see, however, Hieber, W., Sonnekalb, F., and Becker, E. *Ber. Deut. Chem. Ges.* **63**, 977 (1930); as well as references *IV*, 393 (left column) and *V*, *1–3*.
2. Freundlich, H., Cuy, E. J., and Malchow, W., *Ber. Deut. Chem. Ges.* **56**, 2264 (1923); *Z. Anorg. Allgem. Chem.* **141**, 317 (1924).
3. Hieber, W., Sonnekalb, F., and Leutert, F., *Ber. Deut. Chem. Ges.* **61**, 558 (1928); **63**, 973 (1930); **64**, 2832 (1931).
4. Hieber, W., Sonnekalb, F., and Mühlbauer, F., *Ber. Deut. Chem. Ges.* **61**, 555 and 2149 (1928).
5. Behrens, H., Harder, N., and Anders, U., *Chem. Ber.* **97**, 426 (1964); *Z. Naturforsch.* **19b**, 767 (1964).
6. Hieber, W., and Beutner, H., *Angew. Chem.* **74**, 154 (1962); *Z. Anorg. Allgem. Chem.* **317**, 63 (1962); Edgell, W. F., Yang, M. T., Bulkin, B. J., Bayer, R., and Koizumi, N., *J. Am. Chem. Soc.* **87**, 3080 (1965); **88**, 4839 (1966); Schubert, E. H., and Sheline, R. K., *Inorg. Chem.* **5**, 1071 (1966).
7. Reppe, W., and Schweckendiek, W. J., *Ann. Chem.* **560**, 104 and 110 (1948).

8. Hieber, W., and Böckly, E., Z. Naturforsch. **5b**, 129 (1950); Z. Anorg. Allgem. Chem. **262**, 344 (1950).
9. Wilkinson, G., and Irvine, J. W., Jr., Science **113**, 742 (1951); J. Am. Chem. Soc. **73**, 5501 (1951).
10. Behrens, H., Müller, A., and Meyer, K., Z. Anorg. Allgem. Chem. **341**, 124 (1965); Z. Naturforsch. **21b**, 489 (1966); Wilke, G., Müller, E. W., and Kröner, M. Angew. Chem. **73**, 33 (1961).
11. Hieber, W., and von Pigenot, D., Chem. Ber. **89**, 193, 610, and 616 (1956).
12. Malatesta, L., Progr. Inorg. Chem. **1**, 283 (1959).
13. Hieber, W., Kroder, W., and Zahn, E., Z. Naturforsch. **15b**, 325 (1960); Hieber, W., and Ellermann, J., ibid. **18b**, 589 and 595 (1963).
14. For "Trifluorophosphin-Komplexe von Übergangsmetallen," see Kruck, T., Angew. Chem. **79**, 27 (1967).
15. Hieber, W., and Kruck, T., Chem. Ber. **95**, 2027 (1962).
16. Hieber, W., Opavsky, W., and Rohm, W., Chem. Ber. **101**, 2244 (1968).
17. Fischer, E. O., and Werner, H., "Metal-π-Complexes," Vol. 1. Elsevier, Amsterdam, 1966. In the field of π-sandwich compounds, aromatic complexes, etc., my former co-worker, E. O. Fischer, has played a major role. The subject is not, however, relevant to this review.
18. Behrens, H., and Lutz, H., Z. Anorg. Allgem. Chem. **356**, 225 (1968).
19. Hieber, W., and Schropp, W., Jr., Z. Naturforsch. **15b**, 271 (1960); Ziegler, M. L., Haas, H., and Sheline, R. K., Chem. Ber. **98**, 2454 (1965).
20. Hieber, W., Naturwissenschaften **19**, 360 (1931); Hieber, W., and Leutert, F., Ber. Deut. Chem. Ges. **64**, 2832 (1931).
21. Hieber, W., Z. Anorg. Allgem. Chem. **204**, 165 (1932).
22. Hieber, W., and Leutert, F., Z. Anorg. Allgem. Chem. **204**, 145 (1932).
23. Hieber, W., Z. Elektrochem. **40**, 158 (1934); Hieber, W., Schulten, H., and Krämer, K., Angew. Chem. **49**, 463 (1936); Z. Anorg. Allgem. Chem. **232**, 17 (1937).
24. Hieber, W., Mühlbauer, F., and Ehmann, E., Ber. Deut. Chem. Ges. **65**, 1090 (1932).
25. See, for example, LaPlaca, S. J., Hamilton, W. C., and Ibers, J. A., Inorg. Chem. **3**, 1491 (1964).
26. Beck, W., Hieber, W., and Braun, G., Z. Anorg. Allgem. Chem. **308**, 23 (1961).
27. Wilson, W. E., Z. Naturforsch. **13b**, 349 (1958); also Braterman, P. S., Harrill, R. W., and Kaesz, H. D., J. Am. Chem. Soc. **89**, 2851 (1967).
28. Hieber, W., and Lagally, H., Z. Anorg. Allgem. Chem. **245**, 321 (1940); **251**, 96 (1943); Krogmann, K., Binder, W., and Hausen, H. D., Angew. Chem. **80**, 844 (1968).
29. Calderazzo, F., and L'Eplattenier, F., Inorg. Chem. **6**, 1220 and 2092 (1967).
30. Hieber, W., and Stallmann, H., Ber. Deut. Chem. Ges. **75**, 1472 (1942); Z. Elektrochem. **49**, 288 (1943).
31. Bruce, M. I., and Stone, F. G. A., Angew. Chem. **80**, 460 (1968); J. Chem. Soc., A p. 2162 (1968).
32. Blanchard, A. A., and Windsor, M. M., J. Am. Chem. Soc. **56**, 826 (1934); Feigl, F., and Krumholz, P., Z. Anorg. Allgem. Chem. **215**, 242 (1933).
33. Hieber, W., and Hübel, W., Z. Elektrochem. **57**, 235 and 331 (1953); Reppe, W. et al., Ann. Chem. **582**, 116 (1953).
34. Huggins, D. K., Fellmann, W. P., Smith, J. M., and Kaesz, H. D., J. Am. Chem. Soc. **86**, 4841 (1964); Fischer, E. O., and Aumann, R., J. Organometal. Chem. (Amsterdam) **8**, P 1 (1967).
35. Chini, P., Colli, L., and Peraldo, M., Gazz. Chim. Ital. **90**, 1005 (1960).

36. Churchill, M. R., and Bau, R., *Inorg. Chem.* **6**, 2086 (1967).
37. Johnson, B. F. G., Johnston, R. D., Lewis, J., and Robinson, B. H., *Chem. Commun.* p. 851 (1966); *J. Organometal. Chem. (Amsterdam)* **10**, 105 (1967).
38. Hieber, W., and Lindner, E., *Chem. Ber.* **94**, 1417 (1961); Hieber, W., Duchatsch, H., and Muschi, J., *ibid.* **98**, 2933 and 3924 (1965).
39. Hieber, W., Faulhaber, G., and Theubert, F., *Z. Anorg. Allgem. Chem.* **314**, 125 (1962); Hieber, W., Höfler, M., and Muschi, J., *Chem. Ber.* **98**, 311 (1965).
40. Hieber, W., Winter, E., and Schubert, E., *Chem. Ber.* **95**, 3070 (1962).
41. Hieber, W., and Romberg, E., *Z. Anorg. Allgem. Chem.* **221**, 321 (1935).
42. Closson, R. D., Kozikowski, J., and Coffield, T. H., *J. Org. Chem.* **22**, 589 (1957).
43. Hieber, W., Beck, W., and Lindner, E., *Z. Naturforsch.* **16b**, 229 (1961); *Chem. Ber.* **95**, 2042 (1962); see especially, Treichel, P. M., and Stone, F. G. A., *Advan. Organometal. Chem.* **1**, 143 (1964).
44. Hieber, W., and Teller, U., *Z. Anorg. Allgem. Chem.* **249**, 43 (1942).
45. Hieber, W., and Beutner, H., *Z. Anorg. Allgem. Chem.* **320**, 101 (1963).
46. Hieber, W., Fischer, E. O., and Böckly, E., *Z. Anorg. Allgem. Chem.* **269**, 292 and 308 (1952); **271**, 229 (1953).
47. Hieber, W., and Klingshirn, W., *Z. Anorg. Allgem. Chem.* **323**, 292 (1963).
48. Hieber, W., Beck, W., and Nitzschmann, R., unpublished data (1963); Nitzschmann, R., Dissertation, Technische Hochschule, München (1964).
49. Hieber, W., Beutner, H., and Schubert, E., *Z. Naturforsch.* **17b**, 211 (1962); *Z. Anorg. Allgem. Chem.* **338**, 32 and 37 (1965).
50. Hieber, W., and Ellermann, J., *Z. Naturforsch.* **18b**, 595 (1963).
51. Hieber, W., and Muschi, J., *Chem. Ber.* **98**, 3931 (1965).
52. Hieber, W., and Heinicke, K., *Z. Anorg. Allgem. Chem.* **316**, 305 (1967).
53. Fischer, E. O., Fichtel, K., and Öfele, K., *Chem. Ber.* **94**, 1200 (1961).
54. Hieber, W., and Kruck, T., *Angew. Chem.* **73**, 580 (1961); *Z. Naturforsch.* **16b**, 709 (1961); Kruck, T., and Noack, M., *Chem. Ber.* **96**, 3028 (1963); **97**, 1693 (1964); **99**, 1153 (1966); Hieber, W., and Duchatsch, H., *ibid.* **98**, 1744 (1965).
55. Hieber, W., Frey, V., and John, P., *Chem. Ber.* **100**, 1961 (1967).
56. Kruck, T., Höfler, M., Baur, K., Junkes, P., and Glinka, K., *Chem. Ber.* **101**, 3827 (1968).
57. Hieber, W., and Bader, G., *Ber. Deut. Chem. Ges.* **61**, 1717 (1928); *Z. Anorg. Allgem. Chem.* **190**, 193 and 215 (1930).
58. Wojcicki, A., and Basolo, F., *J. Am. Chem. Soc.* **83**, 525 (1961); see also Noack, K., *J. Organometal. Chem. (Amsterdam)* **13**, 411 (1968).
59. Cotton, F. A., and Johnson, B. F. G., *Inorg. Chem.* **6**, 2113 (1967).
60. Hieber, W., and Thalhofer, A., *Angew. Chem.* **68**, 679 (1956).
61. Hieber, W., Appel, H., Woerner, A., and Levy, E., *Z. Elektrochem.* **40**, 262, 287, and 291 (1934).
62. Schützenberger, P., *Compt. Rend.* **70**, 1134 and 1287 (1870); also references quoted in Hieber and Bader (*57*).
63. Hieber, W., Lagally, H., and Wirsching, A., *Z. Anorg. Allgem. Chem.* **245**, 35, 295, and 305 (1940).
64. Pankowski, M., and Bigorgne, M., *Compt. Rend.* **C264**, 1382 (1967).
65. Hieber, W., and Duchatsch, H., *Chem. Ber.* **98**, 2530 (1965); Hieber, W., and Lindner, E., *ibid.* **95**, 273 (1962).
66. Hieber, W., Schulten, H., Schuh, R., and Fuchs, H., *Z. Anorg. Allgem. Chem.* **243**, 164 (1939); **248**, 243 (1941).

67. Hieber, W., and Wagner, G., *Z. Naturforsch.* **12b**, 478 (1957); Abel, E. W., and Wilkinson, G., *J. Chem. Soc.* p. 1501 (1959).
68. Hieber, W., Lux, F., and Herget, C., *Z. Naturforsch.* **20b**, 1159 (1965).
69. Abel, E. W., Hargreaves, G. B., and Wilkinson, G., *J. Chem. Soc.* p. 3149 (1958); p. 1501 (1959).
70. Manchot, W., and König, J., *Ber. Deut. Chem. Ges.* **57**, 2130 (1924); Manchot, W., and Manchot, W. J., *Z. Anorg. Allgem. Chem.* **226**, 385 (1936).
71. Behrens, H., Zizlsperger, H., and Schwab, R., *Z. Naturforsch.* **16b**, 349 (1961); **19b**, 768 (1964).
72. Colton, R., Scollary, G. R., Tomkins, J. B., and Rix, C. J., *Australian J. Chem.* **19**, 1519 (1966); **21**, 15, 1159, and 1427 (1968).
73. Hieber, W., and Schropp, W., Jr., *Z. Naturforsch.* **14b**, 460 (1959); Hieber, W., Fuchs, H., and Schuster, L., *Z. Anorg. Allgem. Chem.* **248**, 269 (1941); **287**, 214 (1956).
74. Hieber, W., and Frey, V., *Chem. Ber.* **99**, 2607 (1966).
75. Hieber, W., Behrens, H., and Teller, U., *Z. Anorg. Allgem. Chem.* **249**, 26 (1942).
76. Hieber, W., and Geisenberger, O., *Z. Anorg. Allgem. Chem.* **262**, 15 (1950).
77. Hieber, W., and Fuchs, H., *Z. Anorg. Allgem. Chem.* **248**, 256 (1941).
78. Hieber, W., and Fischer, H., Deutsches Reichspatent 695589 (1940); *Chem. Abstr.* **35**, 5657 (1941).
79. Corey, E. R., Dahl, L. F., and Beck, W., *J. Am. Chem. Soc.* **85**, 1202 (1963); Dahl, L. F., Martell, C., and Wampler, D. L., *ibid.* **83**, 1761 (1961); Corey, E. R., and Dahl, L. F., *Inorg. Chem.* **1**, 521 (1962).
80. Wei, C. H., and Dahl, L. F., *J. Am. Chem. Soc.* **91**, 1351 (1969).
81. Manchot, W., and Gall, H., *Ber. Deut. Chem. Ges.* **62**, 678 (1929).
82. Blanchard, A. A., and Gilmont, P., *J. Am. Chem. Soc.* **62**, 1192 (1940).
83. Hieber, W., Nast, R., and Bartenstein, C., *Z. Anorg. Allgem. Chem.* **272**, 32 (1953); **276**, 1 and 12 (1954), and references cited therein; Nast, R., von Krakkay, T., and Roos, H., *ibid.* **272**, 234 and 242 (1953).
84. Fackler, J. P., Jr., and Coucouvanis, D., *J. Am. Chem. Soc.* **89**, 1745 (1967); **90**, 2784 (1968).
85. Reppe, W. *et al.*, *Ann. Chem.* **582**, 116 (1953).
86. Behrens, H., Lindner, E., and Schindler, H., *Chem. Ber.* **99**, 2399 (1966).
87. Hieber, W., Ries, K., and Bader, G., *Z. Anorg. Allgem. Chem.* **190**, 215 (1930).
88. Hieber, W., Nast, R., Floss, J. G., and Vohler, O., *Z. Anorg. Allgem. Chem.* **283**, 188 (1956); **294**, 219 (1958).
89. Hieber, W., and Gruber, J., *Z. Anorg. Allgem. Chem.* **296**, 91 (1958).
90. Hieber, W., and Beck, W., *Z. Anorg. Allgem. Chem.* **305**, 265 (1960).
91. Dahl, L. F., and Wei, C. H., *Inorg. Chem.* **2**, 328 (1963).
92. See, for example, Klumpp, E., Bor, G., and Markó, L., *J. Organometal. Chem.* (*Amsterdam*) **11**, 207 (1968).
93. Hieber, W., and Gscheidmeier, M., *Chem. Ber.* **99**, 2313 (1966); *Z. Naturforsch.* **21b**, 1237 (1966).
94. Hieber, W., and Rohm, W., *Chem. Ber.* **102**, 2787 (1969).
95. Hieber, W., and Beutner, H., *Z. Naturforsch.* **15b**, 324 (1960); *Z. Anorg. Allgem. Chem.* **317**, 63 (1962); Frey, V., Hieber, W., and Mills, O. S., *Z. Naturforsch.* **23b**, 105 (1968); Dahl, L. F., Costello, W. R., and King, R. B., *J. Am. Chem. Soc.* **90**, 5422 (1968).

96. For a survey, see Hayter, R. G., *in* "Preparative Inorganic Reactions," Vol. 2, p. 211. Wiley (Interscience), New York (1965).

97. Hieber, W., and Kummer, R., *Z. Naturforsch.* **20b**, 271 (1965); *Chem. Ber.* **100**, 148 (1967); Hieber, W., and Opavsky, W., *ibid.* **101**, 2966 (1968).

98. Anderson, J. S., *Z. Anorg. Allgem. Chem.* **208**, 238 (1932).

99. Barraclough, C. G., and Lewis, J., *Proc. Chem. Soc.* p. 82 (1960).

100. Hieber, W., and Anderson, J. S., *Z. Anorg. Allgem. Chem.* **211**, 132 (1933).

101. Seel, F., *Z. Anorg. Allgem. Chem.* **269**, 40 (1952).

102. Seel, F., *Z. Anorg. Allgem. Chem.* **249**, 308 (1942).

103. Hieber, W., Beutner, H., and Ellermann, J., *Chem. Ber.* **96**, 1659 and 1667 (1963).

104. Nast, R., Pröschel, E., and Gehring, G., *Z. Anorg. Allgem. Chem.* **256**, 145, 159 and 169 (1948).

105. Hieber, W., Beck, W., and Tengler, H., *Z. Naturforsch.* **15b**, 411 (1960); *Z. Anorg. Allgem. Chem.* **318**, 136 (1962).

106. Hieber, W., and Floss, J. G., *Z. Anorg. Allgem. Chem.* **291**, 314 (1957).

107. Hieber, W., and Kramolowsky, R., *Z. Anorg. Allgem. Chem.* **321**, 94 (1963).

108. Weiss, E., *Z. Anorg. Allgem. Chem.* **287**, 223 (1956).

109. Vohler, O., *Chem. Ber.* **91**, 1161 and 1235 (1958); Jahn, A., *Z. Anorg. Allgem. Chem.* **301**, 301 (1959).

110. Beck, W., and Lottes, K., *Z. Anorg. Allgem. Chem.* **335**, 258 (1965); *Chem. Ber.* **98**, 2657 (1965).

111. Beck, W., Nitzschmann, R., Melnikoff, A., and Stahl, R., *Z. Naturforsch.* **17b**, 577 (1962); *Chem. Ber.* **99**, 3721 (1966).

π-Allylnickel Intermediates in Organic Synthesis

P. HEIMBACH, P. W. JOLLY, and G. WILKE

Max-Planck-Institut für Kohlenforschung,
Mülheim-Ruhr, W. Germany

I

INTRODUCTION

It is 5 years since a review devoted to the organometallic chemistry of nickel was published in this series (*1*). In this time much attention has been given to the isolation of the intermediates involved in the catalytic reactions and to the mechanisms of these processes. Notable advances have also been made in using nickel complexes in stoichiometric organic synthesis. The elegance of many of these reactions and the complexities introduced by, what seem at first, minor variations will undoubtedly maintain interest in the use of nickel in organic synthesis.

We have restricted ourselves to discussing stoichiometric and catalytic reactions which clearly involve π-allylnickel complexes and have not considered polymerization reactions, the Reppe synthesis, or "template" reactions. Fortunately there is an abundance of books and reviews which cover these fields as well as the similarities, and dissimilarities, with the other transition metals (*1–8*). The literature up to the end of 1968 has been surveyed but no attempt has been made to include all the material available.

List of Abbreviations

The following abbreviations are used in the text:

THF	Tetrahydrofuran
DMF	Dimethylformamide
BD	1,3-Butadiene
VCH	4-Vinylcyclohexene-1
COD	*cis,cis*-1,5-Cyclooctadiene
DVCB	*cis*-1,2-Divinylcyclobutane
CDT	The sum of all-*trans*-; *trans,trans,cis*-; and *trans,cis,cis*-1,5,9-cyclododecatriene
DT	1,*trans*-4,9-Decatriene
DMCDeT	4,5-Dimethyl-*cis,cis,trans*-1,4,7-cyclodecatriene
CDD	*cis,trans*-1,5-Cyclodecadiene

II

COUPLING REACTIONS

The reaction of allylic halides with nickel tetracarbonyl to form coupled products has been known for over two decades (*9*), but it is only in recent years that an insight into the mechanism has been obtained. Isolation of the intermediate π-allylnickel complexes and the discovery that these react with activated olefins and organic halides in general have led to a considerable increase in the scope of the reaction.

Table I shows the types of reaction that have been reported and is by no means exhaustive.

Understanding of the mechanism of the apparently simple coupling of two allylic groups in the presence of nickel carbonyl is largely due to the investigations of E. J. Corey and his co-workers (*10*). The first step is formation of a π-allylnickel carbonyl halide (I). Corey suggests that (I)

$$CH_2:CHCH_2Br + Ni(CO)_4 \ \rightleftharpoons \ \pi\text{-}C_3H_5NiBr(CO) + 3\ CO \qquad (1)$$

$$(I)$$

TABLE I: COUPLING REACTIONS

Reactants	Solvent	Product (%)	Reference
$CH_2:CHCH_2Cl + Ni(CO)_4$	—	$CH_2:CHCH_2CH_2CH:CH_2$	9
$CH_2:CHCH_2Br + [\pi\text{-}C_3H_5NiBr]_2$	THF, Monoglyme	$CH_2:CHCH_2CH_2CH:CH_2$	10
$CH_3CH:CHCH_2Cl$ or $CH_2:CHC(CH_3)HCl$ $+ Ni(CO)_4$	MeOH	$CH_3CH:CHCH_2CH_2CH:CHCH_3$ $+$ $CH_2:CHCH(CH_3)CH_2CH:CHCH_3$	30
$C_6H_5CH:CHCH_2OAc + Ni(CO)_4$	THF	$C_6H_5CH:CHCH_2CH_2CH:CHC_6H_5$ (31)	31
$CH_3CO_2CH:CHCH_2Br + Ni(CO)_4$	Ether	$CH_3O_2CCH:CHCH_2CH_2CH:CHCO_2CH_3$	32
$BrCH_2CH:CH(CH_2)_nCH:CHCH_2Br + Ni(CO)_4$ ($n = 6, 8, 12$)	DMF	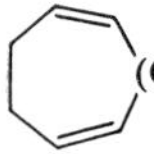 $(CH_2)_n$ (60–80)	16
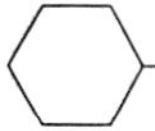 $-I + [\pi\text{-Methallyl NiBr}]_2$	DMF	 $CH_2\text{-}\underset{CH_3}{C}:CH_2$ (91)	22
$Br-\!\!\!-\!\!\!-Br + [\pi\text{-Methallyl NiBr}]_2$ 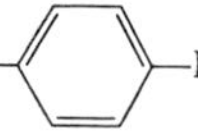	DMF	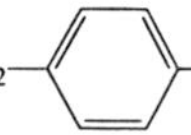 (97)	22
$Cl(CH_2)_3I + \left[EtO_2C\text{-}\!\!\!-\!\!\!-NiBr \right]_2$ 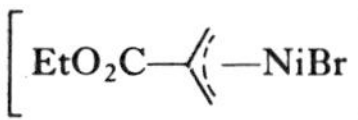	DMF	$CH_2:\underset{CH_3}{C}(CH_2)_4Cl$ CO_2Et	22
$CH_3I + [\pi\text{-}\alpha\alpha'\text{-Dimethallyl NiBr}]_2$	DMF	$CH_3C:CHCH_2CH_3$ (90)	22
$CH_2:CHCHO + [\pi\text{-Methallyl NiBr}]_2$	DMF	$CH_2:\underset{CH_3}{C}CH_2\underset{OH}{C}HCH:CH_2$	22
$CH_2:CHCO_2CH_3 + [\pi\text{-}C_3H_5NiBr]_2$	C_6H_6	$CH_2:CHCH_2CH:CHCO_2CH_3$ $+$ $CH_2:CHCH_2CH_2CH_2CO_2CH_3$	23

reacts with a second molecule of allyl bromide to form an intermediate (II), which then decomposes to liberate bis(allyl).

$$\pi\text{-}C_3H_5NiBr(CO) + CH_2{:}CHCH_2Br \;\rightleftharpoons\; \begin{bmatrix} C_3H_5NiBr(CO) \\ \uparrow \\ C_3H_5Br \\ (II) \end{bmatrix} \tag{2}$$

$$\downarrow$$

$$C_6H_{10} + NiBr_2 + CO$$

Compound (I) readily loses CO to form the known dimer of π-allylnickel bromide.

$$2\,\pi\text{-}C_3H_5NiBr(CO) \;\rightleftharpoons\; [\pi\text{-}C_3H_5NiBr]_2 + 2\,CO \tag{3}$$

The NMR spectrum of the π-methallyl complex (III) similar to (I) has been reported (10), while the complex formed by stabilization with a phosphine ligand (IV) has been known for some years (11, 12). The disposi-

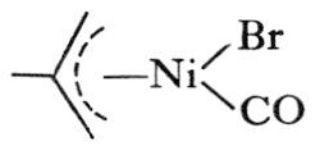

(III) 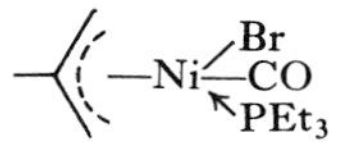 (IV)

tion of the two allyl groups in (II) is probably dependent on the polarity of the solvent used and may vary with the substituent on the allyl group. It is attractive to suppose that the second allyl bromide molecule converts the π-allyl to a σ-bonded group. The influence of solvents on such a π-allyl–σ-allyl equilibrium is well documented and a nickel complex (V) which probably contains both forms has been isolated from the reaction of bis(methallyl)nickel and triethylphosphine (13).

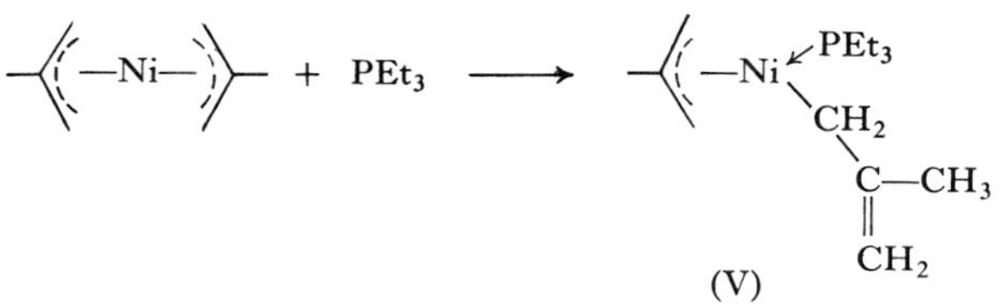

(V) (4)

Furthermore, it has recently been demonstrated (14) that the pentaco-ordinate π-allyl complex similar to (IV) reacts with allyl bromide in methanol to form bis(allyl) and methyl 3-butenoate. This is consistent with coordination of the reacting molecule of allyl bromide and conversion of the π-allyl to a σ-allyl group, which can then undergo either coupling or carbonylation. The carbonylation of π-allyl complexes is discussed in Section III.

The coupling of an allylic halide which can exist as a cis or trans form is, in many cases, stereochemically nonspecific. This interconversion of the allyl groups may be explained by a 1,3-rearrangement through a σ-allyl-nickel complex.

An alternative mechanism for the coupling reaction involving a bis(π-allyl)nickel species must be considered, particularly since it has been shown that a facile equilibrium exists between π-allylnickel bromide and

$$[\pi\text{-}C_3H_5NiBr]_2 \rightleftharpoons \left\langle -Ni- \right\rangle + NiBr_2 \qquad (5)$$

bis(π-allyl)nickel (*15, 18*). It is known that bis(allyl)nickel systems react with carbon monoxide to produce coupling products and nickel tetra-carbonyl (*13*). Bis(π-allyl)nickel itself absorbs one equivalent of CO at $-78°$ C and a further three above $-40°$ C to form 1,5-hexadiene. Similarly, bis(π-methallyl)nickel forms an unstable 1:1 adduct with CO at $-78°$C which takes up further CO at higher temperatures with displacement of 2,5-dimethyl-1,5-hexadiene. A minor side reaction is the insertion of a CO molecule, leading to the formation of the ketone, 2,6-dimethyl-2,5-heptadiene-4-one. Bis(π-crotyl)nickel absorbs 4 moles of CO at room temperature to give the same products as are observed from the reaction of crotyl chloride with nickel carbonyl, i.e., 2,6-octadiene (VI) (38%), 3-methyl-1,5-heptadiene (VII) (38%), and only traces of the third possible

$$2\left\langle -Ni- \right\rangle + 8\,CO \xrightarrow{-2\,Ni(CO)_4} \begin{array}{c} CH_3CH:CHCH_2CH_2CH:CHCH_3 \\ (VI) \\ + CH_3CH:CHCH_2CH(CH_3)CH:CH_2 \\ (VII) \end{array} \qquad (6)$$

isomer, 3,4-dimethyl-1,5-hexadiene. Interestingly, at lower temperature ($-40°$ C) the reaction with CO produces *trans,trans*-2,6-octadiene stereo-specifically.

In the case of cyclooctenyl bromide, a reaction proceeding exclusively through a bis(π-allyl)nickel complex can be excluded, since reaction of bis(π-cyclooctenyl)nickel in dimethylformamide with carbon monoxide gives a 1:1 mixture of bis(2-cyclooctenyl) ketone (VIII) and bis(cyclo-octenyl) (IX), whereas reaction of π-cyclooctenylnickel bromide in the same solvent with CO or of cyclooctenyl bromide with nickel tetracarbonyl

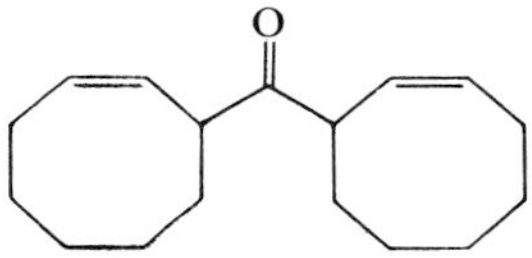

(VIII) 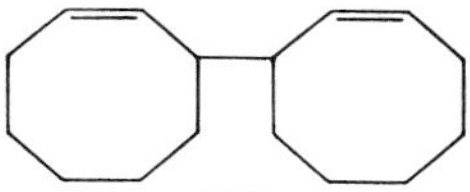   (IX)

produces only bis(cyclooctenyl) (10, 13). In contrast, π-cyclooctenylnickel chloride in pentane absorbs 5 moles of CO at atmospheric pressure to form an acid chloride which, since it readily eliminates HCl, was characterized as its methyl ester (13).

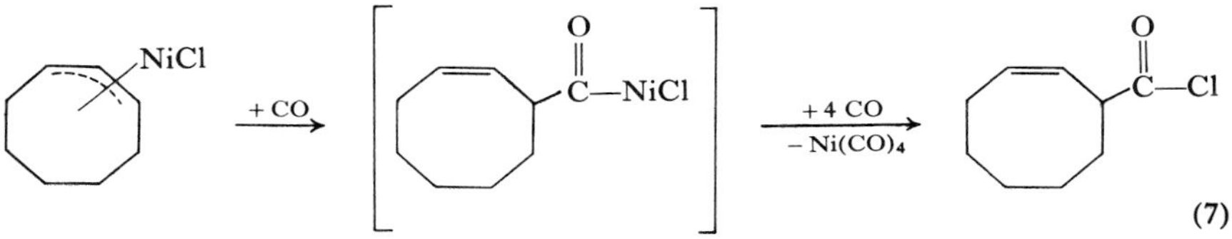

$$\tag{7}$$

The intramolecular coupling reactions of $BrCH_2CH:CH(CH_2)_nCH:CHCH_2Br$, where $n = 6$, 8, and 12 (16), are shown in Table I. When $n = 2$, instead of the expected 1,5-cyclooctadiene, the main product is vinylcyclo-hexene, and when $n = 4$, instead of cyclodecadiene, the product is a mixture of cis- and trans-divinylcyclohexane

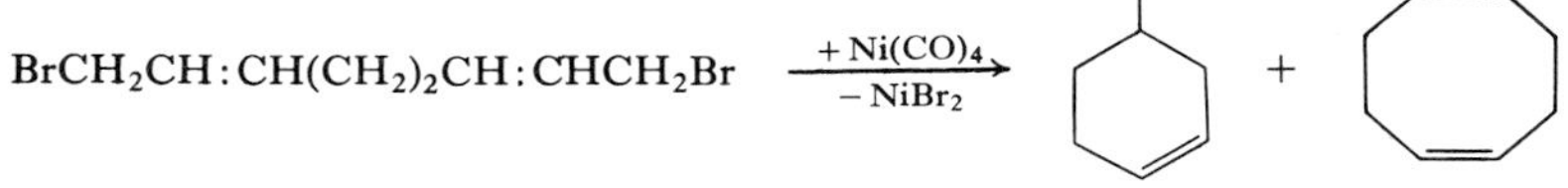

$$BrCH_2CH:CH(CH_2)_2CH:CHCH_2Br \xrightarrow[-NiBr_2]{+Ni(CO)_4}$$

$$\tag{8}$$

(42%)   (5%)

$$BrCH_2CH:CH(CH_2)_4CH:CHCH_2Br \xrightarrow[-NiBr_2]{+Ni(CO)_4}$$

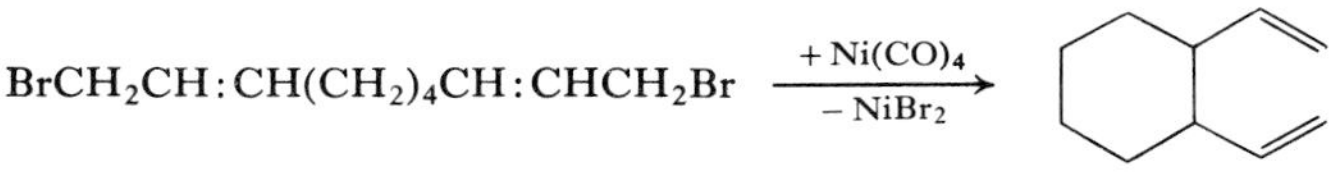

$$\tag{9}$$

The last two examples strongly suggest the possibility that these cyclization reactions proceed through bis(allyl)nickel intermediates,[1] particularly since

[1] The nature of this species may be different in each case. In addition to the σ-allyl–π-allyl equilibrium, the various possible configurations of the bis(allyl) system need to be considered (see Section IV,B).

it has been shown that the bis(π-octadienyl)nickel species (X) reacts with carbon monoxide to produce vinylcyclohexene in over 80% yield (*17*).

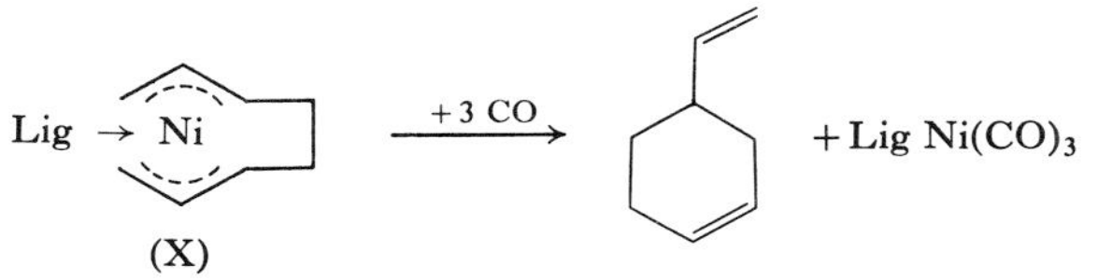

$$\text{Lig} \rightarrow \text{Ni} \quad \xrightarrow{\;+3\;CO\;} \quad \text{vinylcyclohexene} \quad + \text{Lig Ni(CO)}_3 \qquad (10)$$

(X)

It is interesting that the cyclization reactions proceed most efficiently in coordinating solvents (e.g., DMF), and it is possible that the long-chain dibromide first reacts with the nickel tetracarbonyl to form a π-allylnickel halide complex which under the influence of the solvent disproportionates to give a bis(allyl) form stabilized by a solvent molecule, i.e., in (X), Lig = DMF. This species then reacts with carbon monoxide with ring closure. A second role for the solvent molecule may well be in complexing the nickel bromide formed in the disproportionation, thus displacing the equilibrium completely to the right. For example, the reaction of π-allyl-nickel bromide in liquid ammonia at −78°C gives a quantitative yield of bis(π-allyl)nickel which may be sublimed out of the reaction mixture, leaving the hexammoniate of nickel bromide (*18*).

$$[\pi\text{-}C_3H_5NiBr]_2 + 6\,NH_3 \;\rightarrow\; \left(\!\!\left(\!-Ni-\right)\!\!\right) \;+ NiBr_2 \cdot 6\,NH_3 \qquad (11)$$

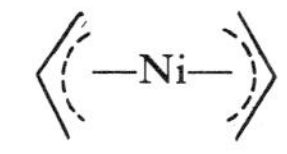

Intramolecular coupling with nickel tetracarbonyl has found application in syntheses of 4,5-*cis*-humulene (XI) (*19*), 1,6-dimethyl-1,5,9- cyclodo-decatriene (XII) (*20*), and 1,4,7-trimethylenecyclononane (XIII) (*21*) from

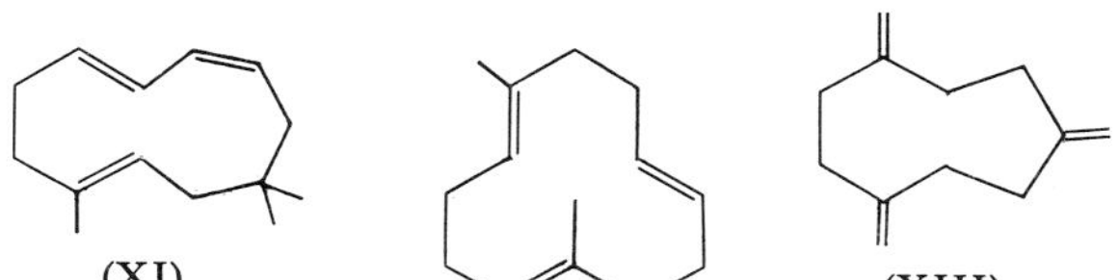

(XI)   (XIII)

(XII)

the appropriate bis(allyl) halides. The catalytic implications of these cyclization processes are developed more fully in Section IV.

The discovery that, in strongly polar media, π-allylnickel halides react with a wide variety of organic halides has introduced what promises to be a most useful method for the selective combination of unlike groups (*22*)

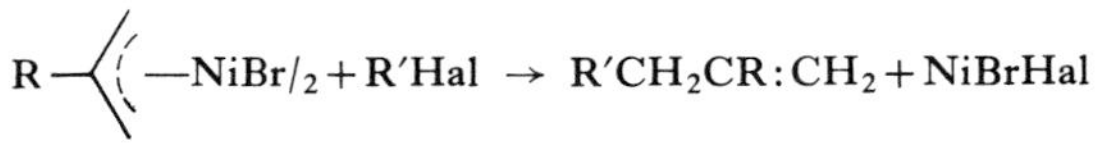

R
—NiBr/2 + R′Hal → R′CH₂CR:CH₂ + NiBrHal

(12)

(see Table I). Dihalides undergo disubstitution, and coupling occurs at the primary rather than at the more substituted end of the allyl group. This last observation has been used in a direct synthesis of α-santalene (XIV) and

I
+
—NiBr/2
− NiBrI
(XIV)
I
(XV)

(13)

epi-β-santalene (XV). The mechanism of this type of reaction has not been investigated but the scheme shown in Eq. (14) has been suggested. This

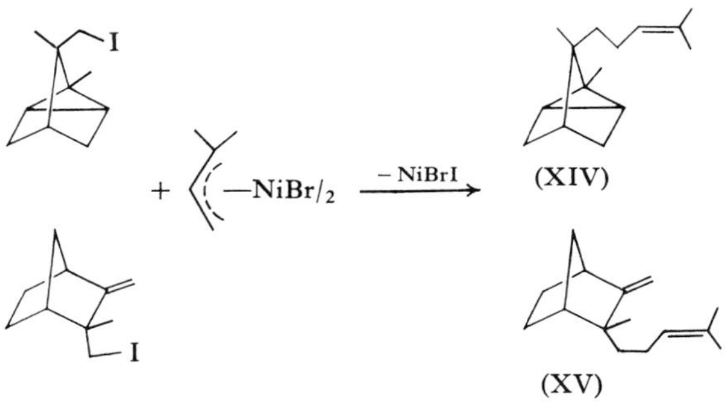

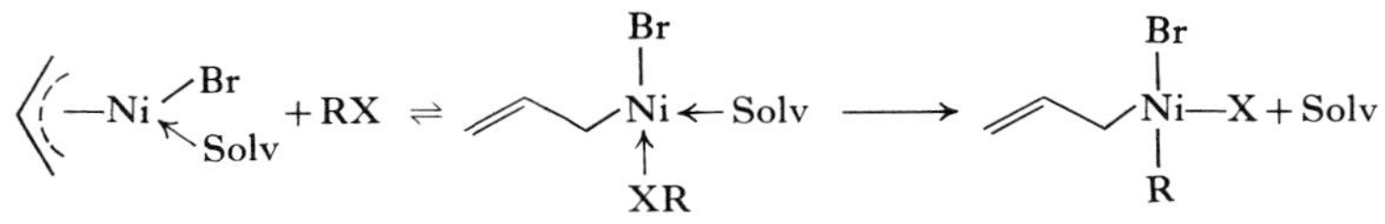

—Ni
Br
Solv
+ RX
Br
Ni ← Solv
XR
Br
Ni—X + Solv
R

(14)

reaction is not limited to organic halides; in addition aldehydes, ketones, or epoxides react with π-allylnickel halides to give alcohols (*22*), while activated olefins undergo coupling with the allyl group (*23*).

From the reaction of acrylonitrile with π-allylnickel bromide an intermediate, $(CH_2{:}CHCH_2CH_2CHCN)_2Ni$, may be isolated which is believed to have a polymeric structure (*24*). This complex is easily decomposed either by water or acids to give 5-hexenonitrile, or, thermally, to a mixture of 2,5-hexadienonitrile and 5-hexenonitrile.

α-Bromo ketones react with nickel tetracarbonyl in dimethylformamide to produce β-epoxy ketones [Eq. (15)], which upon heating eliminate
a molecule of water to give 2,4-disubstituted furans (25). Phenacyl

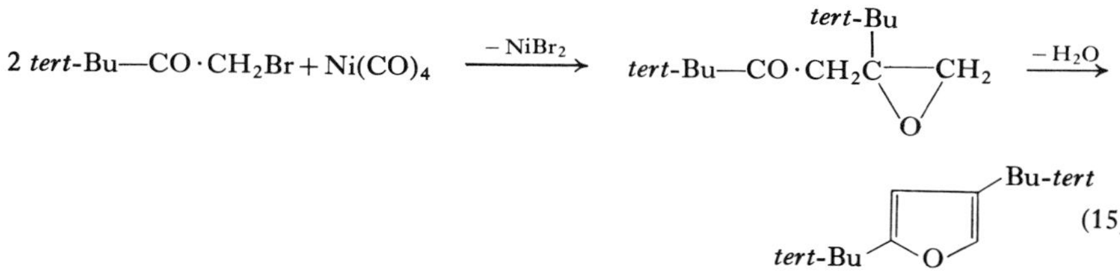

$$2 \text{ } tert\text{-Bu}-\text{CO}\cdot\text{CH}_2\text{Br} + \text{Ni(CO)}_4 \xrightarrow{-\text{NiBr}_2} tert\text{-Bu}-\text{CO}\cdot\text{CH}_2\overset{\overset{\displaystyle tert\text{-Bu}}{|}}{\text{C}}\underset{\text{O}}{\diagdown\diagup}\text{CH}_2 \xrightarrow{-\text{H}_2\text{O}}$$

(15)

bromide in tetrahydrofuran affords the coupled product 1,2-dibenzoyl-
ethane. However, in DMF the product is exclusively 2,4-diphenyl-
furan (26).

Aromatic halides are reported to give only carbonylated products with
nickel tetracarbonyl. In contrast, pentafluorophenyl iodide in DMF gives
decafluorobiphenyl in 70% yield (27). From the other products obtained
(pentafluorobenzene, decafluorobenzophenone) it has been suggested that
a radical mechanism is involved. The reactions of benzyl halides with
nickel carbonyl in various solvents have been reported (28). The main
reaction involves carbonylation, as discussed in Section III. Using benzene
as solvent, a 33% yield of bibenzyl may be obtained. Here again a mechanism involving a π-allylnickel derivative should perhaps be considered,
particularly since such a system is known to exist in (XVI) (29).

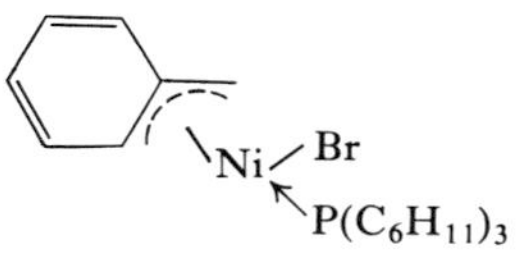

$$\text{Ni}\diagup^{\text{Br}}_{\diagdown\text{P(C}_6\text{H}_{11})_3}$$

(XVI)

This is an appropriate point to mention a method for preparing long-
chain conjugated polyenes that has recently been developed at this Institute
(33). When butadiene is bubbled through a suspension of bis(cyclooctadiene)nickel in acetone a red coloration is immediately produced. Continuous stirring for several hours produces a green gel from which may be
distilled in high yield the coupling product of two molecules of acetone and
one of butadiene, 2,7-dimethyl-trans-octene-4-diol-2,7 (XVII). The

reaction takes a similar course with aldehydes, with the difference that water must be added to decompose the intermediate nickelate. The formation of two molecules of mesityl oxide furnishes the necessary molecules of water in the example described above.

$$(COD)_2Ni + 6\ CH_3COCH_3 + CH_2:CHCH:CH_2 \longrightarrow \tag{16}$$

$$2\ CH_3COCH:C(CH_3)_2 + 2\ COD + CH_3-\overset{\overset{\displaystyle CH_3}{|}}{\underset{\underset{\displaystyle OH}{|}}{C}}-CH_2-\overset{\displaystyle H}{\underset{\displaystyle H}{C}}{=}C-CH_2-\overset{\overset{\displaystyle CH_3}{|}}{\underset{\underset{\displaystyle OH}{|}}{C}}-CH_3 + Ni(OH)_2$$

(XVII)

This reaction seems to be general (Table II), and the diols formed may be easily dehydrated to give the corresponding long-chain conjugated polyene. Substituted butadienes react similarly.

$$2\ C_6H_5CHO + CH_2:C(CH_3)C(CH_3):CH_2 + Ni(COD)_2 + 2\ H_2O$$

$$\rightarrow\ 2\ COD + Ni(OH)_2 + C_6H_5CHOHCH_2C(CH_3):C(CH_3)CH_2CHOHC_6H_5 \tag{17}$$

In certain cases the intermediate complexes involved have been isolated. The reaction of cinnamaldehyde with bis(cyclooctadiene)nickel is typical; in benzene a bis(cinnamaldehyde)nickel derivative may be isolated in which the bonding is believed to be similar to that in bis(acrolein)nickel.

$$(COD)_2Ni + 2\ C_6H_5CH:CHCHO \rightarrow 2\ COD + (C_6H_5CH:CHCHO)_2Ni \tag{18}$$

This intermediate reacts with butadiene to form a paramagnetic compound

$$(C_6H_5CH:CHCHO)_2Ni + CH_2:CHCH:CH_2 \longrightarrow C_6H_5CH:CHCH\underset{\underset{\displaystyle O}{\diagdown}}{\diagup}\underset{Ni}{\overset{\overset{\displaystyle H}{\underset{\displaystyle CH_2C=CCH_2}{\diagup}}}{\cdots}}\underset{\underset{\displaystyle O}{\diagup}}{\overset{\displaystyle H}{\diagdown}}HCCH:CHC_6H_5$$

(XVIII)   (19)

(XVIII) in a yield of over 90%. Spectroscopic evidence leads to the proposal that the structure is as shown. Compound (XVIII) reacts immediately with water to form 1,10-diphenyl-1,5,9-decatrienediol-3,8. Intermediates similar to (XVIII) have been isolated for most of the examples shown in Table II.

TABLE II

REACTION OF $(COD)_2Ni$ AND BUTADIENE WITH ALDEHYDES AND KETONES[a]

Reactant	Product, RCH:CHR (R=)	Yield (%)
CH_3COCH_3	$(CH_3)_2COHCH_2—$	60
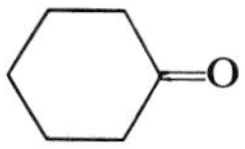		70
$C_6H_5C:O$ with CH_3	$C_6H_5—C(CH_3)(OH)—CH_2—$	68
C_6H_5CHO	$C_6H_5CHOHCH_2—$	72
$C_6H_5CH:CHCHO$	$C_6H_5CH:CHCHOHCH_2—$	51
$C_6H_5CH:CHCH:CHCHO$	$C_6H_5CH:CHCH:CHCHOHCH_2—$	45
Vitamin-A aldehyde	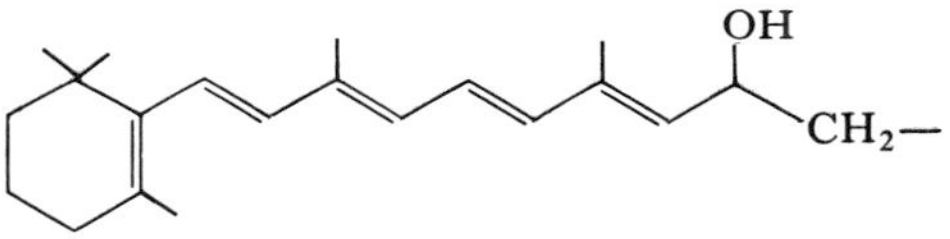	15

[a] de Ortueta Spiegelberg (33).

Another possible intermediate, a π-allylnickelate, is obtained by reacting

$$Ni(COD)_2 + R_2CO + CH_2:CHCH:CH_2 \longrightarrow \text{[structure]} \quad (20)$$

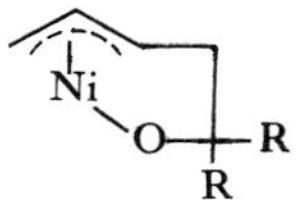

biscyclooctadiene nickel in the presence of butadiene and a carbonyl compound. Bis(π-allyl)nickel reacts similarly. With benzaldehyde, (XIX) is formed, which on hydrolysis gives 4-phenylbut-1-ene,4-ol.

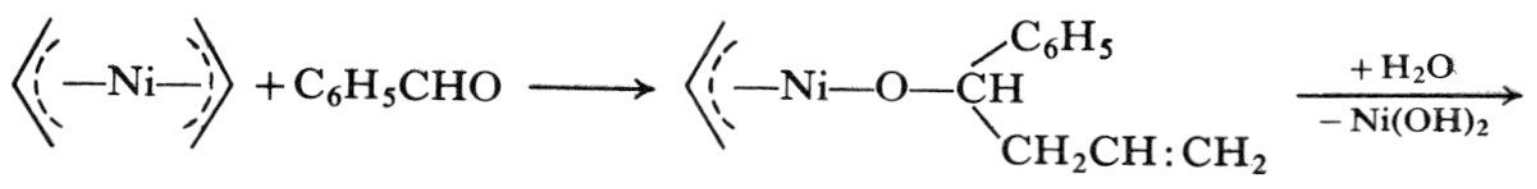

$$\text{[allyl-Ni-allyl]} + C_6H_5CHO \longrightarrow \text{[allyl-Ni-O-CH(C_6H_5)CH_2CH:CH_2]} \xrightarrow[-Ni(OH)_2]{+H_2O}$$

$$\text{(XIX)} \qquad C_6H_5CHOHCH_2CH:CH_2 \qquad (21)$$

III

INSERTION REACTIONS

As discussed in Section I, the reaction of allylic halides with nickel carbonyl at atmospheric pressure leads to coupling products or in some cases, in hydroxylic solvents, to substitutive hydrogenation (34). Under

$$HO_2CCH:CHCH_2Cl + Ni(CO)_4 \xrightarrow{H_2O} HO_2CCH_2CH:CH_2 + NiCl(OH) + 4\ CO \quad (22)$$

slight pressure of carbon monoxide (2–3 atm) the reaction takes a different course (35–38), and insertion of CO occurs, to give unsaturated carboxylic

$$RCH_2:CHCH_2Cl + CO \xrightarrow[R'OH]{Ni(CO)_4} RCH:CHCH_2CO_2R' \quad (23)$$

acid derivatives which are generally isolated as their esters.[2] At higher CO pressure (approx. 300 atm) this reaction practically ceases, presumably because of the increased stability of nickel tetracarbonyl at higher pressures.

The principal investigators in this field (Chiusoli and co-workers) have recently reviewed their contribution (39), and we will here only outline the course of reaction and discuss the most recent developments.

Two groups of workers (40, 41) have demonstrated that the reaction proceeds through the formation of a π-allylnickel intermediate which absorbs CO to form a nickel acyl complex. This then liberates a molecule of acyl halide which is hydrolyzed by the solvent. The presence of the intermediate nickel acyl complex in solution has been demonstrated

$$\tfrac{1}{2}\,[\pi\text{-}C_3H_5NiX]_2 + 2\ CO \rightleftharpoons \pi\text{-}C_3H_5Ni(CO)_2X \underset{-CO}{\overset{+CO}{\rightleftharpoons}} CH_2:CHCH_2Ni(CO)_3X$$
$$(XX) \qquad\qquad\qquad +CO \Big\Updownarrow -CO$$

$$(24)$$

$$CH_2:CHCH_2COX + Ni(CO)_4 \underset{-CO}{\overset{+CO}{\rightleftharpoons}} CH_2:CHCH_2CONi(CO)_3X$$

[2] The complexity of the reactions described in this section does not allow meaningful balanced equations to be written. In the majority of cases the equations are quoted directly from the original publication.

spectroscopically, and the acyl halide formed may be isolated if the reaction is carried out in an inert solvent (36), e.g.,

$$CNCH_2CH:CHCH_2Cl + Ni(CO)_4 \xrightarrow{C_6H_6} CNCH_2CH:CHCH_2COCl \quad (25)$$

The formation of the pentacoordinate species (XX) is also supported by the reaction of the triphenylphosphine derivative (XXI) with CO in methanol to give, besides allyl methyl ether, methyl butenoate (14).

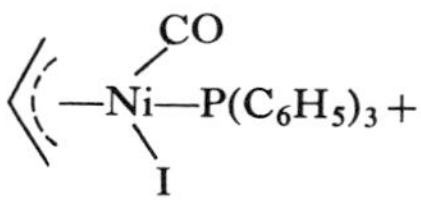

$$\to CH_2:CHCH_2OCH_3 + CH_2:CHCH_2CO_2CH_3$$

$$(26)$$

The reaction may be taken a step further using a mixture of acetylene and carbon monoxide. The products are particularly sensitive to reactant concentration, solvent, and available moisture, and can be visualized as the growth of an organic chain, bonded to a nickel atom, by the successive insertion of CO or acetylene molecules, which is interrupted at various points by the uptake of a proton or hydroxyl ion or by decomposition of the intermediate. This may be followed by rearrangement, or further reaction with molecules of solvent.

The first step is the insertion of a molecule of acetylene and of carbon monoxide to form a cis-2,5-hexadienoyl halide, which may in certain cases

$$RCH:CHCH_2Cl + HC:CH + CO \xrightarrow{Ni(CO)_4} RCH:CHCH_2CH:CHCOCl \quad (27)$$

be isolated (42). If the solvent used contains water, then the cis-2,5-hexadienoic acid (or the corresponding ester from alcoholic solution) is obtained in high yield (39, 43). A similar reaction takes place using allyl alcohols, esters, or ethers in the presence of HCl (44–46).

A particularly simple variation of this reaction has been developed (47, 48) in which the catalytic nickel species is formed in situ by reduction of nickel chloride with a manganese–iron alloy in the presence of thiourea. Allyl halide is added and at the same time acetylene and carbon monoxide are bubbled through the methanolic solution. Conversion is almost complete and yields of cis-methyl-2,5-dienoate of up to 80% have been claimed.

The proposed course of reaction is shown below. The acetylene molecule displaces the π-allyl group from one coordination site. Presumably thiourea plays a similar rôle to carbon monoxide in the reaction mechanism. Such donor-stabilized π-allyl complexes are known to react with acetylene and

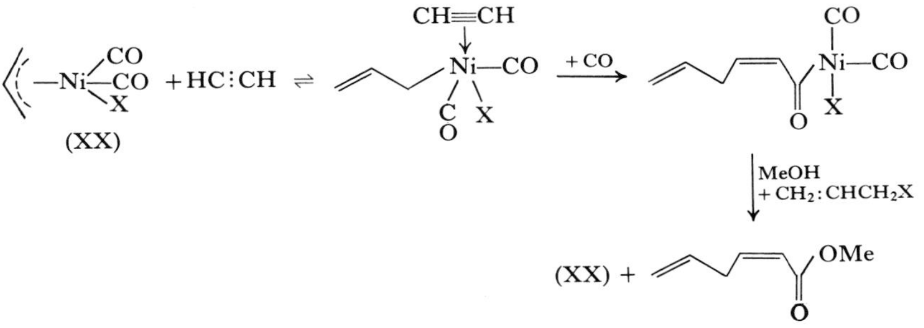

(28)

carbon monoxide. For example, the triphenylphosphine derivative (XXI) reacts in methanol with acetylene and CO to give a 47% yield of *cis*-methyl-2,5-hexadienoate (*14*).

The *cis*-2,5-hexadienoyl halide reacts further with nickel carbonyl in dry diethyl ether, benzene, or heptane, undergoing ring closure followed

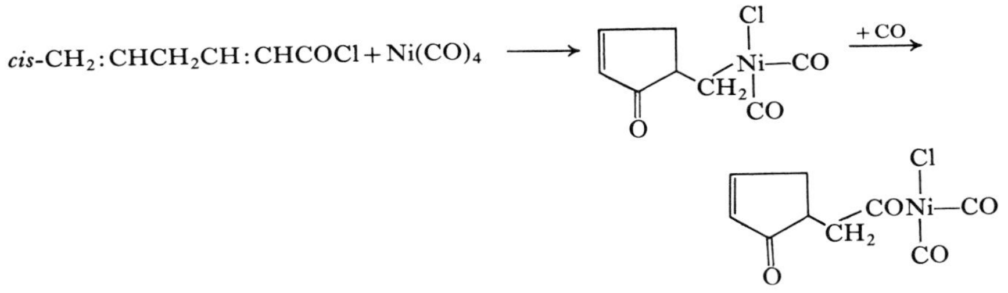

(29)

by insertion of a CO molecule (*42*). If the reaction terminates at this point the main product is an isomeric form of the acyl chloride which is converted, in the presence of water, to 2-oxocyclopent-3-enylacetic acid.

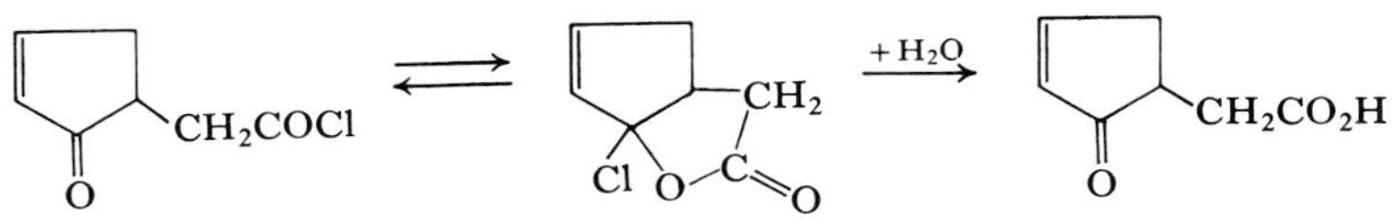

(30)

In the presence of acetylene, the further course of reaction is extremely solvent-dependent. In ketones, and to some extent in esters, the next step is insertion of an acetylene molecule, which may either add a proton to give

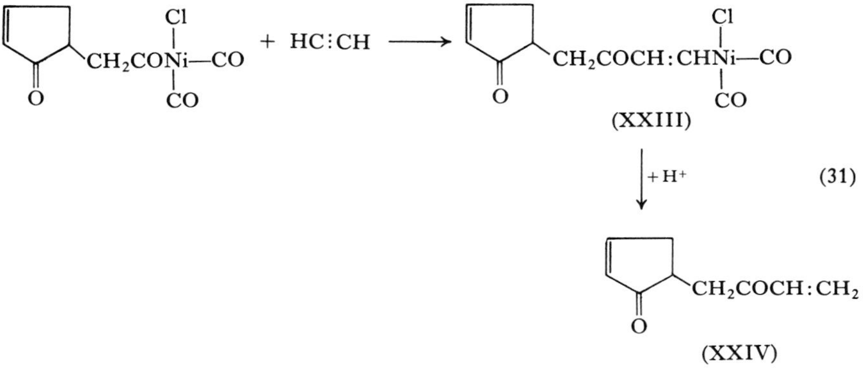

(31)

(XXIV) or be followed by CO insertion and cyclization to give (XXV), which is actually the main product in wet acetone (49, 50). In ketonic

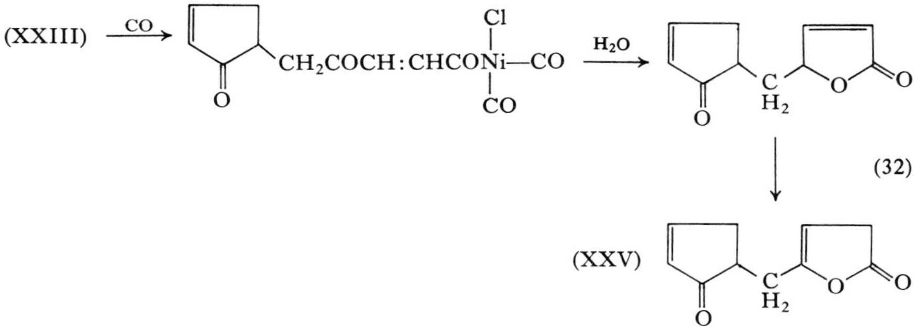

(32)

solvents, in the absence of water, this γ-lactone reacts with the solvent to form (XXVI) and (XXVII), which suggests that a Reformatsky-type reaction has occurred (50, 51).

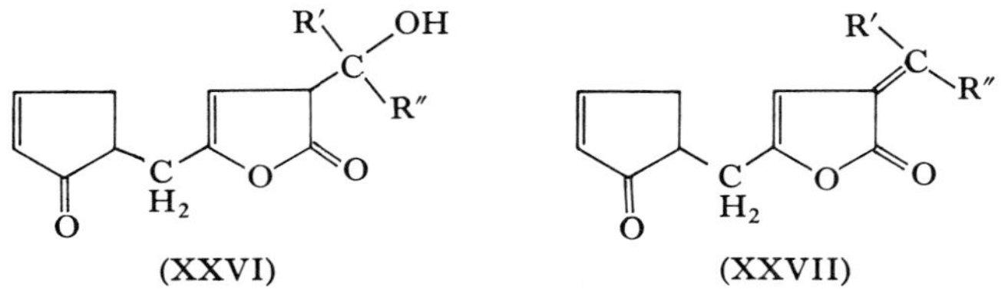

A second mode of reaction is available in which unreacted allyl halide or π-allylnickel halide complex apparently reacts with either (XXII), or the acyl chloride formed from (XXII), and with (XXIII) to give the

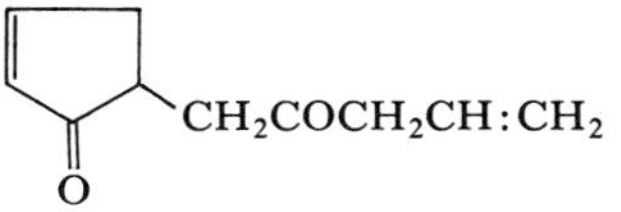

(XXVIII)    (XXIX)

coupling products (XXVIII) and (XXIX). This path is of particular importance in the reaction of allyl bromide in anhydrous diethyl ether when (XXIX) is formed in 20% yield (42).

An interesting by-product is obtained using a high concentration of acetylene in ethers or esters as solvent; an ϵ-lactone (XXX) is formed, probably by insertion into (XXIII) of a second molecule of acetylene, followed by CO insertion and ring closure (52).

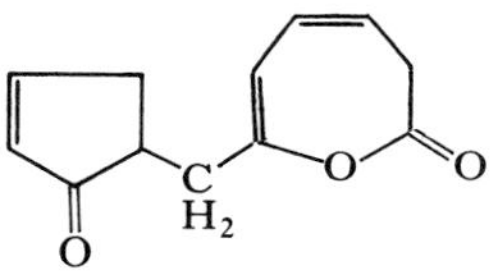

(XXX)

A further subtlety is the cyclization of the initial 2,5-hexadienoyl complex to form a six-membered ring, as an alternative to the cyclopentene system. In the reaction of allyl chloride with acetylene and carbon monoxide in the presence of $Ni(CO)_4$, only traces of six-membered ring systems

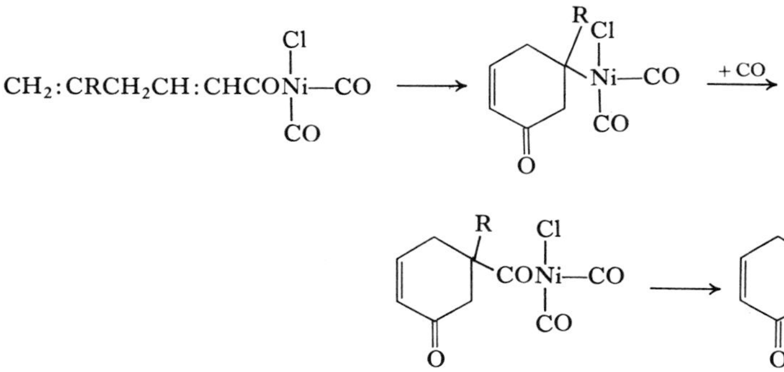

(33)

are formed. However, in the case of methallyl chloride these are the main products (51, 53), substituted phenols being particularly favored. It is not clear whether phenol is formed by further reaction of a cyclohexene derivative or directly from the intermediate 2,5-dienoyl nickel complex.

The reactions described above are not limited to acetylene but appear to be general for monosubstituted acetylenes (54), the allylic group adding predominantly to the unsubstituted carbon atom. Reactions using substituted allyl halides indicate that the acetylene molecule attaches itself to the least-substituted terminal carbon atom of the allyl group. The interested reader is referred to the review by Chiusoli and Cassar (39) for the exact product distribution in these cases.

The rôle of the solvent in practically all of the reactions so far discussed is decisive. For example, allylic halides having electron-attracting substituents, such as methyl bromocrotonate, upon treatment with nickel carbonyl in hydroxylic solvents do not react with CO. Instead substitutive hydrogenation of the halogenated carbon atom occurs (55), while in ketonic solvents the products which might be expected from the carbonylation of normal allylic halides are obtained (50).

The part played by the solvent in such a reaction has received attention and it has been suggested (56) that the ketone group acts as a scavenger of

$$
\begin{array}{c}
R \\
\diagdown \\
C=O \\
\diagup \quad \diagdown \\
O \qquad Ni\begin{array}{c} X \\ CO \end{array} \\
\diagdown \quad \diagup \\
C \\
\diagup \diagdown \\
R' \quad R''
\end{array}
$$

(XXXI)

acyl groups by forming complexes similar to (XXXI). Support for this suggestion comes from the work of Bauld (57), who has shown that in tetrahydrofuran, benzoyl chloride and benzil react in the presence of nickel

$$
2\,C_6H_5COCl + C_6H_5COCOC_6H_5 + Ni(CO)_4 \rightarrow C_6H_5\underset{\underset{\displaystyle}{|}}{\overset{\overset{\displaystyle C_6H_5OCO}{|}}{C}}=\underset{\underset{\displaystyle}{|}}{\overset{\overset{\displaystyle OCOC_6H_5}{|}}{C}}-C_6H_5 + NiCl_2 + 4\,CO
$$

(34)

carbonyl to give an enediol diester. The same reaction in moist acetone gives the benzoate of benzoin (XXXII) as the main product (56). It has been proposed that the benzil coordinates to the nickel atom, and that this complex traps a molecule of acyl halide, forming a C—Ni bond adjacent to the C=O group, and finally adds a proton.

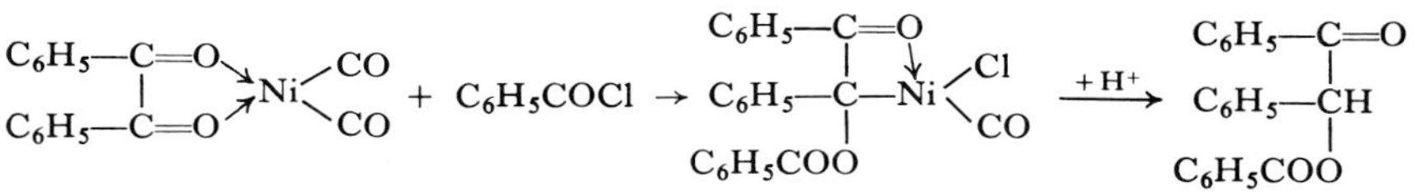

$$
\begin{array}{c}
C_6H_5-C=O \\
| \\
C_6H_5-C=O
\end{array}
Ni\begin{array}{c} CO \\ CO \end{array}
+ C_6H_5COCl \rightarrow
\begin{array}{c}
C_6H_5-C=O \\
| \\
C_6H_5-C-Ni\begin{array}{c}Cl\\CO\end{array} \\
| \\
C_6H_5COO
\end{array}
\xrightarrow{+H^+}
\begin{array}{c}
C_6H_5-C=O \\
| \\
C_6H_5-CH \\
| \\
C_6H_5COO
\end{array}
$$

$$(XXXII) \quad (35)$$

The reaction of an acyl chloride, acrolein, acetylene, and nickel carbonyl, in inert solvents, to give (XXXIII) and (XXXIV) is suggested to proceed by a similar mechanism, the acyl halide and acrolein reacting on the nickel atom to form a substituted allyl system (56).

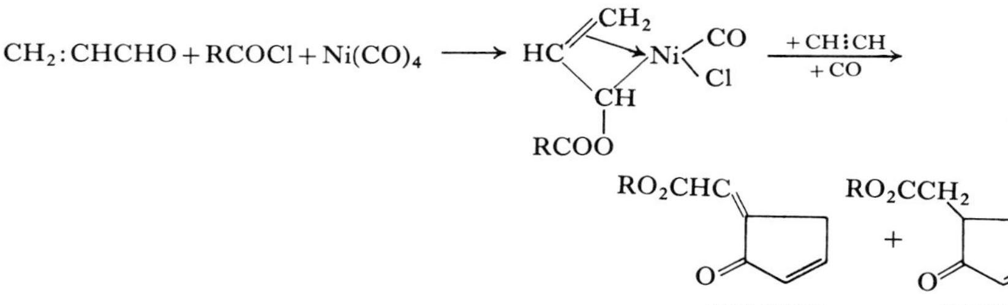

$$
CH_2\text{:}CHCHO + RCOCl + Ni(CO)_4 \longrightarrow
\begin{array}{c}
CH_2 \\
HC \diagdown \\
Ni\begin{array}{c}CO\\Cl\end{array} \\
CH \\
| \\
RCOO
\end{array}
\xrightarrow[+CO]{+CH\text{:}CH}
$$

$$(36)$$

$$
\begin{array}{cc}
RO_2CHC & RO_2CCH_2 \\
\overset{O}{\bigcirc} & + \quad \overset{O}{\bigcirc}
\end{array}
$$

$$(XXXIII) \qquad (XXXIV)$$

We have already mentioned that it is the cis isomer of the 2,5-hexadienoyl halide which in the presence of nickel carbonyl cyclizes to give a cyclopentene derivative. The trans form reacts with acetylene to give (XXXV)

$$
CH_2\text{:}CHCH_2CH\text{:}CH-\overset{}{\underset{O}{\diagup}}\overset{O}{\diagdown}O
$$

$$(XXXV)$$

(42). This is an example of a general reaction that has been observed for aliphatic, alicyclic, and aromatic acyl halides in ketonic solvents (56).

$$
RCOX + HC\text{:}CH + CO + Ni(CO)_4 \rightarrow
RCOCH\text{:}CHCO\overset{X}{\underset{CO}{N i}}-CO
\xrightarrow{+H^+}
\overset{}{R}\diagup_O\diagdown O
$$

$$(37)$$

With high acetylene concentration two molecules of acetylene are inserted to give ε-lactones similar to (XXX) (52).

Under conditions similar to those for allyl halides, 1,4-dichlorobutene reacts with nickel carbonyl to give butadiene. However, a double insertion of acetylene and carbon monoxide can be successfully carried out using 4-chloro-2-buten-1-ol and generating hydrogen halide *in situ* with a weak acid–inorganic halide combination, e.g., $NaBr–H_3PO_4$ (58).

$$ClCH_2CH:CHCH_2OH + 2\ HC:CH + 2\ CO + HBr \xrightarrow[\text{MeOH}]{\text{Ni(CO)}_4}$$

$$CH_3OCOCH:CHCH_2CH:CHCH_2CH:CHCO_2CH_3 \quad (38)$$

Benzyl halides have been reported to react with nickel carbonyl to give both coupling and carbonylation (59). Carbonylation is the principal reaction in polar nonaromatic solvents, giving ethyl phenylacetate in ethanol, and bibenzyl ketone in DMF. The reaction course is probably similar to that of allylic halides. Pentafluorophenyl iodide gives a mixture of coupled product and decafluorobenzophenone. A radical mechanism has been proposed (60). Aromatic iodides are readily carbonylated by nickel carbonyl to give esters in alcoholic solvents or diketones in ethereal solvent (57). Mixtures of carbon monoxide and acetylene react less readily with iodobenzene, and it is only at 320° C and 30 atm pressure that a high yield of benzoyl propionate can be obtained (61). Under the reaction conditions used, the

$$C_6H_5I + HC:CH + CO + R'OH \xrightarrow{\text{Ni(CO)}_4} C_6H_5COCH_2CH_2CO_2R \quad (39)$$

intermediate unsaturated ester is hydrogenated.

Reaction of aryllithium with nickel tetracarbonyl at $-70°$ C gives a black, insoluble, air-sensitive powder which is believed to be lithium aroyltricarbonyl nickelate (62, 63). These lithium salts are highly reactive and on hydrolysis produce α-diketones and acyloins.

$$ArLi + Ni(CO)_4 \rightarrow Li[ArCONi(CO)_3] \quad (40)$$

Reaction with organic halides leads to coupling products (63). From benzyl chloride and lithium *p*-toluoyltricarbonyl nickelate, a 73% yield of α-benzyl *p*-toluoin is obtained. The reaction presumably involves the initial formation of *p*-tolyl benzyl ketone, which reacts with a further molecule of lithium salt. Benzoyl chloride reacts to form a 4,4'-dimethyl-stilbene derivative.

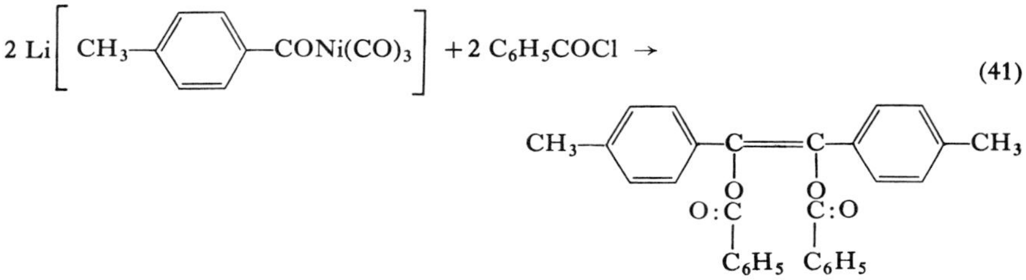

$$2 \, Li\left[CH_3-\bigcirc\!\!\!\!\bigcirc\!\!\!-CONi(CO)_3 \right] + 2 \, C_6H_5COCl \rightarrow \tag{41}$$

$$CH_3-\bigcirc\!\!\!\!\bigcirc\!\!\!-\underset{\underset{C_6H_5}{\overset{O:C}{|}}}{C}\!\!=\!\!\underset{\underset{C_6H_5}{\overset{C:O}{|}}}{C}-\bigcirc\!\!\!\!\bigcirc\!\!\!-CH_3$$

The usefulness of this lithium salt as an intermediate in organic synthesis has been recently extended by the reaction with acetylene and mono-substituted acetylenes (64). At $-70°C$ 2 moles of lithium salt add to 1 mole of acetylene to give, after hydrolysis, a high yield of 1,4-diketone. The same reaction carried out at $-30°C$ produces, in addition to the 1,4-diketone, a smaller yield of a γ-lactone.

$$2 \, Li[RCONi(CO)_3] + R'C\!:\!CH \rightarrow RCOCR'HCH_2COR \tag{42}$$

IV

NICKEL-CATALYZED SYNTHESIS
OF CYCLIC COMPOUNDS

Nickel catalysts for the syntheses of cyclic compounds were first success-fully utilized by Reppe, who was able to prepare cyclooctatetraene from acetylene (65). This eight-membered ring synthesis, and also the prepara-tion of cyclic products from strained olefins (e.g., bicycloheptene and norbornadiene) and acrylonitrile, have been adequately reviewed elsewhere (7) and will therefore not be considered further. A short account of the cyclization reactions of butadiene using nickel-containing catalysts has appeared previously in this series (1). The discovery of new synthetic possibilities and a deeper understanding of the mechanism of these reactions justify a more extensive treatment.

As early as 1954, Reed (66) had shown that butadiene reacts in the presence of nickel-containing catalysts to produce cis,cis-1,5-cyclooctadiene. The catalyst employed was the so-called "Reppe-type catalyst," $(Lig)_2Ni(CO)_2$ (where Lig = phosphine or phosphite), which was previously activated with acetylene. Among others, Reppe himself had shown that this catalytic system was able to convert substituted olefins to six-membered ring compounds (67).

A thorough investigation of the catalytic properties of nickel in its lower valency states could only be undertaken after the school at Mülheim had succeeded in isolating carbonyl-free nickel catalysts containing a stoichiometric ratio of metal to ligand. The extremely active catalysts containing zero-valent nickel may be prepared very easily using the method discovered by Wilke *et al.* (68), whereby nickel acetylacetonate is reduced with an organoaluminum compound (e.g., diethylaluminum ethoxide or triisobutyl-aluminum) in the presence of a suitable ligand. Phosphines, phosphites, and oligo-olefins (e.g., 1,3-butadiene, 1,5-cyclooctadiene, and 1,5,9-cyclododecatriene) have been shown to be particularly effective (17). In the absence, or with a deficiency, of such stabilizing ligands the reduction of nickel acetylacetonate leads to the precipitation of metallic nickel.

Before entering into a detailed discussion of the nickel-catalyzed reactions of olefins, the convention we will use to describe the stereochemistry of the intermediate π-allyl complexes should be mentioned. As an example, consider the bis(π-allyl) C_8 chain postulated as an intermediate in the

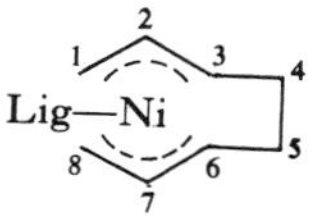

reaction of butadiene with a nickel-ligand catalyst. The reproduction in two dimensions of the configuration of the allyl groups between the C-2 and C-3, as well as the C-6 and C-7, atoms is best described by using the terminology cis and trans.

Thus

trans will be used for *syn*

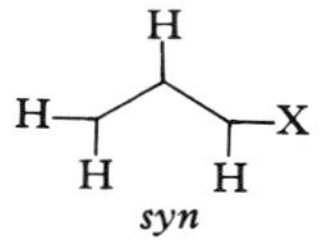

and

cis for *anti*

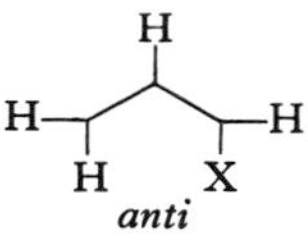

This is preferable, in this case, to the usual syn and anti terminology, since ring closure is accompanied by formation of double bonds having the same configuration as the π-allyl group.

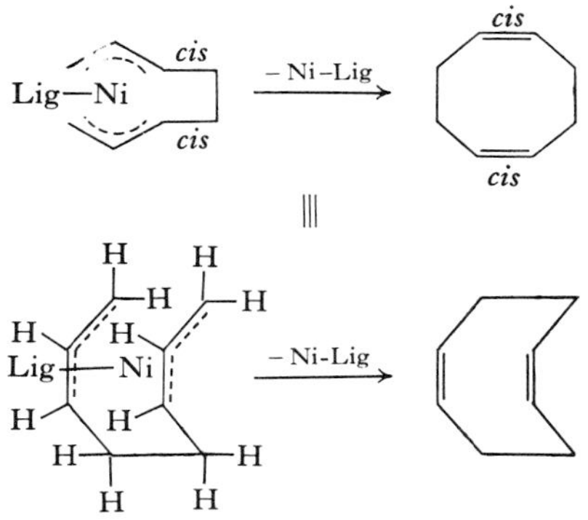

A. Synthesis of 1,5,9-Cyclododecatriene

The reduction of nickel(II) in the presence of butadiene as the only available ligand (i.e., naked-nickel[3]) (69) produces a catalyst which is able to trimerize butadiene to a mixture of all-*trans*-; *trans,trans,cis*-; and *trans,cis,cis*-1,5,9-cyclododecatriene in which the all-trans form predominates.

An understanding of the mechanism of this reaction was obtained by isolation of the intermediate involved. The coupling of three butadiene molecules produces a 12-membered carbon chain bonded to an atom of nickel by two terminal π-allyl groups (70, 71).

The configuration of the C_{12} chain has been determined spectroscopically (72). Bearing in mind that the two terminal π-allyl groups in the C_{12} chain may be mutually cis or trans (71a, 71b) to each other, and also that the molecule contains a trans double bond, the six isomers (XXXVI)–(XLI) are possible. Rotation of one of the π-allyl groups relative to the other gives a further six possibilities; this is illustrated for (XLII). This type of isomerization has been discussed elsewhere in relation to the structure of bis(π-crotyl)nickel (13).

[3] By "naked"-nickel is meant a nickel complex from which all the bonded ligands are easily displaced by butadiene.

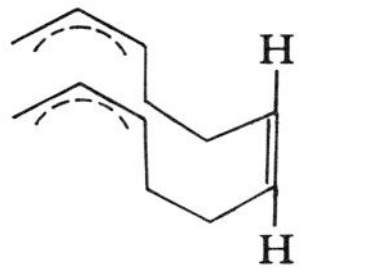

(XXXVI) 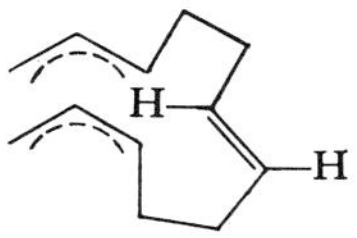 (XXXVII) 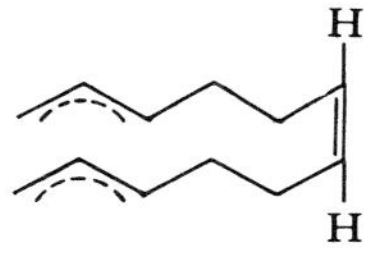 (XXXVIII)

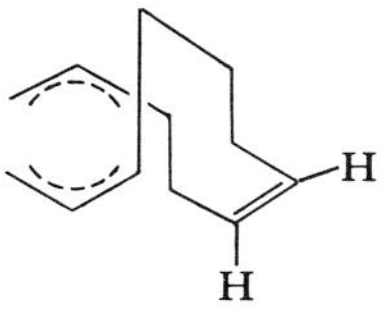

(XXXIX) 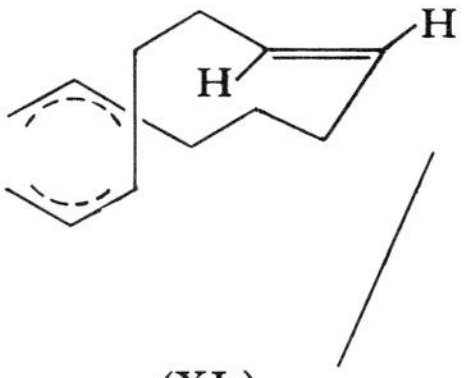 (XL) 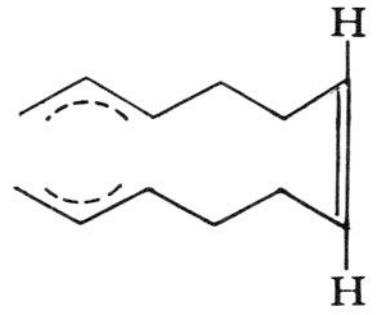 (XLI)

In spite of these possibilities, the NMR spectrum of the C_{12} chain bonded to a nickel atom is relatively simple and indicates that probably only two isomers are present, viz., (XXXVI) and (XXXIX).

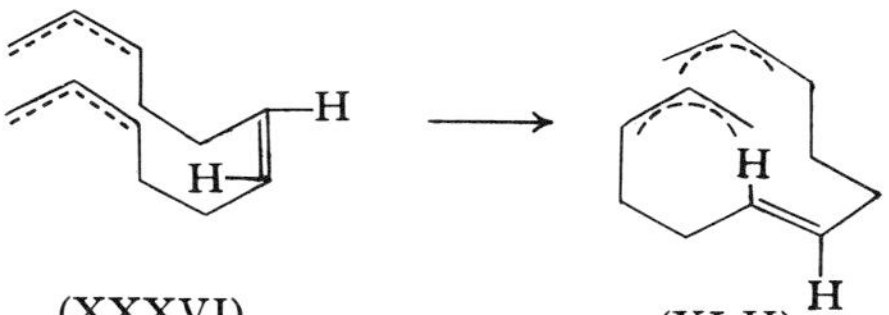

(XXXVI) (XLII)

Interaction of the C_{12} chain with a further molecule of butadiene, or another "accelerating" ligand,[4] leads to closure of the chain to form a 12-membered ring. This final step can be simulated in a stoichiometric reaction using triethylphosphine as ligand (70). In the catalytic reaction the 12-membered ring is displaced by further butadiene with simultaneous

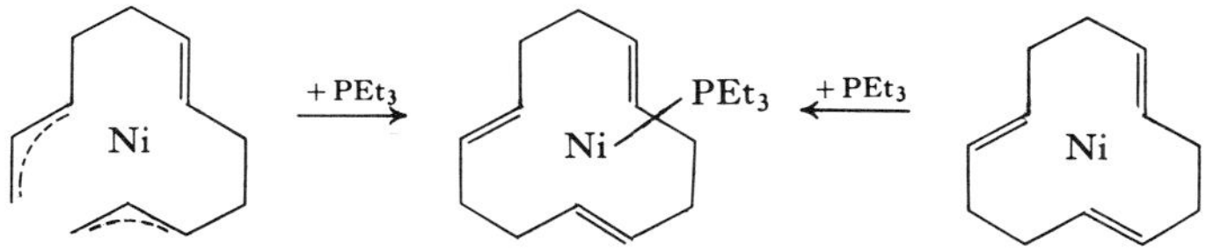

(43)

────────

[4] An "accelerating" ligand is defined as a ligand which causes groups already bonded to a metal atom to react further (e.g., by coupling) without actually being directly involved in the reaction.

reformation of the open-chain species bonded to nickel. Of the four possible isomers of cyclododecatriene, only three are formed catalytically, and none of the all-cis isomer is present in the products.

The distribution of the isomers obtained under various reaction conditions enables the following conclusions to be drawn (72).

(a) The proportion of the trans,trans,cis isomer is dependent only on the temperature of the reaction, and increases with rising temperature at the expense of the sum of the all-trans and trans,cis,cis isomers.

(b) The ratio of the all-trans to trans,cis,cis isomer is mainly dependent on the butadiene concentration. With rising temperature the proportion of the trans,cis,cis isomer also increases at the expense of the all-trans isomer.

(c) Increasing temperature is accompanied by an increase in the proportion of C_8 hydrocarbons formed.

(d) The rate of ring formation decreases with increasing conversion.

The catalytic formation of the various isomers of CDT could occur through the conformers discussed above, viz., (XXXVIII) and (XLI) giving all-*trans*-CDT, (XXXVII) and (XL) giving *trans,trans,cis*-CDT (XXXVI), and (XXXIX) giving *trans,cis,cis*-CDT.

The following hypotheses cannot, however, be ruled out (73).

(a) There exist two types of C_{12} chain bonded to nickel which differ only in the cis or trans configuration of the internal double bond, and which are not in equilibrium with each other.

(b) The ratio of the trans chain (44a) to cis chain (44b) is controlled by the temperature.

(c) Only allyl groups with the same configuration are able to interact with each other to form cyclic products.

The all-cis isomer of cyclododecatriene is not formed. These deductions are schematically summarized in Eqs. (44a) and (44b).

Accelerating ligands capable of reversing the configuration of an allyl group (18) (e.g., pyridine) have the effect, when added to the catalyzed reaction, of depressing the yield of CDT-*tcc*, and the product obtained is mainly CDT-*ttt* (72). At a fast rate of interconversion of the various isomers one would expect preferential reaction through the favored all-trans form.

The isomeric distribution in the catalyzed reaction is apparently not influenced by the stability of the various CDT–nickel complexes. All four of the theoretically possible compounds have been isolated and, remarkably, all-*cis*-CDT (*74*), which is not formed catalytically, forms the most stable nickel complex (*75*).

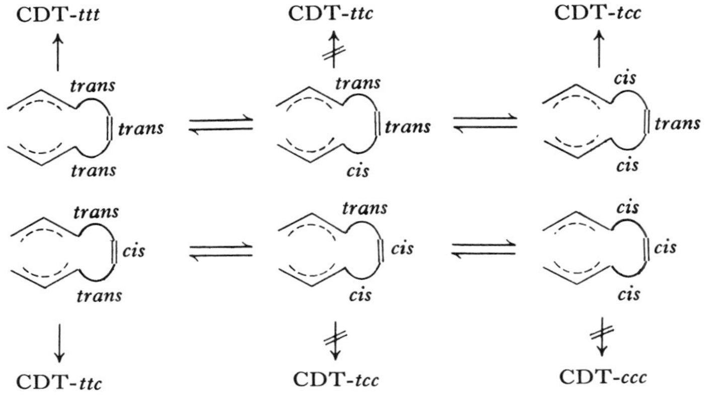

The ratio of the cis internal double bond (from which CDT-*ttc* is formed) to the trans internal double bond (from which CDT-*ttt* is formed) is perhaps controlled by the ratio of single-*trans*-butadiene to single-*cis*-butadiene. This equilibrium is discussed in Reference (*76*).

The suggestion (*73*) that the configuration and concentration of a product may be controlled by the concentration of the conformers of butadiene has precedents (*77, 78*). It has been proposed (*73*), for example, that the ratio of octatriene to 3-methylheptatriene, formed by reaction of butadiene with a cobalt catalyst (*79–82*), is governed by this equilibrium [Eq. (45)].

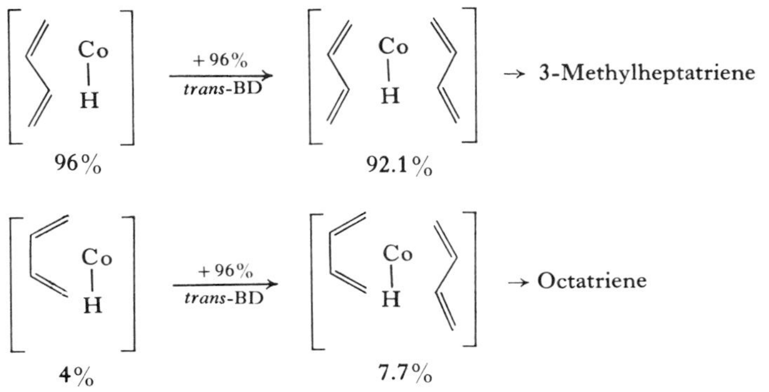

(45)

The 4% single-*cis*-butadiene which is present in butadiene at 20° C reacts to produce almost 8% of octatriene [the formation of a cobalt complex from two single-*cis*-butadiene molecules (0.2%) has been neglected].

B. Cyclodimerization of Butadiene

The cyclotrimerization reaction described above can be converted into a cyclodimerization reaction by blocking one of the vacant coordination positions around the nickel atom (*17, 66, 83*). Phosphines and phosphites in a ligand-to-nickel ratio of 1:1 have been found to be particularly effective.

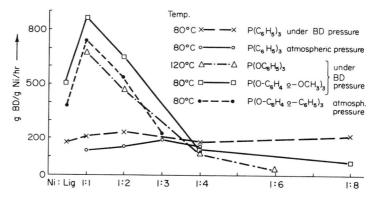

FIG. 1. Rate of reaction of butadiene with varying ligand:nickel concentration (*84*).

As Fig. 1 shows (*84*) systems containing phosphines are independent, to a first approximation, of the nature or concentration of the ligand. It may be supposed that butadiene easily displaces the phosphine molecules. This is confirmed by the addition of 1 mole each of triphenylphosphine and tri(*o*-phenylphenyl)phosphite to a catalyst containing a gram-atom of nickel. The system shows the characteristics of a nickel phosphite catalyst (Fig. 1).

Complete conversion of butadiene (e.g., at 80° C and atmospheric pressure) gives as the main products COD and VCH. The ratio of COD to VCH is dependent on the ligand attached to the nickel atom (Table III).

TABLE III

CYCLODIMERIZATION OF BUTADIENE WITH NICKEL-LIGAND CATALYSTS [a,b]

	1	2	3	4	5	6
VCH(%)	5.8	39.6	26.8	7.4	5.8	3.1
COD(%)	17.3	41.2	64.3	81.1	88.4	96.5
CDT(%)	60.4	14.4	6.0	9.2	4.4	0.2
$> C_{12}$(%)	14.1	4.8	2.8	2.3	2.4	0.2
Gm C_4H_6/gm Ni/hr	20	35	180	100	190	780

[a] Brenner et al. (84).
[b] Ligand: Nickel = 1:1; atmospheric pressure, 80° C, 3 hours.

$1 = As(C_6H_5)_3$ $4 = P(OC_6H_5)_3$
$2 = P(C_6H_{11})_3$ $5 = P(OC_6H_5)_2[O—C_6H_4\text{-}o\text{-}C_6H_5]$
$3 = P(C_6H_5)_3$ $6 = P(O—C_6H_4\text{-}o\text{-}C_6H_5)_3$

The relative donor–acceptor strength of various ligands in $Lig_2Ni(CO)_2$ complexes has been studied by Meriwether and Fiene (85) using IR techniques. In Table IV the CO stretching frequencies are correlated with

TABLE IV

CORRELATION BETWEEN CO STRETCHING FREQUENCY AND COD-TO-VCH RATIO [a,b]

Ligand in $Lig_2Ni(CO)_2$	ν_{CO}		COD:VCH
$P(OC_6H_5)_3$	1987	2040	11.0
$P(OC_6H_4\text{-}p\text{-}C_6H_5)_3$	1987	2035	9.4
$P(OC_6H_4\text{-}p\text{-}OCH_3)_3$	1980	2035	7.8
$P(C_6H_5)_3$	1937	2000	2.4
$P(C_6H_{11})_3$	1910	1950	1.04
$P(OC_6H_4\text{-}o\text{-}C_6H_5)_3$	1987	2035	31
$P(OC_6H_4\text{-}o\text{-}OCH_3)_3$	1980	2035	15.6

[a] Brenner et al. (84) and Meriwether and Fiene (85).
[b] Ni:Lig = 1:1; atmospheric pressure, 80° C.

the selectivity[5] of formation of COD for those ligands that have been found to be particularly useful in the catalytic reaction (84). The CO frequency (i.e. the acceptor ability of the ligand) decreases in the series

$$P(OC_6H_5)_3 > P(OC_6H_4-p-C_6H_5)_3 > P(C_6H_5)_3 > P(C_6H_{11})_3$$

At the same time the selectivity of COD formation decreases. Too much reliance cannot, of course, be placed on this qualitative correlation, and it is not to be assumed that the transition state of the cyclization reaction has the same geometry as a $Lig_2Ni(CO)_2$ complex. It can be seen from Table IV that the selective formation of COD using ligands with o-substituents is not associated with electronic factors. In these cases steric effects play an important role. This is illustrated clearly in Table V.

TABLE V

RELATION BETWEEN PRODUCT DISTRIBUTION AND POSITION OF SUBSTITUTION OF TRIPHENYLPHOSPHITE[a,b]

	Substituent				
	o-OCH$_3$	m-OCH$_3$	p-OCH$_3$	o-C$_6$H$_5$	p-C$_6$H$_5$
VCH(%)	5.8	7.5	10.1	3.1	6.9
COD(%)	90.3	83.3	78.6	96.5	65.2
CDT(%)	3.7	9.3	9.6	0.2	18.2
C$_{12}$(%)	0.3	—	1.8	0.2	9.6
Gm BD/gm Ni/hr	325	140	90	780	75

[a] Brenner et al. (84).
[b] Ni:Ligand = 1:1; atmospheric pressure, 80° C.

Using the nickel-tri(o-phenylphenyl)phosphite catalyst, the composition of the reaction product is markedly dependent on the extent of conversion of the butadiene. With a conversion of less than 85%, 1,2-divinylcyclo-butane (DVCB) is obtained in a yield of up to 40%. At higher conversions DVCB catalytically rearranges to COD and VCH (Table VI)

The catalytic Cope rearrangement of DVCB to COD is illustrated in Fig. 2. The reaction is zero order with respect to the four-membered ring up to a conversion of around 97%. An additional condition which favors

[5] Selectivity is here defined as the preferred formation of a product catalytically.

the formation of DVCB, by limiting the thermal Cope rearrangement, is a relatively low reaction temperature or a reaction of short duration at higher temperature.

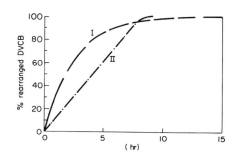

FIG. 2. The Cope rearrangement of DVCB (*84*). I, Thermal rearrangement at 80°C; II, catalytic rearrangement with nickel-tri(*o*-phenylphenyl)phosphite (0.2 *M* solution), 24° C.

TABLE VI

RELATIONSHIP BETWEEN BD CONVERSION AND YIELD OF DVCB[a,b]

	BD Conversion (%)				
	4	33	85	95	100
DVCB[c](%)	36.3	39.3	38.4	14.0	—
VCH[c](%)	8.7	2.5	1.8	2.2	2.0
COD[c](%)	55.0	58.2	59.8	83.8	98.0
> C$_8$[c](%)	($\sim$1% in all cases, 60% of which CDT)				

[a] Brenner *et al.* (*84*).
[b] Nickel:tri(*o*-phenylphenyl)phosphite = 1:1, Benzene:BD = 1:1. Rate $\sim$6.8 gm BD/gm Ni/hr.
[c] $\sum$ C$_8$ made 100%.

The catalytic formation of COD could occur by two alternative routes: either directly from butadiene, or indirectly from DVCB. The present evidence does not allow a distinction between these possibilities.

The formation of both DVCB and COD directly from butadiene requires that the intermediate nickel complex exist in at least two isomeric forms.

The scheme shown in Eq. (46) summarizes and explains mechanistically the results so far obtained. Two butadiene molecules may be supposed to react together to form a bis(π-allyl) C_8 chain bonded to the Ni-ligand moiety. This bis(π-allyl) system exists in two configurations, (XLIII) and (XLIV), from which DVCB and COD are formed. The nature of the ligand in these intermediates seems to be relatively unimportant.

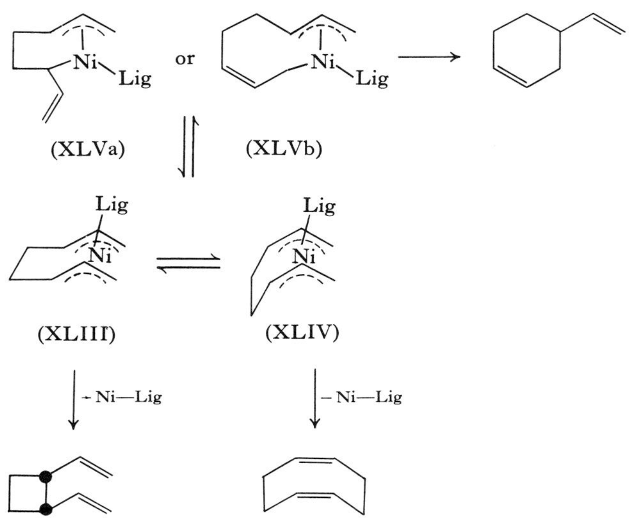

(46)

In contrast, the formation of a third conformer (XLV) which leads to VCH is markedly dependent on the donor–acceptor character of the ligand (see Table III). The interconversion of (XLIII) into (XLV) is shown in simplified form in Eq. (47).

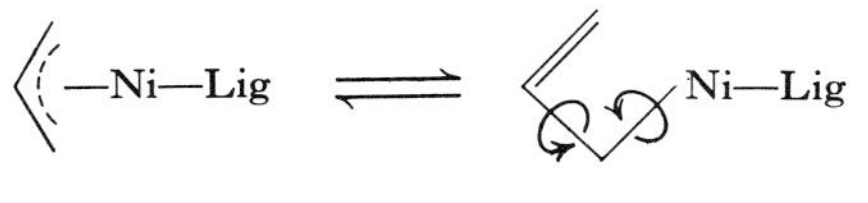

(47)

Phosphites as ligand will displace the equilibrium in Eq. (47) to the left while phosphines should favor the σ-allyl form, and hence promote the formation of VCH. This is indeed observed (Table III).

Models show that the formation of VCH can proceed more easily through a complex containing a "terminal" σ-alkyl, π-allyl arrangement (XLVb) than via an "internal" σ-alkyl, π-allyl arrangement (XLVa). Coupling at the

C-1 and C-6 carbon atoms in form (XLVb) would result in the formation of a terminal double bond complexed to the nickel atom and might be supposed to favor reaction through this isomer.

The transformation of (XLIII) into (XLIV) possibly occurs through a σ-allyl group. However, it could proceed through DVCB or the generation of free butadiene. The isolation of free butadiene in up to 30% yield during the catalytic conversion of DVCB to COD and VCH at 80 mm makes this

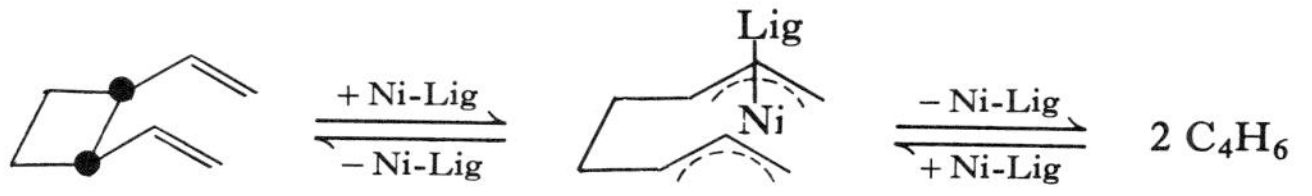

$$\text{Lig} = \text{P(O—C}_6\text{H}_4\text{O-C}_6\text{H}_5)_3$$

particularly feasible (*86, 87*). The formation of the individual carbon–carbon bonds at the nickel atom is discussed at greater length in Section IV,H.

The catalytic rearrangement of DVCB to COD (Fig. 2) supports the alternative possibility that the formation of COD occurs exclusively through the four-membered ring [this has been postulated for the thermal synthesis of COD from butadiene (*88*) but has since been discredited by Benson (*89*)]. The results of investigations with piperylene, discussed in Section IV,E,2, are also in accord with this view. [See Eq. (57), p. 72.]

C. Synthesis of cis, trans-1,5-Cyclodecadiene

The formation of CDT is suppressed if ethylene as well as butadiene is brought into contact with a naked-nickel catalyst. Depending on the reaction conditions, the product is a mixture of *cis,trans*-1,5-cyclodecadiene (CDD) and 1,*trans*-4,9-decatriene (DT) (*90*). With equal concentration of butadiene and ethylene the co-oligomerization occurs some six times faster than the cyclotrimerization of butadiene to CDT.

The dependence of the ratio of DT to CDD on reaction temperature and concentration of ethylene is shown in Tables VII and VIII. At low temperatures practically only CDD is formed, but the reaction takes place very slowly. At higher temperatures the product is mainly DT. The yield of CDD and rate of reaction reach a maximum at a $C_2H_4 : C_4H_6$ ratio of between 1.5 and 3; at the same time, the formation of COD and VCH and CDT is drastically reduced. The following reaction conditions appear to be optimal for normal laboratory preparations: An autoclave is charged with butadiene, ethylene (20–30 atm), and the catalyst [(COD)$_2$Ni] and left at 20° C for 3–4 weeks. Depending on the ratio of the reactants, more ethylene is added if necessary. In this way one obtains a product consisting of about 80% CDD, which, after destruction of the catalyst, may be purified by distillation (22° C/0.3 mm) through a column. This

TABLE VII

VARIATION OF DT:CDD WITH TEMPERATURE[a,b]

	Reaction temperature (°C)				
	0°	20°	40°	60°	80°
DT(%)	4	12	30	50	69
CDD(%)	96	88	70	50	31
$\sum C_{10}$[c](%)	78.9	88.9	82.4	78.2	78.4
Gm product/gm Ni/hr	0.2	2.2	12.7	71	370

[a] Heimbach and Wilke (90).
[b] Catalyst, "naked"-nickel. C$_2$H$_4$:C$_4$H$_6$ = 1:1.
[c] Remainder CDT with traces of VCH and COD.

TABLE VIII

DEPENDENCE OF PRODUCT DISTRIBUTION ON C$_2$H$_4$:C$_4$H$_6$[a], USING "NAKED"-NICKEL CATALYST AT 39°–41° C

	C$_2$H$_4$:C$_4$H$_6$[b]			
	0.29	0.76	2.9	13
DT(%)	37	31	27	40
CDD(%)	63	69	73	60
$\sum C_{10}$[c](%)	52.9	79.4	89.1	93.6
Gm product/gm Ni/hr	9.4	13.4	16.4	5.3

[a] Heimbach and Wilke (90).
[b] Average value calculated from value before and after reaction.
[c] Rest CDT and higher oligomers.

extremely simple synthesis allows preparation of relatively large quantities of a system which was previously obtainable only by fairly lengthy classical synthetic methods.

A variation, which has the advantage that the rate of reaction may be increased, is to use a catalyst which normally converts butadiene into COD, i.e., the Ni-ligand system. At the same time this introduces the disadvantage that COD and VCH are also produced. The effect of varying the ligand on the co-oligomerization of butadiene and ethylene is summarized in Table IX.

TABLE IX

CO-OLIGOMERIZATION OF BUTADIENE AND ETHYLENE USING A NICKEL-LIGAND CATALYST[a,b]

	Ligand[c]					
	I	—[d]	II	III	IV	V
VCH (%)	4.7	0.9	5.7	13.6	5.6	2.3
COD (%)	1.8	1.0	19.6	33.6	51.8	92.1
DT (%)	26.9	24.6	13.3	2.1	0.03	—
CDD (%)	51.2	57.8	48.2	34.0	35.9	3.7
CDT (%)	14.0	14.5	11.7	16.0	5.8	1.5
Higher olig. (%)	1.5	1.2	1.5	0.7	0.8	0.5
CDD:DT	1.9	2.4	3.6	16	1000	—
Gm product/gm Ni/hr	6.3	17.7	2.2	11	23	19

[a] Heimbach and Wilke (90).
[b] Ligand:nickel = 1:1; $C_2H_4:C_4H_6 = 1:1$ (initially); 40° C. BD conversion 60–70%.
[c] I = As(C_6H_5)$_3$; II = P(C_6H_{11})$_3$; III = P(C_6H_5)$_3$; IV = P(OC_6H_5)$_3$; V = P(O—C_6H_4-o-C_6H_5)$_3$.
[d] No ligand.

The donor–acceptor character of the ligand as well as steric effects parallel those found for the cyclodimerization of butadiene. Particularly characteristic is the variation in the CDD-to-DT ratio, which (exactly as found for the COD to VCH ratio) is controlled by the extent of charge donation from the ligand to the metal.

The product distribution using a Ni-ligand catalyst is also dependent on the ratio of ethylene to butadiene. With increasing ethylene concentration the proportion of C_{10} products increases while at the same time the rate of reaction decreases (90).

A reasonable mechanism for the co-oligomerization of butadiene with ethylene on a naked-nickel catalyst is shown in Eq. (49). Interaction of an ethylene molecule with the bis(π-allyl) C_8 chain produces a C_{10} chain, containing both an alkyl- and a π-allylnickel group (XLVI). Coupling of the alkyl bond with the terminal atom of a *cis*-π-allyl group or the terminal

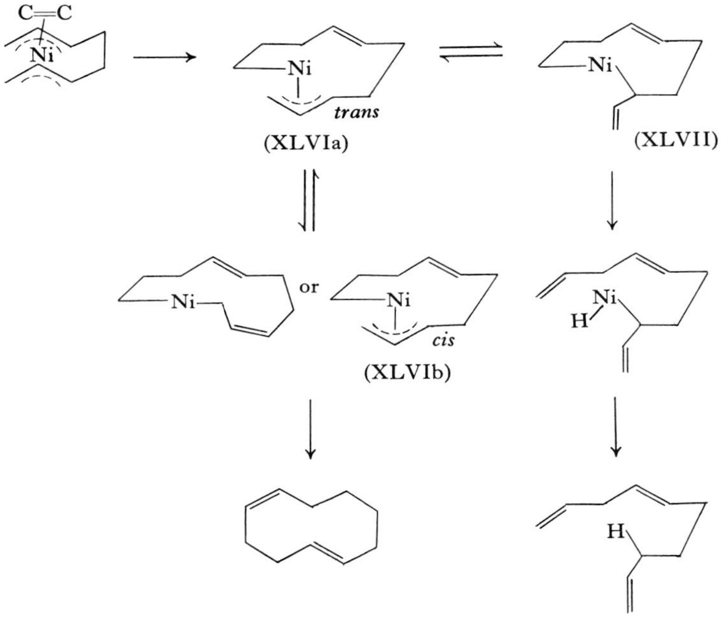

$$(49)$$

atom of a *cis*-σ-allyl group leads to the formation of CDD. Alternatively, the π-allyl group can rearrange to an "internal" σ-allyl group (XLVII). Elimination of the hydrogen β to the least stable nickel–carbon bond and addition to the σ-allyl group gives 1,*trans*,4,9-decatriene. This process of β-elimination is discussed in greater detail in Section IV,F.

Ligands with good acceptor character (e.g., triphenylphosphite, Table IX) stabilize the *cis*-π-allyl, σ-alkyl intermediate (XLVIb) and only CDD is formed. The stabilization introduced by triphenylphosphite is so effective that even at elevated temperatures (60° C) practically no DT results, while a naked-nickel catalyst at the same temperature produces DT and CDD in about equal proportions.

The steric hindrance introduced by bulky ligands [e.g., tri(o-phenyl-phenyl)phosphite] has the effect that the ethylene molecule is unable to coordinate to the nickel atom and mainly COD is formed.

The thermal Cope rearrangement of CDD to cis-1,2-divinylcyclohexane (*91*) becomes significant above 80° C and effectively limits the temperature at which the catalytic reaction can be carried out.

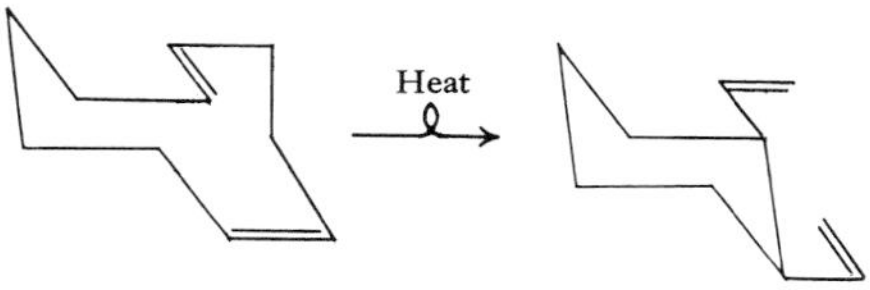

$$\text{(structure)} \xrightarrow[\text{Heat}]{} \text{(structure)} \qquad (50)$$

D. Synthesis of 4,5-Dimethyl-cis,cis,trans-1,4,7-cyclodecatriene

Immediately after the discovery of the cyclodecadiene synthesis by the co-oligomerization of butadiene and ethylene, the question of whether other unsaturated systems could be used in place of ethylene was investigated. Of the many variations on this theme which have been studied (*92, 93*), we will limit ourselves to discussing the co-oligomerization of 2-butyne, since this system became a model for all the other combinations.

Butadiene and butyne, in a 2-to-1 ratio, react with both the naked-nickel and nickel-ligand catalyst to form 4,5-dimethyl-*cis,cis,trans*-1,4,7-cyclo-decatriene (DMCDeT) (*94*). The yield with naked-nickel, however, never exceeds 25% and will not be discussed further.

DMCDeT undergoes an extremely facile Cope rearrangement (*91*) to give 1,2-dimethyl-*cis*-4,5-divinylcyclohexene [Eq. (51)], and hence the temperature of the catalytic reaction must be kept as low as possible

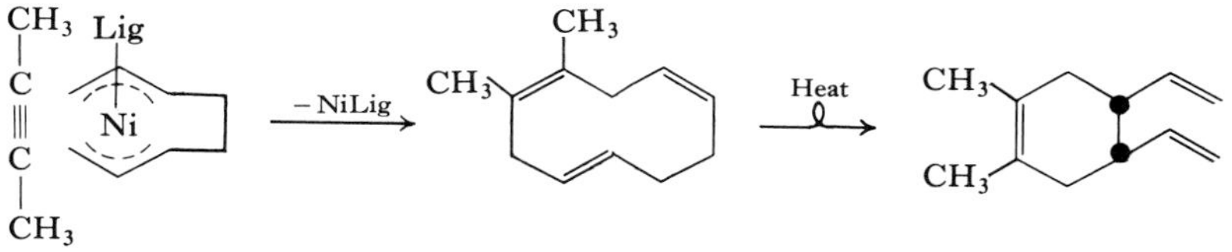

$$\text{(structure)} \xrightarrow[]{-\text{NiLig}} \text{(structure)} \xrightarrow[\text{Heat}]{} \text{(structure)} \qquad (51)$$

($< 60°$ C). No particular configuration of the π-allyl groups about the nickel is intended in Eq. (51).

The effect of various ligands on the yield of DMCDeT is illustrated in Table X and should be compared with the cyclodimerization of butadiene and the co-oligomerization of butadiene with ethylene (Tables III and IX).

If one bases the yield of DMCDeT on reacted butyne, then the system containing tri(o-phenylphenyl)phosphite produces satisfactory results. Although large quantities of butadiene dimers are formed, only a trace of CDT is produced and hardly any higher oligomers, which facilitates the work-up. This combination, therefore, has definite advantages for the synthesis of cyclodecatriene derivatives from expensive higher alkynes.

TABLE X

CO-OLIGOMERIZATION OF BUTADIENE AND 2-BUTYNE[a,b]

	Ligand[c]			
	I	II	III	IV
C_8 (%)	77.4	44.1	7.6	18.0	13.3
DMCDeT (%)	19.9	42.1	87.5	71.4	54.1
CDT + higher olig. (%)	2.7	13.8	4.9	10.7	32.5
Yield of C_{10} (%) based on BD reacted	14	33	80	63	47
Yield of C_{10} (%) based on butyne reacted	85	79	95	86	64
Gm product/gm Ni/hr[d]	36	28	40	27	4.3
Length of reaction (hr)	8.25	11.5	2.3	8	48

[a] Brenner et al. (94).
[b] Ligand:Nickel = 1:1; BD:Butyne = 5:1 (initially); 40° C.
[c] I = P(O—C_6H_4-o-C_6H_5)$_3$; II = P(OC$_6$H$_5$)$_3$; III = P(C$_6$H$_5$)$_3$; IV = P(C$_6$H$_{11}$)$_3$.
[d] Average value.

Table X, however, shows clearly that, in contrast to the butadiene–ethylene system, triphenylphosphine is the most successful ligand. We will return to this point later. The preparation of DMCDeT on a laboratory scale can be conveniently carried out by dissolving the nickel-ligand catalyst [which may be prepared by reduction of nickel acetylacetonate or directly from bis(cyclooctadiene)nickel and triphenylphosphine] in a solution of butadiene in toluene. Butyne is then added to give a butadiene-to-butyne ratio of 5–10:1. The reaction is conducted at 20° C and the contraction in volume is observed. The reaction is terminated at the break in the contraction curve (Fig. 3).

Excess of butadiene is necessary in order to minimize the formation of co-oligomers containing more than one molecule of butyne [see Eq. (52)], and the reaction must be interrupted when all the alkyne is consumed to prevent further reaction of DMCDeT with butadiene to give higher oligomers.

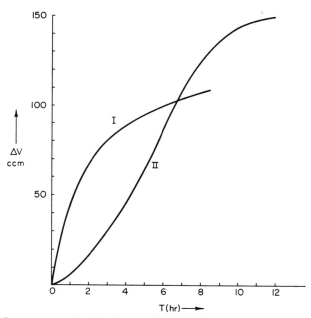

Fig. 3. Volume contraction in the co-oligomerization of butadiene with 2-butyne (nickel-ligand catalyst) (94); I=P(C₆H₅)₃, II=P(OC₆H₅)₃.

A particularly intriguing phenomenon is the difference in the ligand that is necessary to obtain a maximum yield of the ten-membered ring from butadiene and ethylene or from butadiene and butyne. The incorporation of ethylene is most successful using the ligand having the better acceptor character (triphenylphosphite) while incorporation of butyne occurs most smoothly with the ligand having the better donor character (triphenylphosphine). Here we intend to refer only to the donor–acceptor properties of triphenylphosphine relative to triphenylphosphite, i.e., since triphenylphosphite is the better acceptor, triphenylphosphine is the better donor A study of the effect of varying the ligand on the behavior of ethylene or butyne in the catalytic co-oligomerization (Tables IX and X) reveals that ethylene, relative to butyne, is the better donor. The conclusion can be drawn that the two reactions operate under optimal conditions when the electronic density at the nickel atom is the same, i.e., with triphenylphosphite and ethylene on the one hand and triphenylphosphine and butyne on the other, the metal atom here operating both as a matrix for the reaction and as a relay of electronic charge. Similar effects in the preparation of metal olefin and acetylene complexes are described in the literature (95).

The mechanism of formation of DMCDeT may be supposed to be stepwise and similar to the previously described CDD synthesis. In favor of this type of mechanism it may be pointed out that during the course of reaction the coordination number around the nickel atom does not drastically vary as it would if several carbon–carbon bonds were formed simultaneously. A study of the products obtained by reaction of alkynes with a deficiency of butadiene (96) supports this stepwise mechanism and is schematically summarized in Eq. (52). It is highly improbable that four or more carbon–carbon bonds would be formed simultaneously.

Cyclic and open-chain oligomers

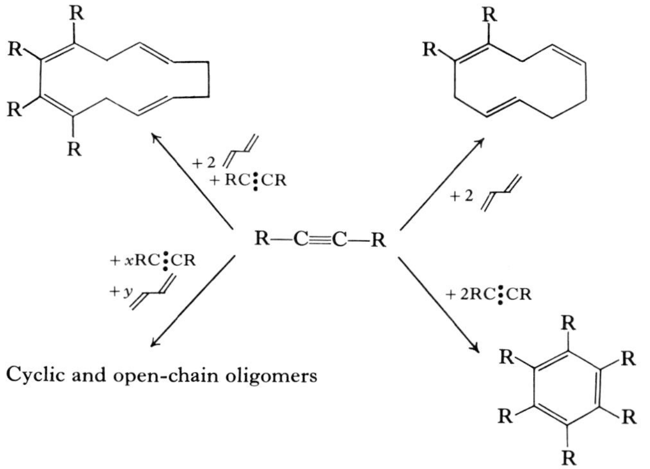

(52)

The formation of the different co-oligomers is not only controlled by the butadiene and alkyne concentration or the choice of catalyst, but is also dependent on the nature of the substituents on the alkyne or on the butadiene (96). It has been found that the formation of a ten-membered ring from a substituted alkyne and two butadiene molecules with a nickel-triphenylphosphine catalyst occurs most readily if the functional groups are separated from the alkyne group by at least two methylene groups (e.g., $ROCH_2CH_2CH_2C\vdots CCH_2CH_2CH_2OR$).

E. Syntheses of Substituted Ring Systems

Extension of the study of the cyclization reactions to substituted buta-diene was restricted initially to methyl substituents for three reasons:

(a) The starting materials (isoprene, *cis*- and *trans*-piperylene, and 2,3-dimethylbutadiene) are available in large quantities.

(b) The stereochemistry of the reaction products might be expected to give valuable information on the reaction mechanism.

(c) The constitution and configuration of the products could be relatively easily determined by gas chromatography, and the usual physical and chemical methods.

1. *Synthesis of Methyl-Substituted 1,5-Cyclooctadienes*

We will first consider the formation of substituted eight-membered ring systems using the dimerization catalyst (nickel-ligand). Two routes are available: either two molecules of substituted butadiene may dimerize

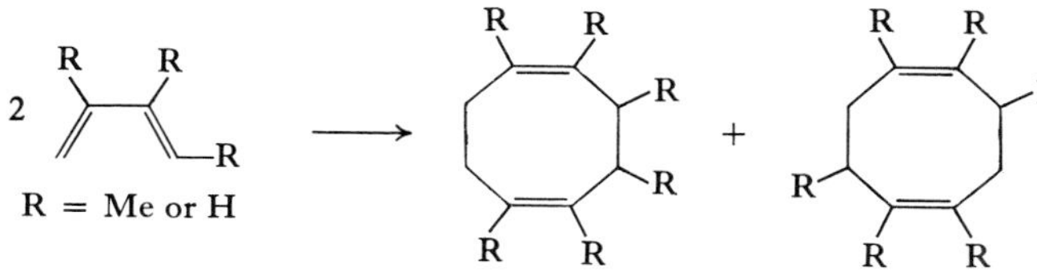

$$\text{(53)}$$

or one molecule of substituted butadiene may codimerize with an un-substituted molecule

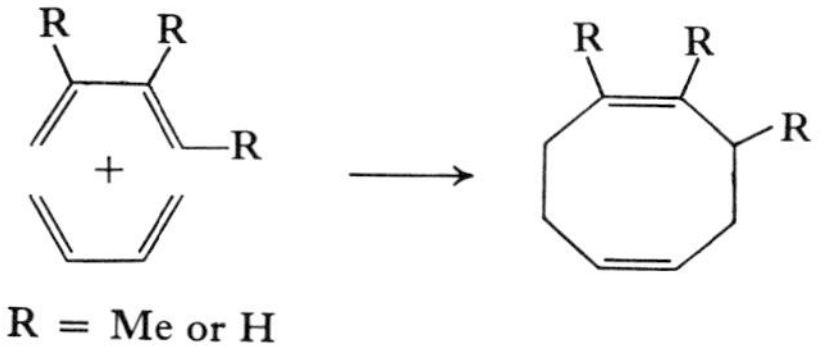

$$\text{(54)}$$

The method used (97, 98) to identify the various isomers formed warrants a short digression. Several groups (99, 100, 100a) have made use of the fact that by using a deficiency of methylene groups (generated by irradiation of diazomethane), insertion occurs, to give the corresponding methyl compound, to an extent which reflects quite accurately the number and ratio of the different types of C—H groups present in the molecule. The products are therefore relatively easy to identify gas chromatographically. Furthermore, the dimerization of methyl-substituted butadiene can only give a restricted number of products. Equation (55) illustrates how both pieces of information complement each other in the simplest case of the codimerization of butadiene with isoprene and *trans*-piperylene. The

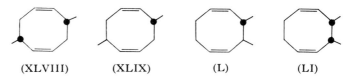

(55)

extension of this technique to the reaction of diazomethane in the presence of 1- and 3-methylcyclooctadiene enables all the possible dimethylcyclo-octadiene derivatives which could result from the dimerization of two molecules of monomethylbutadiene or from the dimerization of butadiene with a molecule of dimethylbutadiene, to be rapidly identified.

We have already discussed the effect of the ligand on the selectivity of eight-membered ring formation (Section IV,B). Table XI illustrates that this is also strongly dependent on the number of methyl substituents.

From the cyclodimerization of *cis*-piperylene all four possible isomers of dimethylcyclooctadiene, (XLVIII)–(LI), may be isolated. *trans*-Piperylene,

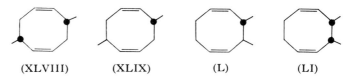

(XLVIII)   (XLIX)   (L)   (LI)

however, gives principally (XLIX) and (L) (*86*). With increasing methyl substitution, not only the yield of the eight-membered ring but also the reaction rate sharply decrease. One may suppose that the methyl groups stabilize the intermediate complexes involved. In addition, the higher electron density at the nickel atom should facilitate the isomerization of a π-allyl group to a σ-allyl group [Eq. (46)] and favor the formation of six-membered ring.

TABLE XI

CYCLODIMERIZATION OF METHYL-SUBSTITUTED 1,3-DIENES [a, b]

	1,3-Diene			
	Butadiene	Piperylene	Isoprene	2,3-Dimethyl-BD
C_6 Ring (%)	2.3	5.3	34.8	86.3
C_8 Ring (%)	97.2	90.9	55.1	6.1
C_8 Ring:C_6 ring	42	14.4	1.6	0.07
Gm Product/gm Ni/hr	220	31	14	0.6

[a] Heimbach *et al.* (*97, 98*).
[b] Catalyst, Ni–P(OC$_6$H$_4$-*o*-C$_6$H$_5$)$_3$ (1:1), 60° C. 1,3-Diene totally converted.

The codimerization of methyl-substituted 1,3-dienes with butadiene can lead not only to "mixed" dimers but also to "pure" dimers. The undesired cyclodimerization of butadiene with itself can be kept to a minimum by using as low a concentration of butadiene as possible. The yield of substituted eight-membered ring is then generally 10–45% based on the converted butadiene. After completion of the reaction, excess 1,3-diene is distilled off, with the result that any four-membered ring products are catalytically rearranged to substituted cyclooctadiene. Table XII shows the effect of increasing substitution on the yield of codimer (*97, 98*). In contrast to Table (XI), the "electivity"[6] of formation of codimer is practically independent of the extent of substitution, whereas the yield of "pure" dimer of the 1,3-diene decreases with increasing substitution. The two effects of "selectivity" and "electivity" operate here in opposition to each other; consequently, the yield of codimer remains approximately constant.

[6] The preferential formation of a codimer is defined as the "electivity" of the catalyst.

TABLE XII

CODIMERIZATION OF METHYL-SUBSTITUTED 1,3-DIENES WITH BUTADIENE[a,b]

1,3-Diene	Piperylene	Isoprene	2,3-Dimethyl-BD
C_8 Ring (%)	86.4	84.0	92.3
C_6 Ring (%)	1.6	5.5	7.6
Dimer of subst. 1,3-diene	11.5	9.8	Trace

[a] Heimbach et al. (97, 98).
[b] Catalyst = Ni–P(OC$_6$H$_4$-o-C$_6$H$_5$)$_3$ (1:1). Yield based on converted subst. 1,3-diene.

2. Preparation of Methyl-Substituted Divinylcyclobutanes

Insight into the stereochemistry of the dimerization process is provided by terminating the reaction of cis- and trans-piperylene with a nickel-tri(o-phenylphenyl)phosphite catalyst after only 10% of the 1,3-diene has been consumed. The product from cis-piperylene is the DVCB derivative (LII), and from trans-piperylene compounds (LIII) and (LIV) (86, 87).

(LII)

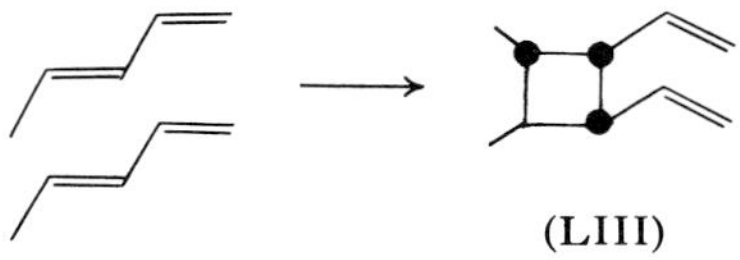

(LIII) 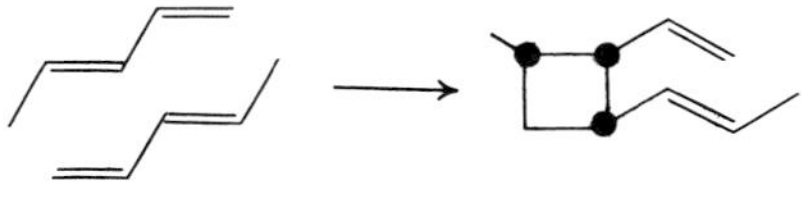   (LIV)

A mixture of cis- and trans-piperylene leads to two additional isomers (LV) and (LVI), as well as (LII)–(LIV). The yield of cyclobutane derivative is around 90%, with a 1,3-diene conversion of 10%.

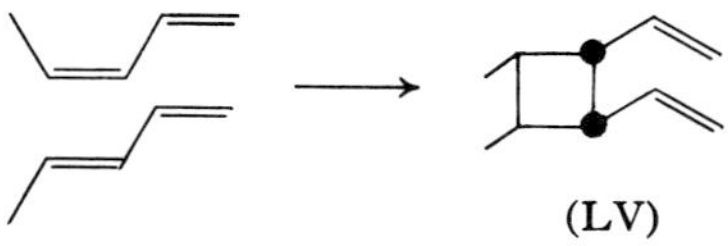

(LV) 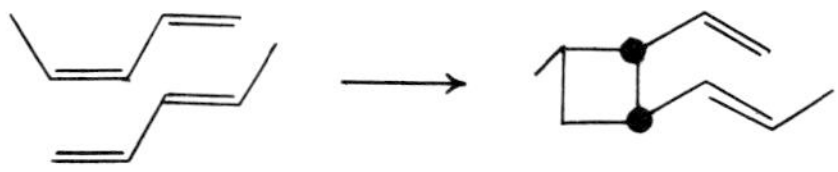   (LVI)

The four isomers produced indicate that formation of the second carbon–carbon bond (the ring closure) proceeds from a complex having a plane of symmetry and the first (formation of the C_8 chain) from a complex having a center of inversion.

Moreover, the reorientation of the carbon skeleton which necessarily occurs between these two steps is stereoselective, since the isomers (LIIa),

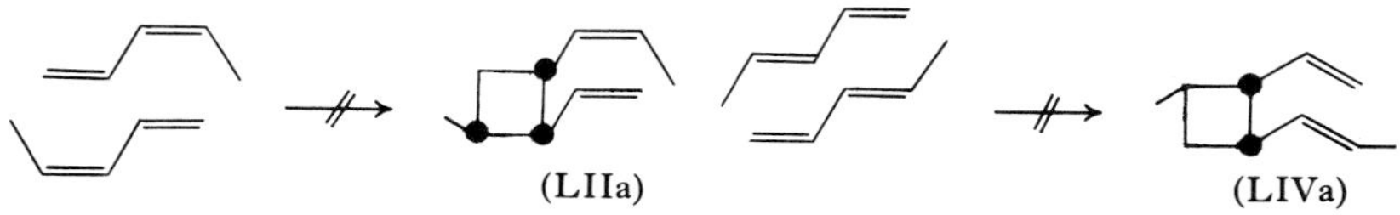

(LIIa) (LIVa)

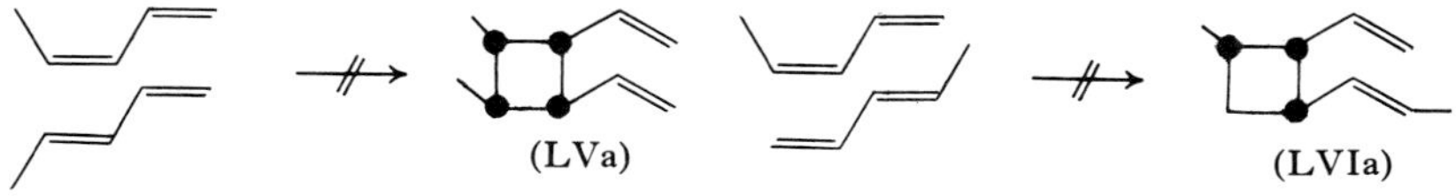

(LVa) (LVIa)

(LIVa), (LVa), and (LVIa) are not formed. Formally these would result from rotation of the unsubstituted methylene group instead of the methyl-substituted methylene group, or in the case of (LVa), of the trans- and not the cis-methyl-substituted methylene group.

Formation of the DVCB derivatives is complicated by the nickel-catalyzed isomerization of cis-piperylene into trans-piperylene which proceeds through the formation of the cyclobutane (LIII). The rate of isomerization is dependent on the nature of the ligand attached to the metal, and increases in the series tri(o-phenylphenyl)phosphite < triphenyl-phosphine < tricyclohexylphosphine. In the case of the nickel-tricyclo-hexylphosphine catalyst, the rate of isomerization is faster than the cyclization reaction.

Another factor must be considered. With increasing conversion, the rate of the nickel-catalyzed cleavage of methyl-substituted DVCB to piperyl-ene [Eq. (56)] increases. This reverse reaction is dependent on the ligand in the same sense as the isomerization of cis-piperylene. If the diene is removed as soon as it is formed (by working in a partial vacuum), no

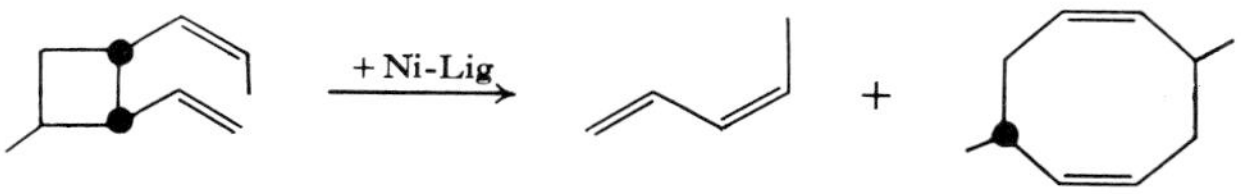

$$+ \text{Ni-Lig} \qquad\qquad + \qquad\qquad (56)$$

isomerization is observed. In addition to the diene, only dimethylcyclo-octadiene is formed. The ratio of the eight-membered ring to piperylene depends on the ligand in the opposite sense to the isomerization reaction.

In certain cases the formation of dimethylcyclooctadiene occurs only after an induction period of approximately an hour, while the cyclobutane is formed immediately. This strongly suggests that the cyclooctadiene is formed by the nickel-catalyzed rearrangement of the cyclobutane derivative, the necessary reorientation of a trans,trans into a cis,cis configuration occurring by conformational changes in the free cyclobutane [Eq. (57)].

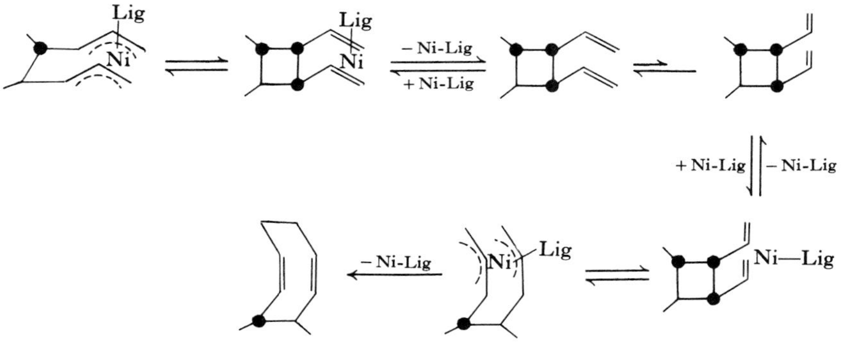

$$(57)$$

The implications of this proposal are discussed in greater detail elsewhere (*101*).

3. *Synthesis of Other Substituted Cyclooctadiene Compounds*

In Table XIII we have summarized the results obtained by codimeriza-tion of various substituted 1,3-dienes with butadiene. In all cases the highest yields were obtained using the nickel-tri(*o*-phenylphenyl)phosphite catalyst.

In contrast to the examples shown in Table XIII, the codimerization of sorbic ester with butadiene is most favored using triphenylphosphine as the additional ligand (Table XIV). Sorbic ester is a better electron donor than alkyl-substituted 1,3-dienes and hence, in order to obtain optimal results, a better electron-accepting ligand is necessary. As in the previous examples of the co-oligomerization of butadiene with butyne or ethylene, the "relay" function of the nickel atom, in transferring electronic charge from the ligand to the substrate, is of fundamental importance.

TABLE XIII

CODIMERIZATION OF SUBSTITUTED 1,3-DIENES WITH BUTADIENE[a,b]

Diene ($R_1HC:CR_2CR_3:CR_4H$)				Codimeric C_8 ring		Yield[c] (%) (Based on subst. 1,3-diene)
R_1	R_2	R_3	R_4	B.p.(°C)/mm	n_D^{20}	
CH_3	H	H	H	166.5°/760	1.4859	86
C_2H_5	H	H	H	61°/8	1.4869	82
H	C_2H_5	H	H	76°/16	1.4900	92
cis-$CH_2CH:CHCH_3$	H	H	H	47°/0.1	1.5058	93
$CH(CH_3)CH:CH_2$	H	H	H	48°–50°/0.2	1.5001	96
OCH_3	H	H	H	86°/20	1.4887	94
H	CH_3	H	H	59.5°/14	1.4910	84
CH_3	CH_3	H	H	72°/14	1.4861	86
H	CH_3	CH_3	H	78.5°/18	1.4941	93
CH_3	H	CH_3	H	67.5°/13.5	1.4858	85

[a] Heimbach et al. (97, 98).
[b] Catalyst = Nickel–$P(OC_6H_4$-o-$C_6H_5)_3$ (1:1).
[c] Yield based on reacted butadiene = 10–45% (remainder COD). Conversion of subst. 1,3-diene = 20–80%, depending on rate of addition of BD.

TABLE XIV

CODIMERIZATION OF SORBIC ESTER WITH BUTADIENE[a,b]

	Ligand[c]					
	I(a)	I(b)	II(a)	II(b)	III(a)	III(b)
Temperature (°C)	60°	40°	60°	40°	60°	40°
Total codimer (%)	2	—	10	11	40	49
C_8 Ring (%)	76	—	79	87	84	90
C_6 Ring (%)	24	—	21	13	16	10

[a] Delliehausen (102).
[b] Nickel-Ligand = 1:1. Butadiene bubbled into sorbic ester at atmospheric pressure.
[c] I = $P(OC_6H_4$-o-$C_6H_5)_3$; II = $P(OC_6H_5)_3$; III = $P(C_6H_5)_3$.

It is instructive, from a mechanistic standpoint, to compare the six-membered ring obtained from sorbic ester and butadiene (LVII) with that from 2,3-dimethylbutadiene (LVIII). Using models and bearing in mind

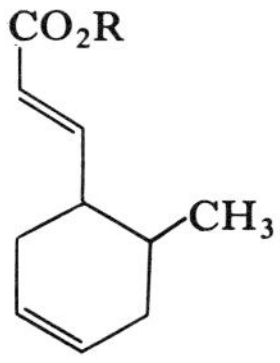

(LVII) 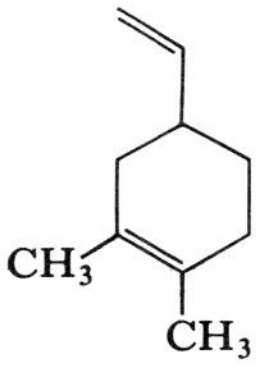   (LVIII)

the configuration of the product, the probable intermediates are (LIX) and (LX). In both of these intermediates the most stable terminal π-allyl system is adopted.

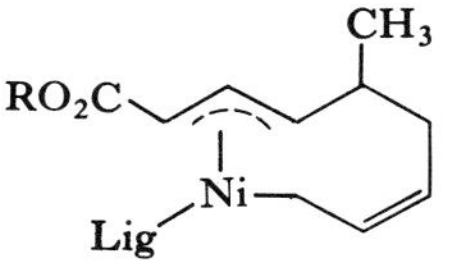

(LIX) 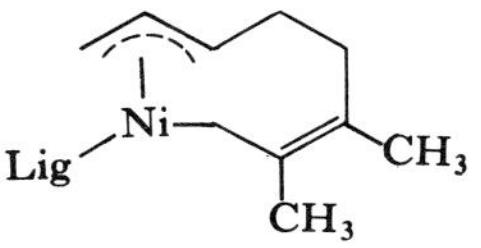   (LX)

The structure of the product (LXI) obtained by co-oligomerization of butadiene, sorbic ester, and ethylene also supports the suggestion that it is the methyl-substituted carbon atom of sorbic ester which couples to the butadiene. Compound (LXI) is formed in over 80% yield (102).

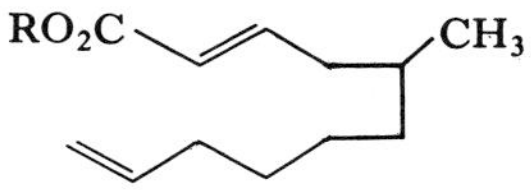

(LXI)

Confirmation of the preferred coupling of the primary carbon atom of the alkyl-substituted 1,3-diene has been referred to earlier, e.g., in the formation of cyclobutane derivatives from *trans*-piperylene.

F. Hydrogen-Transfer Reactions

We have already mentioned that the co-oligomerization of butadiene with ethylene leads to the formation of decatriene (DT) by a hydrogen-transfer process. The ratio of cyclized to open-chain product depends on the temperature and the nature of the ligand bonded to the nickel. An additional factor which affects the product distribution is the presence and nature of substituents on the olefin. Aryl and ester groups are particularly effective in promoting a hydrogen-transfer reaction, and are treated in detail below.

1. Co-oligomerization of Butadiene with Styrene

In contrast to the reaction of ethylene, styrene reacts with butadiene to give a mixture of isomers [Eq. (58), Table XV] (103, 104). It is not clear

TABLE XV

Co-oligomerization of Butadiene with Styrene[a]

	I (20°)[b]	II (110°)[c]
$C_8 + C_{12}$ (%)	51.3	46.4
C_{16} Hydrocarbons (%)	35.1	42.9
>C_{16} (%)	13.5	10.7
Gm product/gm Ni/hr	0.7	700
Cyclic C_{16}-compounds (%)	39.3	4.2
1-Phenyl-t,t,t-1,4,8-DT (%)	25.4	31.9
1-Phenyl-t,t,c-1,4,8-DT (%)	14.2	21.0
1-Phenyl-1,4,9-DT (%)	21.0	42.9

[a] Heimbach and Wilke (104).
[b] I. Butadiene:Styrene = 2:1; 20° C in autoclave, naked-nickel catalyst.
[c] II. Butadiene:Styrene = 2:1; 110° C continuous process; naked-nickel catalyst.

whether the 1,4,8-isomers are formed by hydrogen transfer from the terminal carbon atom of a *cis*- or *trans*-π-allyl group [i.e., (LXIIa) and (LXIIb)] or from the corresponding terminal σ-allyl group. Treatment of the mixture of 1,4,8- and 1,4,9-derivatives with the C_3H_5NiX—AlX_3 catalyst system (105) causes isomerization exclusively to the isomers of 1-phenyl-1,4,8-DT, greatly facilitating their identification. The main product using a naked-nickel catalyst is the straight-chain compound. Using the nickel-ligand system, the hydrogen-transfer reaction is not completely suppressed, but with tri(o-phenylphenyl)phosphite as ligand, 75% of the styrene is converted into phenylcyclodecadiene.

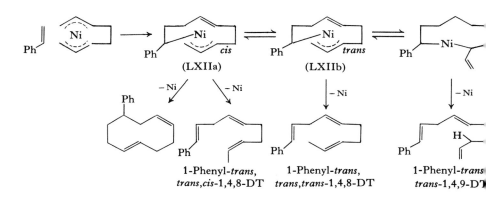

| 1-Phenyl-*trans*, | 1-Phenyl-*trans*, | 1-Phenyl-*trans* |
| *trans,cis*-1,4,8-DT | *trans,trans*-1,4,8-DT | *trans*-1,4,9-DT |

2. Co-oligomerization of Butadiene with Acrylic Esters

The products formed by the co-oligomerization of acrylic esters with butadiene (102, 106) provide useful information concerning the nature and configuration of the intermediates involved. Naked-nickel, methyl acrylate, and butadiene do not react together.[7] However, reaction does occur if the nickel-ligand system is used. The formation of the Diels-Alder adduct between the diene and olefin (a cyclohexene derivative) can be suppressed by adding the reactants dropwise to the catalyst (Table XVI footnote C).

The structure of the products and a reasonable mechanism of formation are shown in Eq. (59). The ester function is found only in the C-1 position of the open-chain product, and it is therefore probable that only one cyclodecadiene isomer, *cis,trans*-3,7-CDD-1-carboxylic acid ester (LXIIIa), is formed. However, the thermal Cope rearrangement of (LXIII) leads to

[7] At normal temperatures methyl crotonate does not react with butadiene in the presence of either naked-nickel or the nickel-ligand catalyst. Moreover, since no oligomerization of the butadiene occurs, it is probable that the formation of a stable nickel complex renders the catalyst inactive.

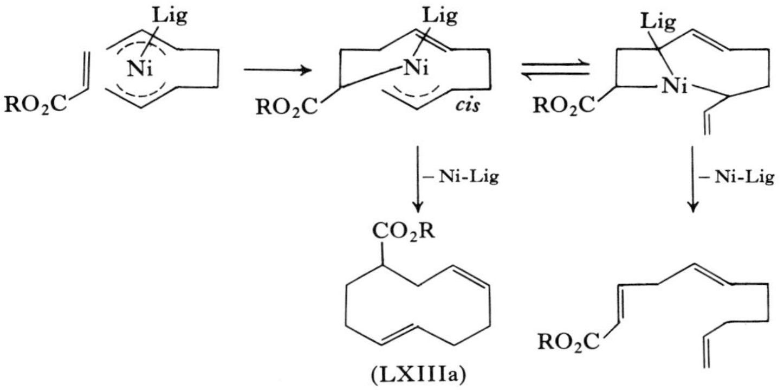

(59)

(LXIIIa)

the formation of two isomeric divinylcyclohexane derivatives which may be separated gas chromatographically. This does not conclusively demonstrate the presence of more than one isomer in the original cyclodecadiene, since the transition state of the Cope rearrangement could lead to two products differing only in the equatorial or axial position of the ester

TABLE XVI

CO-OLIGOMERIZATION OF METHYL ACRYLATE AND BUTADIENE[a, b]

	Ligand					
	P(OC$_6$H$_5$)$_3$			P(OC$_6$H$_4$-o-C$_6$H$_5$)$_3$		
Temperature	20°	60°	70°	40°	60°	60°[c]
BD-oligomer	4.8	8.5	29.1	23.2	12.1	16.0
Cyclohexene deriv. (%)	—	4.5	24.8	25.0	35.1	—
DT deriv. (%)	1.0	3.7	5.5	4.1	1.7	5.1
CDD deriv. (%)	2.1	—	—	15.8	11.1	62.7
Dicarboxylic acid ester (%)	92.1	83.3	40.6	31.9	40.0	16.2
Rate of reaction, gm ester/gm Ni/hr	0.1	1.4	1.8	~0.08	0.12	0.8
Yield based on converted ester						
Dicarboxylic acid ester (%)	97.4	92.3	58.6	45.1	47.4	24.8
CDD deriv. (%)	1.6	—	—	15.8	9.6	69.7

[a] Delliehausen (102).
[b] Nickel:Ligand = 1:1.
[c] A mixture of BD, benzene, and methyl acrylate added dropwise to the catalyst.

function (LXIV and LXV). The alternative structure, *trans,cis*-3,7-CDD-1-carboxylic acid ester, results if the first carbon–carbon bond is formed by coupling a *cis*-π-allyl group of the bis(π-allyl) C_8 chain with the unsubstituted carbon atom of the acrylic ester, or by coupling of a *trans*-π-allyl group with the substituted carbon atom.

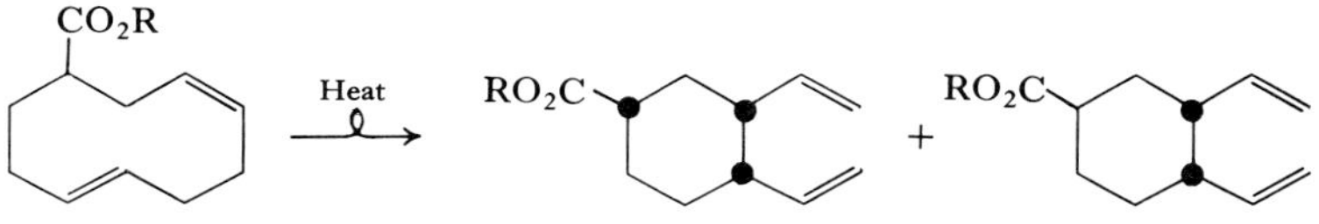

$$(60)$$

Which isomer of (LXIII) is actually formed remains an open question.

In addition to the expected cyclic and open-chain products, two isomeric dicarboxylic acid esters are produced. These result from the reaction of the σ-allyl intermediate with a second molecule of acrylic ester, followed by

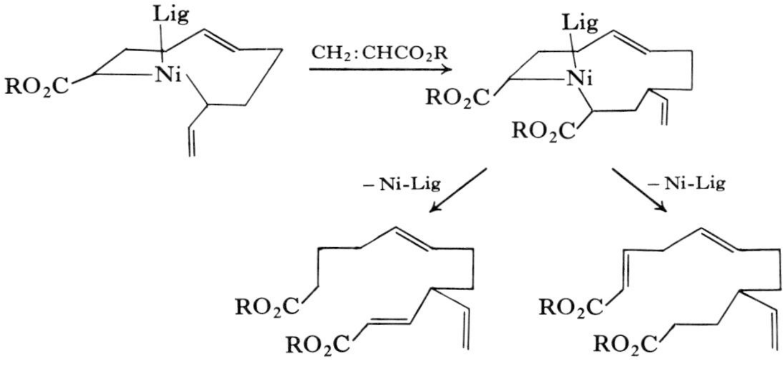

$$(61)$$

hydrogen transfer [Eq. (61)]. The yield of these dicarboxylic acid derivatives may be significantly influenced by decreasing the concentration of acrylic ester or by increasing the steric hindrance of the ligand (Table XVI).

The co-oligomerization of methacrylic ester with butadiene using a naked-nickel catalyst is particularly interesting. Surprisingly, the main product is a 1,5,10-undecatriene derivative (*102*). The formation of a cyclodecadiene derivative has been neither substantiated nor disproved, and is not considered in Eq. (62). A hydrogen atom in (LXVI) can be abstracted from either the methyl group β to the nickel atom or from the methylene group. Abstraction from the methylene group, which would lead to the thermodynamically favored tertiary olefin, is completely suppressed. The exclusive formation of a terminal olefin indicates that the preferred complexation of a primary olefin to the nickel atom plays a significant role in the hydrogen-transfer process. The stability of $(R_3P)_2Ni$-olefin complexes is known to increase with decreasing alkyl substitution (107–109).

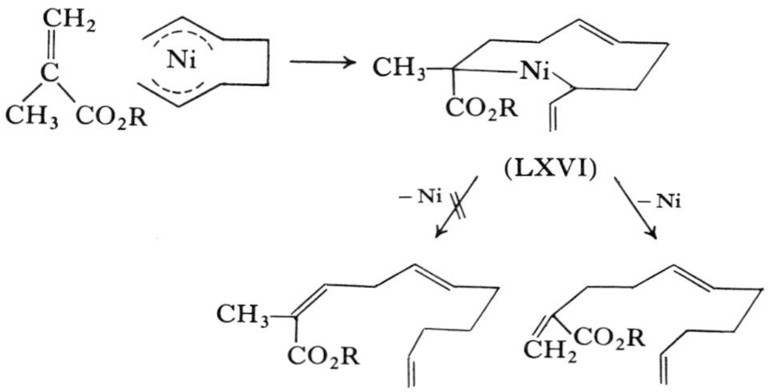

(62)

The undecatriene formed can react further with the C_8 nickel intermediate to form a nonadecapentaene derivative (LXVII) (102).

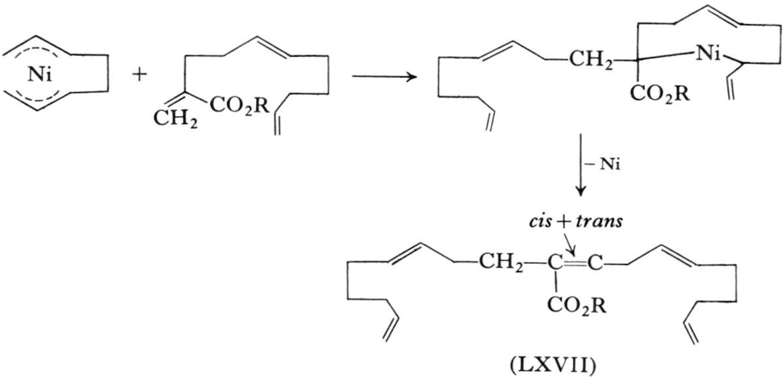

(63)

The coupling of the unsubstituted carbon atom of the mono-olefin with the C_8 chain, which was observed in the co-oligomerization of styrene with butadiene, and of acrylic esters with butadiene, is not, however, a general phenomenon. For example, the co-oligomerization of 1-decene with butadiene using nickel-tricyclohexylphosphine as catalyst leads (after hydrogenation) to the saturated hydrocarbons (LXVIII)–(LXX). The total yield is only 15% based on converted butadiene, the remainder being co-oligomers of butadiene.

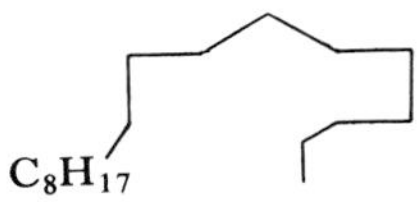

(LXVIII)	(LXIX)	(LXX)
Octadecane (34%)	9-Methylheptadecane (46%)	Octylcyclodecane (20%)

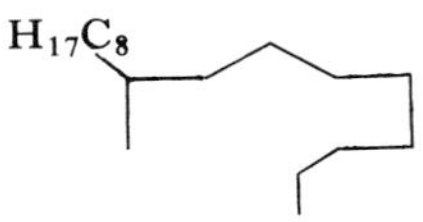

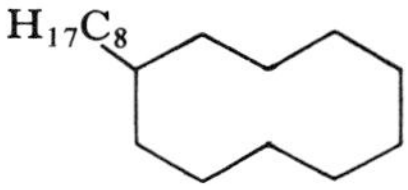

3. Other Hydrogen-Transfer Reactions

The co-oligomerization of butadiene with ethylene on naked-nickel leads to the formation of traces (2–4%) of higher oligomers, of which about 75% consists of two isomeric C_{18} tetraenes in a 60-to-40 ratio (90). Hydrogenation converts these isomers into octylcyclodecane. It is suggested that the trans double bond of cis,trans-1,5-cyclodecadiene is alkylated by a further C_8 chain. This alkylation of CDD can be carried out by dissolving the catalyst [e.g., Ni(COD)$_2$] in CDD and then slowly passing butadiene in at 40° C. The alkylated ten-membered ring is formed in a yield of over 90% based on converted CDD. CDT-t,t,t can be smoothly alkylated in a similar way (98).

G. Syntheses of Open-Chain Oligomers of Butadiene

In the presence of alcohols, butadiene is oligomerized by the nickel ligand catalyst to open chain hydrocarbons (110–112). The product, an isomer of octatriene, is dependent on the ligand attached to the nickel (113).

Ligand	Product
Triethylphosphite	1,3,6-Octatriene
Tributylphosphine	1,3,7-Octatriene
Phosphoric acid trimorpholide	2,4,6-Octatriene

Amines have the same effect as alcohols, except that here the product is

independent of the nature of the ligand and is almost exclusively *trans,trans-*
and *cis,trans*-1,3,6-octatriene *(114)*. The probable reaction mechanism is

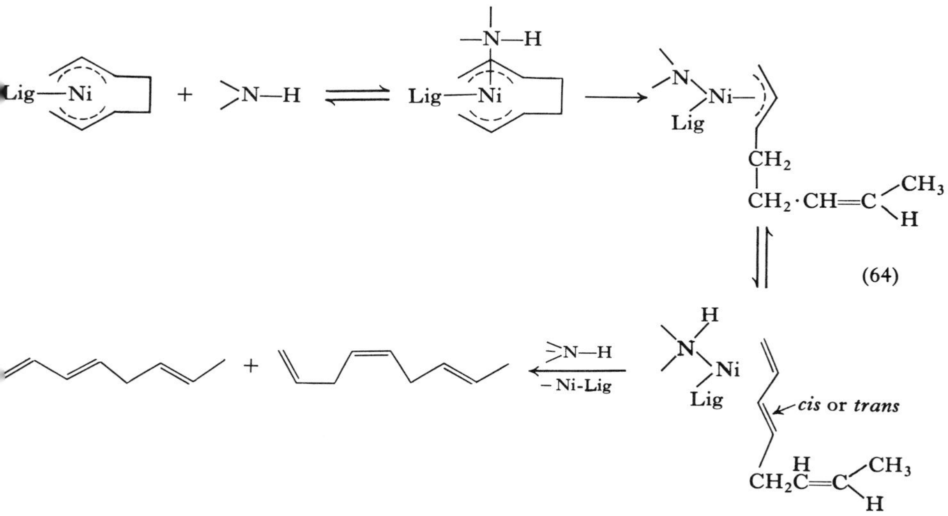

$$(64)$$

outlined in Eq. (64). This mechanism is supported by the formation, in a
side reaction, of a trialkylamine [Eq. (65)]. At 20° C aminoalkylation occurs
in greater than 75% yield based on converted amine.

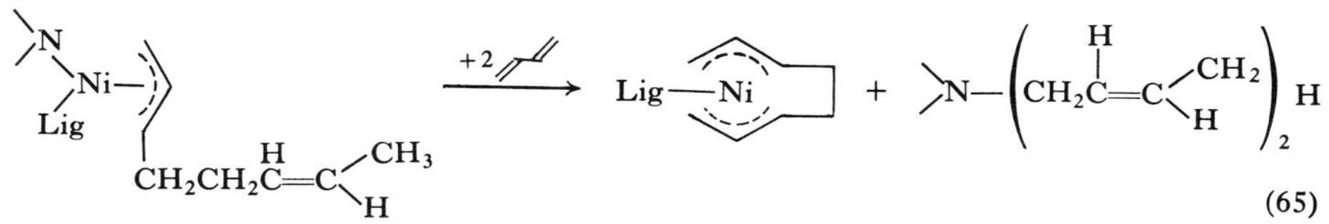

$$(65)$$

A parallel may be drawn between the formation of CDD and DT from
ethylene and butadiene using a naked-nickel catalyst and the formation of
trialkylamine and octatriene. In both cases low temperature favors the
formation of a C—C or C—N bond (i.e., formation of CDD or R_3N),
whereas at higher temperatures ($< 60°$) the hydrogen-transfer reaction
becomes predominant (i.e., formation of DT and *n*-octatriene).

The octatriene formed reacts further, with excess of butadiene, to form
butenyloctatriene [Eq. (66)]. Similar hydrogen-transfer reactions have
been reported for other catalytic systems *(115–117)*.

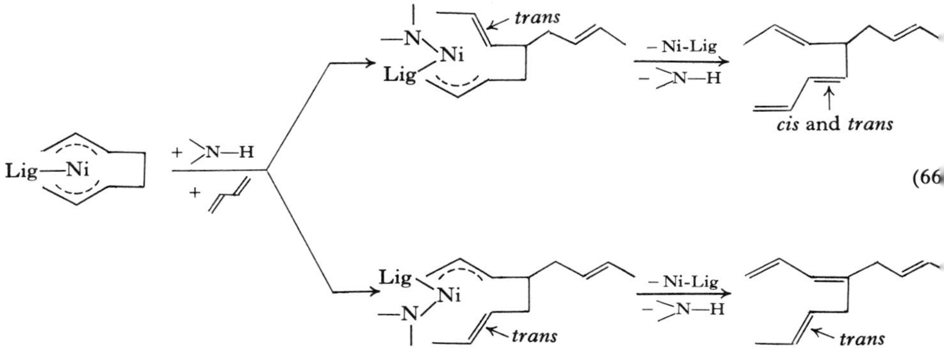

$$(66)$$

H. Mechanism of the Coupling and Cleavage Reactions

In the discussion of the mechanism of the cyclization and hydrogen-transfer processes we have not touched on the question of the origin of the reactions which occur at the metal. It is our opinion that the coupling and cleavage reactions which occur are the result of an inherent "coordinative lability" in the intermediate nickel complexes. Coordinative unsaturation results in a cleavage reaction, e.g., a transition metal alkyl compound will undergo β-elimination to give a hydride-olefin complex, while overfilling of the coordination sphere leads to decomplexation or coupling.[8] The first step in the reaction of the bis(π-allyl)nickel complex to form cis-1,2-DVCB, COD, or CDT is the occupation of an additional coordination position by an "accelerating" ligand which causes ring closure.

Two extreme cases are possible: (a) An entering olefin (this may be the second double bond of an already bonded 1,3-diene) labilizes the complex, and coupling results. In this case the entering olefin is involved in the reaction. (b) An accelerating ligand labilizes the complex, and coupling occurs between groups already bonded to the metal. A clear example of this is to be found in Eq. (43).

A second important aspect to the mechanism of the reactions discussed in this review is that a step-by-step mechanism is assumed, individual bond-making steps being a result of a minimum alteration in the coordination number around the nickel atom.

We have already shown that the formation of a four-membered ring from cis- and trans-piperylene with a *nickel-ligand* catalyst is not consistent with

[8] A more detailed treatment of this view is to be found in Heimbach and Traunmüller (*118*).

the synchronous process proposed by Mango and Schachtschneider (119). Essentially they have suggested "that certain metal systems containing orbital configurations of the prerequisite energy are capable of rendering, otherwise forbidden cycloaddition reactions, allowed, by providing a template of atomic orbitals through which electron pairs of transforming hydrocarbon ligands and metal system can interchange and flow in the required regions of space" (119). Only cycloaddition reactions are discussed from this point of view. Based on the reactions discussed in this article we have adopted a different approach (120), that "by the bonding to the central metal the originally separated ligands and the metal atom form a unique electron system. In the case of π-ligands a conjugated system involving the metal as a heteroatom is produced, to which the Woodward-Hoffmann rule is applicable. Thus the formation of a single σ-bond corresponds to the ring closure in a conjugated system. To distinguish the two different bond formations we call the first one an "electronic hetero ring closure." If, after such a step, an electronic rearrangement occurs, a certain course of reaction is conserved for the next step. In this way multistep reactions may be controlled stereoelectronically" (120).

This general approach has been tested on the synthesis of CDT with the help of the SCCC-MO method. The extension to other systems has also been undertaken (120, 121).

REFERENCES

1. Schrauzer, G. N., *Advan. Organometal. Chem.* 2, 2 (1964).
2. Collman, J. P., *Transition Metal Chem.* 2, 1 (1966).
3. Candlin, J. P., Taylor, K. A., and Thompson, D. T., "Reactions of Transition Metal Complexes." Elsevier, Amsterdam, 1968.
4. Chalk, A. J., and Harrod, J. F., *Advan. Organometal. Chem.* 6, 119 (1968).
5. Ugo, R., *Coord. Chem. Rev.* 3, 319 (1968).
6. Ochiai, E-I., *Coord. Chem. Rev.* 3, 49 (1968).
7. Schrauzer, G. N., *Advan. Catalysis* 18, 373 (1968).
8. Bird, C. W., "Transition, Metal Intermediates in Organic Synthesis." Logos, London, 1967.
9. I. G. Farbenind A. G., Belgian Patent 448,884 (1943); *Chem. Abstr.* 41, 6576a (1945).
10. Corey, E. J., Semmelhack, M. F., and Hegedus, L. S., *J. Am. Chem. Soc.* 90, 2416 (1968).
11. Walter, D., MPI für Kohlenforschung, Dissertation Tech. Hochschule Aachen (1965).
12. Guerrieri, F., and Chiusoli, G. P., *Chem. Commun.* p. 781 (1967).
13. Keim, W., MPI für Kohlenforschung, Dissertation Tech. Hochschule Aachen (1963); see Wilke, G. et al., *Angew. Chem.* 78, 157 (1966).
14. Guerrieri, F., and Chiusoli, G. P., *J. Organometal. Chem.* (*Amsterdam*) 15, 209 (1968).
15. Corey, E. J., Hegedus, L. S., and Semmelhack, M. F., *J. Am. Chem. Soc.* 90, 2417 (1968).

16. Corey, E. J., and Wat, E. K. W., *J. Am. Chem. Soc.* **89**, 2757 (1967).
17. Wilke, G., *et al.*, *Angew. Chem.* **75**, 10 (1963).
18. Birkenstock, U., MPI für Kohlenforschung, Dissertation Tech. Hochschule Aachen (1966). Studiengesellschaft Kohle M.D.H., Belg. Pat. 702682 (1968).
19. Corey, E. J., and Hamanaka, E., *J. Am. Chem. Soc.* **89**, 2758 (1967).
20. Corey, E. J., and Hamanaka, E., *J. Am. Chem. Soc.* **86**, 1641 (1964).
21. Corey, E. J., and Semmelhack, M. F., *Tetrahedron Letters* p. 6237 (1966).
22. Corey, E. J., and Semmelhack, M. F., *J. Am. Chem. Soc.* **89**, 2755 (1967).
23. Dubini, M., Montino, F., and Chiusoli, G. P., *Chim. Ind.* (*Milan*) **47**, 839 (1965).
24. Dubini, M., and Montino, F., *J. Organometal. Chem.* (*Amsterdam*) **6**, 188 (1966).
25. Yoshisato, E., and Tsutsumi, S., *J. Am. Chem. Soc.* **90**, 4488 (1968).
26. Yoshisato, E., and Tsutsumi, S., *Chem. Commun.* p. 33 (1968).
27. Beckert, W. F., and Lowe, J. U., *J. Org. Chem.* **32**, 1215 (1967).
28. Yoshisato, E., and Tsutsumi, S., *J. Org. Chem.* **33**, 869 (1968).
29. Walter D., and Wilke, G., unpublished (1967).
30. Webb, I. D., and Borchert, G. T., *J. Am. Chem. Soc.* **73**, 2654 (1951).
31. Bauld, N. L., *Tetrahedron Letters* p. 859 (1962).
32. Chiusoli, G. P., and Cometti, G., *Chim. Ind.* (*Milan*) **45**, 461 (1963).
33. de Ortueta Spiegelberg, C., MPI für Kohlenforschung, Dissertation Tech. Hochschule Aachen (1965); see also Wilke, G., *J. Appl. Chem.* **17**, 179 (1968).
34. Chiusoli, G. P., Bottacio, G., and Cameroni, A., *Chim. Ind.* (*Milan*) **44**, 131 (1962).
35. Chiusoli, G. P., *Gazz. Chim. Ital.* **89**, 1332 (1959).
36. Chiusoli, G. P., *Chim. Ind.* (*Milan*) **41**, 503 (1959).
37. Chiusoli, G. P., *Chim. Ind.* (*Milan*) **43**, 363 (1961).
38. Chiusoli, G. P., *Angew. Chem.* **72**, 74 (1960).
39. Chiusoli, G. P., and Cassar, L., *Angew. Chem.* **79**, 177 (1967).
40. Chiusoli, G. P., and Merzoni, S., *Z. Naturforsch.* **17b**, 850 (1962).
41. Heck R. F., *J. Am. Chem. Soc.* **85**, 2013 (1963).
42. Cassar, L., Chiusoli, G. P., and Foa, M., *Tetrahedron Letters* p. 285 (1967).
43. Chiusoli, G. P., *Chim. Ind.* (*Milan*) **41**, 506 (1959).
44. Chiusoli, G. P., *Chim. Ind.* (*Milan*) **41**, 762 (1959).
45. Chiusoli, G. P., *Chim. Ind.* (*Milan*) **45**, 6 (1963).
46. Chiusoli, G. P., and Merzoni, S., *Chim. Ind.* (*Milan*) **42**, 6 (1963).
47. Chiusoli, G. P., Dubini, M., Ferraris, M., Guerrieri, F., Merzoni, S., and Mondelli, G., *J. Chem. Soc.*, *C* p. 2890 (1968).
48. Guerrieri, F., *Chem. Commun.* p. 983 (1968).
49. Chiusoli, G. P., and Bottacio, G., *Chim. Ind.* (*Milan*) **47**, 165 (1965).
50. Cassar, L., and Chiusoli, G. P., *Tetrahedron Letters* p. 3295 (1965).
51. Cassar, L., and Chiusoli, G. P., *Chim. Ind.* (*Milan*) **48**, 323 (1966).
52. Foa, M., Cassar, L., and Venturi, M. T., *Tetrahedron Letters* p. 1357 (1968).
53. Chiusoli, G. P., Bottacio, G., and Venturello, C., *Tetrahedron Letters* p. 2875 (1965).
54. Chiusoli, G. P., and Bottacio, G., *Chim. Ind.* (*Milan*) **47**, 165 (1965).
55. Chiusoli, G. P., and Cometti, G., *Chim. Ind.* (*Milan*) **45**, 401 (1963).
56. Cassar, L., and Chiusoli, G. P., *Tetrahedron Letters* p. 2805 (1966).
57. Bauld, N. L., *Tetrahedron Letters* p. 1841 (1963).
58. Mettalia, G. B., and Specht, E. H., *J. Org. Chem.* **32**, 3941 (1967).
59. Yoshisato, E., and Tsutsumi, S., *J. Org. Chem.* **33**, 869 (1968).
60. Beckert, W. F., and Lowe, J. U., *J. Org. Chem.* **32**, 1215 (1967).
61. Chiusoli, G. P., Merzoni, S., and Mondelli, G., *Tetrahedron Letters* p. 2777 (1964).

62. Myeong, S. K., Sawa, Y., Ryang, M., and Tsutsumi, S. S., *Bull. Chem. Soc. Japan* **38**, 330 (1965).
63. Ryang, M., Myeong, S. K., Sawa, Y., and Tsutsumi, S., *J. Organometal. Chem. (Amsterdam)* **2**, 305 (1966).
64. Sawa, Y., Hashimoto, I., Ryang, M., and Tsutsumi, S., *J. Org. Chem.* **33**, 2159 (1968).
65. Reppe, W., Schlichting, O., Klager, K., and Toepel, T., *Ann. Chem.* **560**, 1 (1948).
66. Reed, H. W. B., *J. Chem. Soc.* p. 1931 (1954).
67. Reppe, W., and Schweckendiek, W. J., *Ann. Chem.* **560**, 104 (1948).
68. Wilke, G., Müller, E. W., and Kröner, M., *Angew. Chem.* **73**, 33 (1961).
69. Breil, H., Heimbach, P., Kröner, M., Müller, H., and Wilke, G., *Makromol. Chem.* **69**, 18 (1963).
70. Wilke, G., Kröner, M., and Bogdanovič, B., *Angew. Chem.* **73**, 755 (1961).
71. Bogdanovič, B., MPI für Kohlenforschung, Dissertation Technische Hochschule Aachen (1962).
71a. Bönnemann, H., Bogdanovič, B., and Wilke, G., *Angew. Chem.* **79**, 817 (1967).
71b. Bönnemann, H., MPI für Kohlenforschung, Dissertation, Technische Hochschule Aachen (1967).
72. Bogdanovič, B., Heimbach, P., Kröner, H., Wilke, G., Hoffmann, E. G., and Brandt, J., *Ann. Chem.* **727**, 143 (1969).
73. Heimbach, P., *Symp. Allyl- Olefin-Complexes Metals, Sheffield, Gt. Brit.*, 1967.
74. Untch, K. G., and Martin, D. J., *J. Am. Chem. Soc.* **87**, 3518 (1965).
75. Jonas, K., Heimbach, P., and Wilke, G., *Angew. Chem.* **80**, 1033 (1968).
76. For a summary of the literature, see Hoffmann, R., *Tetrahedron* **22**, 521 (1966).
77. Smith, W. B., and Massingill, J. L., *J. Am. Chem. Soc.* **83**, 4301 (1961).
78. Hammond, G. S., and Liu, R. S. H., *J. Am. Chem. Soc.* **85**, 477 (1963).
79. Studiengesellschaft Kohle m.b.H., Austrian Patent 219,580 (1959/60).
80. Wittenberg, D., *Angew. Chem.* **75**, 1124 (1963).
81. Otsuka, S., Kikuchi, T., and Taketomi, T., *J. Am. Chem. Soc.* **85**, 3709 (1963).
82. Natta, G., Giannini, U., Pino, P., and Cassata, A., *Chim. Ind. (Milan)* **47**, 524 (1965).
83. Müller, H., Wittenberg, D., Seibt, H., and Scharf, E., *Angew. Chem.* **77**, 318 (1965).
84. Brenner, W., Heimbach, P., Hey, H., Müller, E. W., and Wilke, G., *Ann. Chem.* **727**, 161 (1969).
85. Meriwether, L. S., and Fiene, M. L., *J. Am. Chem. Soc.* **81**, 4200 (1960).
86. Heimbach, P., and Hey, H., *Angew. Chem.* (1969) (in press).
87. Hey, H., MPI für Kohlenforschung, Dissertation, Universität Bochum (1969).
88. Vogel, E., *Ann. Chem.* **615**, 2 (1958).
89. Benson, S. W., *J. Chem. Phys.* **46**, 4920 (1967).
90. Heimbach, P., and Wilke, G., *Ann. Chem.* **727**, 183 (1969).
91. Heimbach, P., *Angew. Chem.* **76**, 859 (1964).
92. Heimbach, P., and Brenner, W., *Angew. Chem.* **78**, 983 (1966).
93. Ploner, K., MPI für Kohlenforschung, Dissertation, Universität Bochum (1969).
94. Brenner, W., Heimbach, P., and Wilke, G., *Ann. Chem.* **727**, 194 (1969).
95. Chatt, J., Rowe, G. A., and Williams, A. A., *Proc. Chem. Soc.* p. 208 (1957).
96. Brenner, W., Heimbach, P., Ploner, K., and Thömel, F., *Angew. Chem.* **81**, 744 (1969).
97. Heimbach, P., Schomburg, G., and Wilke, G., unpublished results.
98. Heimbach, P., and Wilke, G., unpublished results.
99. Doering, W. von E., Buttery, R. G., Laughlin, R. G., and Chaudhuri, N., *J. Am. Chem. Soc.* **78**, 3224 (1956).

100. Simmons, M. C., Richardson, D. B., and Durett, C. R., *in* "Gas Chromatography" (R. P. W. Scott, ed.), p. 211. Butterworth, London and Washington, D.C., 1960.

100a. Dvoretsky, T., Richardson, D. B., and Durett, C. R., *Anal. Chem.* **35**, 545 (1963).

101. Heimbach, P., and Hey, H., in preparation.

102. Delliehausen, C., MPI für Kohlenforschung, Dissertation, Univ. Bochum (1968).

103. Lautenschlager, H., Scharf, E., Wittenberg, D., and Müller, H., BASF, Belgian Patent 622,195 (1961/63).

104. Heimbach, P., and Wilke, G., unpublished results.

105. Bogdanovič, B., and Wilke, G., *Brennstoff-Chem.* **49**, 323 (1968).

106. Rhone-Poulenc, French Patent 1,433,409 (1965).

107. Wilke, G., and Herrmann, G., *Angew. Chem.* **78**, 591 (1966).

108. Herrmann, G., MPI für Kohlenforschung, Dissertation, Technische Hochschule Aachen (1963).

109. Jonas, K., MPI für Kohlenforschung, Dissertation, Universität Bochum (1968).

110. Seibt, H., and Kutepour, N. V., BASF, Belgian Patent 635,483 (1962/63).

111. Feldmann, J., Frampton, O., Saffer, B., and Thomas, M., *Am. Chem. Soc., Div. Petrol. Chem., Preprints* **9**, No. 4, A55–A64 (1964).

112. Smutny, E. J., *J. Am. Chem. Soc.* **89**, 6793 (1967).

113. Müller, H., Wittenberg, D., Seibt, H., and Scharf, E., *Angew. Chem.* **77**, 320 (1965)

114. Heimbach, P., *Angew. Chem.* **80**, 967 (1968).

115. Carbonaro, A., Greco, A., and Dall'Asta, G., *Tetrahedron Letters* p. 2037 (1967).

116. Miller, R. G., Kealy, T. J., and Barney, A. L., *J. Am. Chem. Soc.* **89**, 3756 (1967).

117. Miyake, A., Hata, G., Iwamoto, M., and Yuguchi, S., *Proc. 7th World Petrol. Conf.,* **5**, 317 (1967).

118. Heimbach, P., and Traunmüller, R., "Metall-Olefin-Komplex Chemie." Verlag Chemie, Weinheim.

119. Mango, F. D., and Schachtschneider, J. H., *J. Am. Chem. Soc.* **89**, 2486 (1967).

120. Traunmüller, R., Polansky, O. E., Heimbach, P., and Wilke, G., *Chem. Phys. Letters* **3**, 300 (1969).

121. Traunmüller, R., Dissertation, Universität Wien (1969).

Transition Metal–
Carborane Complexes

LEE J. TODD

Department of Chemistry,
Indiana University,
Bloomington, Indiana

I

INTRODUCTION

A. Scope

The chemistry of compounds having boron groups connected to transition metals has expanded very rapidly in the past few years. One class of compounds involves bonding between boron hydride fragments (BH_3, $BH_4{}^-$, $B_3H_8{}^-$, $B_{10}H_{12}{}^{2-}$) and metals. Examples of this class are (a) Lewis acid–base adducts (*32*) like $BH_3 \cdot Re(CO)_5{}^-$; (b) compounds containing metal—H—B bridge bonding such as $Zr(BH_4)_4$ (*2*), $Cr(CO)_4B_3H_8{}^-$ (*24*), and the unusually complex polynuclear system, $HMn_3(CO)_{10}(B_2H_6)$

(23); and (c) transition metal, $M(B_{10}H_{12})^{2-}$ ions (8, 25) which appear to have multicenter bonding between the borane fragment and the metal.

A second class of compounds having boron–metal σ-bonds is exemplified by species such as $(C_6H_5)_2BMn(CO)_4P(C_6H_5)_3$ (31).

The third and largest class of compounds comprises closed-cage polyhedral molecules in which the metal atom is held in the cage by π-bonding to the borane fragment in a manner formally analogous to ferrocene. The first members of this class were reported in 1965 by Hawthorne et al. (10). This diverse, chemically stable class of compounds will comprise the subject of this review.

B. Nomenclature

The nomenclature problems in this emerging area of chemistry will be very complex judging from the variety of structures reported thus far. Tentative nomenclature rules for boron compounds approved by the Council of the American Chemical Society have recently appeared.[1] Fortunately to date most authors have elected to use formulas instead of names in their papers. This review will attempt to follow their example. However, for brevity and convenience, the 11-atom ligands will be described by the specific nomenclature suggested by Hawthorne (18). The hypothetical $B_{11}H_{11}^{4-}$ ion is given the trivial name "ollide" after the Spanish noun "olla" meaning water jar. The isoelectronic heteroatom ions, $B_{10}CH_{11}^{3-}$, $B_9C_2H_{11}^{2-}$, and $B_9CPH_{10}^{2-}$ are then named the "carbollide," "dicarbollide," and "phosphacarbollide" ions, respectively. Rules for numbering the cage atoms are necessary to locate the positions of the heteroatoms (1). To date all 11-atom ligands have the heteroatoms in the open five-member face of the cage. The numbering of the heteroatoms of the ollide ion is derived from the numbering system of the parent icosahedral molecule. Thus, reaction of $1,2-B_{10}C_2H_{12}$ with base generates the 1,2-dicarbollide ion (see Section II,A). The transition metal complexes are then named as the appropriate "ollyl" complexes, in accordance with the cyclopentadienyl metal nomenclature. The molecule shown in Fig. 5 would be named π-cyclopentadienyl-π-1,2-dicarbollyl-iron(III).

---

[1] These rules are presented in *Inorg. Chem.* **7**, 1945 (1968).

II

COMPLEXES WITH ELEVEN-ATOM LIGANDS

A. 1,2-$B_9C_2H_{11}^{2-}$ and C-Substituted Ligands

The icosahedral molecules 1,2-$B_{10}C_2H_{12}$, 1,2-$B_{10}CPH_{11}$, and 1,2-$B_{10}CAsH_{11}$ can be thermally rearranged at 400°–600° C to the corresponding 1,7-isomers in good yield (see Fig. 1 for numbering of icosahedral cage).

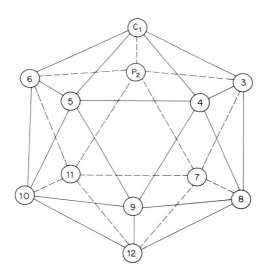

Fig. 1. Proposed structure of 1,2-$B_{10}H_{10}CHP$.

The 1,2- and 1,7-isomers are susceptible to degradation with strong base, which removes one boron atom from each of the cages. In the 1,2-isomers either the 3- or 6-boron atom is removed while for the 1,7-isomers either the 2- or 3-boron atom is removed. The boron atoms capable of extraction in each isomer are adjacent to both heteroatoms. These boron atoms are the most electron-deficient and the most susceptible to nucleophilic base attack. The 11-atom fragments are not normally isolated as the ollide ions themselves but in a protonated form (i.e., 1,2-$B_9C_2H_{12}^-$). The extra protons are easily removed with base prior to metal complexation. The ollide ions discussed in this section may well exist as open-face icosahedral fragments (Fig. 2). This point has not been definitely determined yet and an alternative

TABLE I

THE VARIOUS TYPES OF ELEVEN-ATOM LIGANDS

Ligand	Synthesis references
1,2- and 1,7-$B_9C_2H_{11}{}^{2-}$	18
1,2- and 1,7-$B_9CPH_{10}{}^{2-}$	36
1,2- and 1,7-$B_9H_9CHPCH_3{}^{-}$	36
1,2- and 1,7-$B_9CAsH_{10}{}^{2-}$	38
1,7-$B_9H_9CHAsCH_3{}^{-}$	38
$B_{10}CH_{11}{}^{3-}$	22, 26
$B_{10}H_{10}CNH_3{}^{2-}$	26
$B_{10}H_{10}S^{2-}$	21

closed-cage 11-atom polyhedral structure is a possibility. The basic types of 11-atom ligands for which metal complexes have been reported to date are presented in Table I. The "thiollide" ion, $B_{10}SH_{10}{}^{2-}$, does not in a strict sense belong in this review but is added for the sake of completeness because it is of the same class.

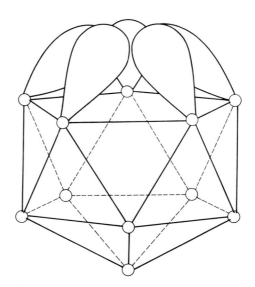

FIG. 2. Proposed structure for the ollide ions.

If the ollide ions are or can rearrange to open icosahedral species (Fig. 2), the open face of these structures would have the characteristics of the cyclopentadienide ion. The five nearly equivalent sp^3 atomic orbitals pointing toward the missing apex of the ollide ion should yield five π-type molecular orbitals which correspond to the A_1, the two E_1, and the two E_2 molecular orbitals of the $C_5H_5^-$ ion. This type of ligand would then be expected to form sandwich-bonded complexes similar to the metallocenes.

The properties of the π-ollyl metal compounds are summarized in Table II.

1. *Chromium Group*

Treatment of chromium(III) chloride with the $1,2$-$B_9C_2H_{11}^{2-}$ ion or various carbon-substituted derivatives produced bis(1,2-dicarbollyl)-chromium complexes which contain formal Cr(III) (*33*). The ($1,2$-$B_9C_2H_{11})_2Cr^-$ salts are stable in hot aqueous solution, in contrast to chromicinium salts which are easily hydrolyzed. Under basic conditions these bis(dicarbollide) complexes are decomposed and the $B_9C_2H_{12}^-$ ion can be recovered in good yield. The ^1H and ^{11}B NMR spectra were not obtained on these complexes because they are strongly paramagnetic (three unpaired electrons—see Table III). An X-ray diffraction study of the cesium salt of $[1,2$-$B_9H_9C_2(CH_3)_2]_2Cr^-$ indicated a symmetrical sandwich structure similar to that shown in Fig. 6 (*33*). Neutral mixed ligand complexes such as $C_5H_5Cr(1,2$-$B_9C_2H_{11})$ have also been obtained. These are sublimable red compounds which are subject to air oxidation in solution.

Exposure of a tetrahydrofuran solution of $W(CO)_6$ and the $1,2$-$B_9C_2H_{11}^{2-}$ ion to ultraviolet radiation produced immediate carbon monoxide evolution. Ultimately the air-sensitive $(1,2$-$B_9C_2H_{11})W(CO)_3^{2-}$ ion was obtained as the tetramethylammonium salt (*18*). The corresponding chromium and molybdenum complexes have been obtained in the same manner (*17, 18*). These dianions undergo nucleophilic reactions characteristic of the analogous π-$C_5H_5Mo(CO)_3^-$ ion

$$(1,2\text{-}B_9C_2H_{11})Mo(CO)_3^{2-} \quad \begin{array}{l} \xrightarrow{H^+} (1,2\text{-}B_9C_2H_{11})Mo(CO)_3H^- \\ \xrightarrow{CH_3I} (1,2\text{-}B_9C_2H_{11})Mo(CO)_3CH_3^- \\ \xrightarrow{M(CO)_6} (1,2\text{-}B_9C_2H_{11})Mo(CO)_3M(CO)_5^{2-} \end{array} \qquad (1)$$

$$M = W \text{ or } Mo$$

TABLE II

π-OLLYL METAL COMPOUNDS

Compound	Color	Melting point (°C)	References of synthesis
Chromium Group			
$[1,2\text{-}B_9C_2H_{11}]W(CO)_3{}^{2-}$	Pale yellow	—	18
$[1,2\text{-}B_9C_2H_{11}]Mo(CO)_3{}^{2-}$	Gray	—	18
$[1,2\text{-}B_9H_9C_2(CH_3)_2]Mo(CO)_3{}^{2-}$	Pale yellow	—	18
$[1,2\text{-}B_9C_2H_{11}]W_2(CO)_8{}^{2-}$	Orange	—	18
$[1,2\text{-}B_9H_9C_2(CH_3)_2]Mo_2(CO)_8{}^{2-}$	Yellow	—	18
$[1,2\text{-}B_9C_2H_{11}]Mo(CO)_3W(CO)_5{}^{2-}$	—	—	18
$[1,2\text{-}B_9C_2H_{11})W(CO)_3Mo(CO)_5{}^{2-}$	Orange–yellow	—	18
$[1,2\text{-}B_9C_2H_{11})W(CO)_3CH_3{}^{-}$	Green–yellow	—	18
$[1,2\text{-}B_9C_2H_{11}]Mo(CO)_3CH_3{}^{-}$	Tan	—	18
$[1,2\text{-}B_9C_2H_{11}]Mo(CO)_3H^{-}$	Red	—	18
$[1,2\text{-}B_9C_2H_{11}]Cr(CO)_3{}^{2-}$	—	—	17
$[1,2\text{-}B_9C_2H_{11}]_2Cr^{-}$	Dark violet	—	33
$[1,2\text{-}B_9H_9CHC(CH_3)]_2Cr^{-}$	Dark violet	—	33
$[1,2\text{-}B_9H_9C_2(CH_3)_2]_2Cr^{-}$	Dark violet	—	33
$[1,2\text{-}B_9H_9CHC(C_6H_5)]_2Cr^{-}$	Dark violet	—	33
$[1,7\text{-}B_9C_2H_{11}]_2Cr^{-}$	Brown	—	33
$(C_5H_5)Cr[1,2\text{-}B_9C_2H_{11}]$	Dark red	248°–249°	33
$(C_5H_5)Cr[1,2\text{-}B_9H_9CHC(CH_3)]$	Dark red	219°–220°	33
$(C_5H_5)Cr[1,2\text{-}B_9H_9C_2(CH_3)_2]$	Dark red	261°–262°	33
$(C_5H_5)Cr[1,2\text{-}B_9H_9CHC(C_6H_5)]$	Dark red	208°–209°	33
$(C_5H_5)Cr[1,7\text{-}B_9C_2H_{11}]$	Dark red	217°–218°	33
Manganese Group			
$[1,2\text{-}B_9C_2H_{11}]Mn(CO)_3{}^{-}$	Pale yellow	—	18
$[1,2\text{-}B_9C_2H_{11}]Re(CO)_3{}^{-}$	Pale yellow	—	18
$[B_{10}CH_{11}]Mn(CO)_3{}^{2-}$	Pale yellow	—	22
$[1,7\text{-}B_9H_9CHPCH_3]Mn(CO)_3$	Yellow	99°–100°	36
$[B_{10}H_{10}S]Re(CO)_3{}^{-}$	White	—	21
$[B_{10}CH_{11}]_2Mn^{2-}$	—	—	26
$[1,7\text{-}B_9H_9CHAs]Mn(CO)_3{}^{-}$	Pale yellow	—	38
Iron Group			
$[1,2\text{-}B_9C_2H_{11}]_2Fe^{-}$	Red	—	18
$[1,2\text{-}B_9H_9C_2(CH_3)_2]_2Fe^{-}$	Red	247°–249° [N(CH_3)_4 salt]	18
$[1,2\text{-}B_9H_9CHC(C_6H_5)]_2Fe^{-}$	Red	—	18
$(C_5H_5)Fe[1,2\text{-}B_9C_2H_{11}]$	Deep red	181°–182°	18
$(C_5H_5)Fe[1,2\text{-}B_9C_2H_{11}]^{-}$	Orange	158° [N(CH_3)_4 salt]	18

TABLE II—*continued*

Compound	Color	Melting point (°C)	References of synthesis
$[1,2\text{-}B_9C_2H_{11}]_2Fe^{2-}$	Pink	—	18
$[B_{10}CH_{11}]_2Fe^{3-}$	Red	—	22
$[B_{10}C(NH_3)H_{10}]_2Fe^-$	—	—	26
$[1,7\text{-}B_9H_9CHP]_2Fe^{2-}$	Lavender	—	36
$[1,7\text{-}B_9H_9CHPCH_3]_2Fe$	Brown	239.5°–240.5°	36
$[1,7\text{-}B_9H_9CHPCH_3]_2Fe$	Red	233°–234°	36
$[1,2\text{-}B_9H_9CHPCH_3]_2Fe$	Pink	325° (dec)	36
$(C_5H_5)Fe[1,7\text{-}B_9H_9CHPCH_3]$	Orange	165°–167°	36
$[1,7\text{-}B_9H_9CHPCH_3]Fe[1,7\text{-}B_9H_9CHP]^-$	Red–orange	—	37
$[B_{10}SH_{10}]_2Fe^{2-}$	Dark red	—	21
$[B_9H_9B(C_6H_5)S]_2Fe^{2-}$	—	—	21
$[1,2\text{-}B_9H_9C_2(CH_3)_2]_2Fe^{2-}$	Blue	—	18
$[1,2\text{-}B_9H_9CHC(C_6H_5)]_2Fe^{2-}$	Blue	—	18
$C_5H_5Fe[1,7\text{-}B_9H_9CHAsCH_3]$	Red	—	38
$[1,7\text{-}B_9H_9CHAs]_2Fe^{2-}$	Red	—	38
$[1,7\text{-}B_9H_9CHAs]_2Fe^-$	Green	—	38

Cobalt Group

Compound	Color	Melting point (°C)	References of synthesis
$[1,2\text{-}B_9C_2H_{11}]_2Co^-$	Yellow	—	18
$[1,2\text{-}B_9H_9C_2(CH_3)_2]_2Co^-$	Red	273°–275° $[N(CH_3)_4salt]$	18
$[1,2\text{-}B_9H_9CHC(C_6H_5)]_2Co^-$	Red	290°–293° $[N(CH_3)_4salt]$	18
$[1,2\text{-}B_9C_2H_8Br_3]_2Co^-$	Orange	—	18
$(C_5H_5)Co[1,2\text{-}B_9C_2H_{11}]$	Yellow	246°–248°	18
$[1,7\text{-}B_9C_2H_{11}]_2Co^-$	Tan	—	18
$[B_{10}CH_{11}]_2Co^{3-}$	Yellow	—	22, 26
$[B_{10}C(NH_2C_2H_5)H_{10}]_2Co^-$	Orange	—	22
$[B_{10}CH_{11}]_2Co^{2-}$	Dark blue	—	26
$[B_{10}C(NH_2)H_{10}]Co[B_{10}C(NH_3)H_{10}]^{2-}$	Yellow	—	26
$[1,7\text{-}B_9H_9CHP]_2Co^-$	Yellow	—	36
$[1,7\text{-}B_9H_9CHPCH_3]_2Co$	Orange	278°–279°	36
$[1,7\text{-}B_9H_9CHPCH_3]_2Co$	Orange	231°–233°	36
$C_5H_5Co[B_{10}SH_{10}]$	Orange	267.5°–268.5°	21
$[B_{10}SH_{10}]_2Co^-$	Orange	287°–287.5° $[(C_5H_5)_2Co^+salt]$	21
$[B_{10}SH_{10}]Co[P(C_2H_5)_3]_2$	Brown	—	21
$[1,7\text{-}B_9H_9CHAs]_2Co^-$	Orange	—	38
$[1,2\text{-}B_9H_9CHPCH_3]_2Co$	Brown	276°–278°	39
$(6\text{-}C_6H_5\text{-}1,2\text{-}B_9H_8C_2H_2)_2Co^-$	Yellow	275°–277°	19

Continued

TABLE II—*continued*

Compound	Color	Melting point (°C)	References of synthesis
	Nickel Group		
$[1,2\text{-}B_9C_2H_{11}]_2Ni^{2-}$	Pale brown	—	18
$[1,2\text{-}B_9C_2H_{11}]_2Ni^{-}$	Black	—	18
$[1,2\text{-}B_9C_2H_{11}]_2Ni$	Orange	—	18
$[1,7\text{-}B_9C_2H_{11}]_2Ni^{-}$	—	—	18
$[1,7\text{-}B_9C_2H_{11}]_2Ni$	Orange	—	18
$(B_{10}CH_{11})_2Ni^{2-}$	Yellow	—	22, 26
$[B_{10}C(NH_2C_3H_7)H_{10}]_2Ni$	Orange	—	22
$[B_{10}C(NH(CH_3)C_3H_7)H_{10}]_2Ni$	Dark orange	—	22
$B_{10}C(OH)H_{10}]_2Ni$	—	—	26
$[B_{10}C(N(CH_3)_2H)H_{10}]_2Ni$	—	—	26
$[1,7\text{-}B_9H_9CHPCH_3]_2Ni$	Red brown	229°–231°	36
$[B_{10}SH_{10}]Pt[P(C_2H_5)_3]_2$	Yellow	—	21
$[C_4(C_6H_5)_4]Pd[1,2\text{-}B_9H_9C_2(CH_3)_2]$	Red	> 325°(dec)	43
$[C_4(C_6H_5)_4]Pd[1,2\text{-}B_9C_2H_{11}]$	Red	308°–309°	43
$[1,2\text{-}B_9C_2H_{11}]_2Pd^{2-}$	Brown	—	42
$[1,2\text{-}B_9C_2H_{11}]_2Pd^{-}$	Brown	—	42
$[1,2\text{-}B_9C_2H_{11}]_2Pd$	Yellow	—	42
	Copper Group		
$[1,2\text{-}B_9C_2H_{11}]_2Cu^{2-}$	Blue	—	18
$[1,2\text{-}B_9C_2H_{11}]_2Cu^{-}$	Red	—	42
$[1,2\text{-}B_9C_2H_{11}]_2Au^{2-}$	Blue	—	42
$[1,2\text{-}B_9C_2H_{11}]_2Au^{-}$	Red	—	42

Preliminary results of an X-ray diffraction study of $[(CH_3)_4N]_2(1,2\text{-}B_9C_2H_{11})Mo(CO)_3W(CO)_6$ give the tentative structure shown in Fig. 3 (*18*).

2. *Manganese Group*

Both bromomanganese pentacarbonyl and bromorhenium pentacarbonyl rapidly precipitated sodium bromide at room temperature when added to a tetrahydrofuran solution of $Na_2[1,2\text{-}B_9C_2H_{11}]$. Two equivalents of CO were evolved only after the mixture was refluxed, producing the corresponding 1,2-dicarbollyl metal tricarbonyl anion (*11, 18*).

$$1,2\text{-}B_9C_2H_{11}{}^{2-} + BrM(CO)_5 \xrightarrow[\text{reflux}]{\text{THF}} (1,2\text{-}B_9C_2H_{11})M(CO)_3{}^{-} + Br^{-} + 2\ CO \qquad (2)$$

$$(M = Mn \text{ or } Re)$$

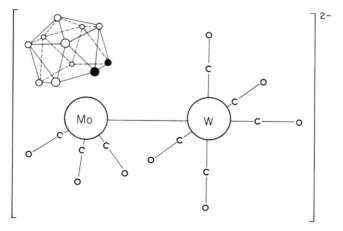

FIG. 3. Probable structure of the [(1,2-B$_9$C$_2$H$_{11}$)Mo(CO)$_3$W(CO)$_5$]$^{2-}$ ion.

The molecular structure of the rhenium derivative has been determined by an X-ray diffraction study and is reproduced in Fig. 4 (47).

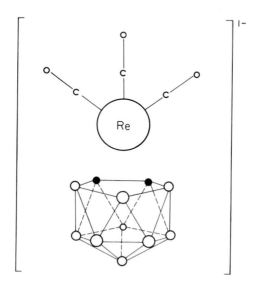

FIG. 4. Structure of the [(1,2-B$_9$C$_2$H$_{11}$)Re(CO)$_3$]$^-$ ion.

3. Iron Group

The complexes in this group have significance in the history of organo-metallic and boron chemistry. The first reported examples of π-bonded carborane–transition metal compounds were the 1,2-dicarbollyl analogs of ferrocene and the ferricinium ion discovered by Hawthorne and co-workers (10). These were initially prepared by an anhydrous route. The 1,2-$B_9C_2H_{12}^-$ ion was deprotonated with sodium hydride in tetrahydrofuran and then $FeCl_2$ added to the mixture. Subsequent air oxidation produced the stable red $(1,2-B_9C_2H_{11})_2Fe^-$ ion, isoelectronic with the ferricinium ion. Later it was discovered that the Fe(III) species could also be made satis-factorily from a mixture of $FeCl_2$ and the 1,2-$B_9C_2H_{12}^-$ ion in 40% aqueous sodium hydroxide solution. In this reaction the 1,2-dicarbollide ion is generated *in situ* (16). Reduction of the Fe(III) species with sodium amalgam in acetonitrile produced the air-sensitive $(1,2-B_9C_2H_{11})_2Fe^{2-}$ ion, isoelectronic with ferrocene. This complex, like ferrocene, was diamagnetic. The mixed ligand complex, π-C_5H_5—$Fe(1,2-B_9C_2H_{11})$ was prepared by reaction of an equimolar mixture of $C_5H_5^-$ and 1,2-dicarbollide ion with $FeCl_2$ followed by air oxidation (13). This was the first neutral, sublimable complex of the dicarbollide series. The crystal and molecular structure of the cyclopentadienyl complex obtained by X-ray diffraction afforded the first confirmation of the sandwich bonding of the carborane ligand to the iron atom (Fig. 5) (46). The π-$C_5H_5Fe(1,2-B_9C_2H_{11})^-$ ion could be obtained by sodium amalgam reduction of the neutral, formally Fe(III) species.

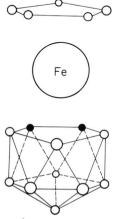

Fig. 5. Structure of $(\pi$-$C_5H_5)Fe(1,2-B_9C_2H_{11})$.

4. *Cobalt Group*

High yields of bis(dicarbollyl)cobalt(III) derivatives in which the 1,2-dicarbollide ion is unsubstituted, 1,2-dimethyl or 1-phenyl have been obtained by the anhydrous route using cobalt(II) chloride (*12, 18*). An internal redox process occurs in this reaction and cobalt metal is formed.

$$1.5 \, CoCl_2 + 2 \, [1,2\text{-}B_9C_2H_{11}{}^{2-}] \rightarrow (1,2\text{-}B_9C_2H_{11})_2Co^- + 3 \, Cl^- + 0.5 \, Co \qquad (3)$$

Preparation of $(1,2\text{-}B_9C_2H_{11})_2Co^-$ has also been accomplished by the aqueous route. The neutral complex, $C_5H_5Co(1,2\text{-}B_9C_2H_{11})$ was obtained in low yield by reaction of an equimolar mixture of $C_5H_5^-$ and $1,2\text{-}B_9C_2H_{11}{}^{2-}$ with cobalt(II) chloride. The proton NMR spectrum of this compound is consistent with a sandwich-bonded structure similar to the iron analog (see Fig. 5).

Exhaustive bromination of $Rb[(1,2\text{-}B_9C_2H_{11})_2Co]$ in refluxing glacial acetic acid produced a hexabromo derivative. The molecular structure of this compound has been determined by an X-ray diffraction study of a single crystal (Fig. 6) (*4*). Each dicarbollide ligand has three bromine atoms

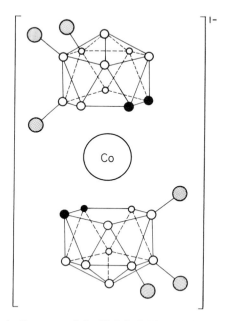

FIG. 6. Structure of the $[(1,2\text{-}B_9C_2H_8Br_3)_2Co]^-$ ion.

bonded to boron atoms in a triangular array. The positions of substitution are as far as possible from the carbon atoms. If this reaction is electrophilic in nature, this result pinpoints the region of highest electron density in this type of molecule.

The 6-phenyl-1,2-dicarbollide ion derived from 3-phenyl-1,2-dicarba-closododecaborane (12) by boron atom removal with base also forms a cobalt complex $[1,2-B_8H_8B(C_6H_5)C_2H_2]_2Co^-$ under the normal conditions (19).

5. Nickel Group

Formation of the $(1,2-B_9C_2H_{11})_2Ni^-$ ion was accomplished by reaction of nickel(II) salts with the 1,2-dicarbollide ion generated by the anhydrous method or in strong aqueous base followed by O_2 oxidation (18, 41). Reaction of the Ni(III) species with one equivalent of ferric ion produced the sublimable complex, $(1,2-B_9C_2H_{11})_2Ni$. Reduction of the nickel(III) species with sodium amalgam afforded the air-sensitive, paramagnetic $(1,2-B_9C_2H_{11})_2Ni^{2-}$. Preliminary X-ray studies (45) suggest that the Ni(III) compound has a normal sandwich-bonded structure but that the Ni(II) analog has a slipped structure (Fig. 7).

Addition of palladium acetylacetonate in glyme to a cold glyme solution of excess dicarbollide ion produced the air-sensitive $(1,2-B_9C_2H_{11})_2Pd^{2-}$ ion (42). Solutions of this ion deposited palladium metal on standing. The Pd(II) compound was oxidized with iodine to produce the neutral species, $(1,2-B_9C_2H_{11})_2Pd$.

$$[(C_2H_5)_4N]_2Pd(1,2-B_9C_2H_{11})_2 + 3\ I_2 \rightarrow (1,2-B_9C_2H_{11})_2Pd + 2\ (C_2H_5)_4NI_3 \quad (4)$$

Reduction of the neutral complex in nonaqueous media with cadmium amalgam produced $(1,2-B_9C_2H_{11})_2Pd^-$, an air-sensitive and water-sensitive compound. In contrast, the analogous Ni(III) species is water stable.

The Ni(IV) and Pd(IV) species appear to be Lewis acids, since crystalline adducts have been obtained with π-bases and Lewis bases such as pyrene, phenanthrene, N,N-dimethylaniline halide ions, and SCN^- (18, 42). It has been briefly mentioned (42) that the substituted complexes $[B_9H_9C_2(CH_3)_2]_2Ni^n$, $n = 0$, -1, -2, exhibit a series of molecular rearrangements to form isomeric systems.

Mixed ligand compounds containing a tetraphenylcyclobutadiene ring have been formed in low yield according to Eq. (5) (43).

$$2 \ Na_2[1,2\text{-}B_9H_9C_2(CH_3)_2] + [(C_6H_5)_4C_4PdCl_2]_2$$
$$\rightarrow 2 \ \pi\text{-}(C_6H_5)_4C_4Pd[(1,2\text{-}B_9H_9C_2(CH_3)_2)] + 4 \ NaCl \qquad (5)$$

This complex has a structure in which both ligands are symmetrically π-bonded to the palladium atom (43).

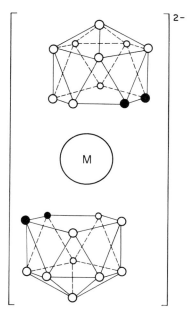

FIG. 7. General "π-slipped" structure of $(1,2\text{-}B_9C_2H_{11})_2M$ (M=Ni(II), Cu(II), Cu(III), and Au(III).

6. Copper Group

Preparation of the deep blue cupricene analog, $(1,2\text{-}B_9C_2H_{11})_2Cu^{2-}$, involved reaction at $0°C$ of $CuSO_4 \cdot 5 \ H_2O$ with $1,2\text{-}B_9C_2H_{11}^{2-}$ generated in 40% aqueous sodium hydroxide (18). The blue ion which has copper formally in the $+2$ oxidation state could be oxidized with air to the red ion, $(1,2\text{-}B_9C_2H_{11})_2Cu^-$ (42). The structures of both the Cu(II) and Cu(III) species have been determined by Wing (44, 45). They have the unusual slipped configuration shown in Fig. 7.

Addition of anhydrous $AuCl_3$ to excess $1,2\text{-}B_9C_2H_{11}^{2-}$ in 1,2-dimethoxy-ethane produced a deep blue species which could be oxidized with acidic hydrogen peroxide to red $(1,2\text{-}B_9C_2H_{11})_2Au^-$ (42). Sodium amalgam reduced the Au(III) compound to the deep blue-green ion, $(1,2\text{-}B_9C_2H_{11})_2$ Au^{2-} (42). An X-ray study by Wing indicated that the triphenylmethyl-phosphonium salt of the Au(III) compound was isomorphous with its Cu(III) analog (45).

The bonding in these slipped structures has been described (45) as π-allyl in nature with the distortion attributed to the electron-rich nature of the system and the heterocyclic character of the bonding face of the carborane ligand. An alternative viewpoint of the bonding suggests that the $3d$ orbitals in the d^8 and d^9 systems decrease in energy and become less important in metal–ligand bonding. The primary metal acceptor orbitals are then $4s$ and $4p$ in character. Maximum overlap of the metal orbitals with a cyclopentadienyl-type filled E_1 molecular orbital would require distortion away from the pseudo-fivefold axis of the ligand (42).

B. $1,7\text{-}B_9C_2H_{11}^{2-}$ Ligand

Curiously, only chromium (33), cobalt, and nickel complexes of this 11-atom ligand have been prepared thus far. The complex $(1,7\text{-}B_9C_2H_{11})_2Co^-$ has been prepared by both the anhydrous and aqueous methods (18). An X-ray diffraction study of the cesium salt of the cobalt(III) complex confirmed the symmetrical sandwich bonding of the carborane cages but failed to locate the carbon atoms because of disorder in the crystal (48).

The 1,7-dicarbollide ion and nickel acetylacetonate reacted in tetrahydrofuran to yield a red complex which was air oxidized to $(1,7\text{-}B_9C_2H_{11})_2Ni^-$ (18). Further oxidation of the Ni(III) species with ferric ion produced the sublimable $(1,7\text{-}B_9C_2H_{11})_2Ni$. The 1,2-dicarbollide complexes appear to be more stable than their 1,7-analogs.

C. $B_{10}CH_{11}^{3-}$ and C-Substituted Ligands

The syntheses of the carbollide–transition metal complexes thus far reported are based on three general routes closely related to the preparative methods used for the dicarbollyl sandwich compounds. Method A consists of adding n-butyllithium to a mixture of a metal(II) halide and $B_{10}H_{12}CNH_3$ or the $B_{10}H_{12}CH^-$ ion in tetrahydrofuran (26). Method B involves reaction of $B_{10}H_{12}CN(CH_3)_3$ with sodium in refluxing tetrahydrofuran to form a precipitate of the composition $Na_3B_{10}H_{10}CH\cdot(THF)_2$ and subsequent reaction of this solid with a transition metal halide (22). Method C consists of reaction of a metal halide and $B_{10}H_{12}CH^-$ or $B_{10}H_{12}CNH_2R$ (R = H or alkyl) in concentrated aqueous sodium hydroxide (22, 26, 27).

The preferred formal oxidation states in the chromium, manganese, iron, cobalt, and nickel complexes of the type $(B_{10}H_{10}CH)_2M^{n-}$ are III, IV, III, III, IV, respectively. All but the manganese complex are air stable and not readily decomposed by acids. The manganese complex decomposes slowly in acetonitrile solution (26). The $B_{10}H_{10}CH^{3-}$ ligand appears to be even better in stabilizing formal high oxidation states of metal atoms than the dicarbollide ligand. Thus Ni(IV) carbollide complexes are isolated directly without the need for $M^{3+} \rightarrow M^{4+}$ oxidation by ferric ion. In addition, the Co(III) complex can be oxidized by ceric ion to $(B_{10}H_{10}CH)_2Co^{2-}$ containing the metal formally in the +4 oxidation state (26).

Manganese(IV), iron(III), cobalt(III), and nickel(IV) complexes of the type $(B_{10}H_{10}CNH_2R)_2M$ (R = H or alkyl) have been made from $B_{10}H_{12}CNH_2R$ by either method A or C (22, 26). Further chemistry of the amine group has been reported [Eq. (6)] (26).

$$(B_{10}H_{10}CNH_3)_2Ni \Big\langle \begin{array}{l} \xrightarrow[\text{2. (CH}_3)_2SO_4]{\text{1. NaOH}} [B_{10}H_{10}CNH(CH_3)_2]_2Ni \\ \xrightarrow{\text{HONO}} (B_{10}H_{10}COH)_2Ni^{2-} \end{array} \tag{6}$$

The only metal carbonyl compound thus far reported for this class of compounds is $(B_{10}H_{10}CH)Mn(CO)_3^{2-}$, prepared from $BrMn(CO)_5$ and $Na_3B_{10}H_{10}CH\cdot(THF)_2$ (22).

D. 1,2- and 1,7-$B_9H_9CHP^{2-}$ and P-Substituted Ligands

Removal of one boron atom from the recently synthesized icosahedral cage molecules 1,2- and 1,7-$B_{10}H_{10}CHP$ (29) can be accomplished cleanly by reflux in excess piperidine for several hours (36). This forms a piperidinium salt of the 1,2- or 1,7-$B_9H_{10}CHP^-$ ion isoelectronic with the previously described $B_9C_2H_{12}^-$ ion (see Section II,A). Treatment of the phosphacarba ions with methyl iodide produced sublimable 1,2- and 1,7-$B_9H_{10}CHPCH_3$ in which the methyl group is attached to the phosphorus atom (36). Each of these 11-atom cage phosphacarbaboranes can be deprotonated with sodium hydride or other bases to produce "phosphacarbollide" ions, and these react with transition metal halides to produce sandwich-bonded complexes.

Reaction of $1,7\text{-}B_9H_{10}CHP^-$ with sodium hydride in tetrahydrofuran followed by addition of one equivalent of ferrous chloride formed the red-violet complex $(1,7\text{-}B_9H_9CHP)_2Fe^{2-}$ (36). Alkylation of this dianion with methyl iodide produced two isomeric neutral compounds with the composition $(1,7\text{-}B_9H_9CHPCH_3)_2Fe$. This same pair of isomers could be generated by reaction of $1,7\text{-}B_9H_9CHPCH_3^-$ with ferrous chloride. One of the isomers (m.p. 239.5°–240.5° C) was partially resolved by chromatography on a lactose hydrate column to give a rotation of $[\alpha]_{5000}^{25} + 69°$. A d,l pair of $B_9H_{10}CHP^-$ ions would be generated by base abstraction of either of two equivalent boron atoms adjacent to both the carbon and phosphorus atoms in $1,7\text{-}B_{10}H_{10}CHP$. The two isomeric iron complexes would then be a $d,d\text{-}l,l$ racemate and a d,l meso form. A cocrystallite containing both isomers has been analyzed by X-ray methods (36). The molecular structure is shown in Fig. 8. The sandwich-bonded nature of

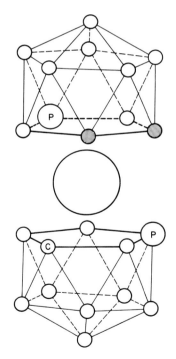

Fig. 8. Structure of $(1,7\text{-}B_9H_9CHPCH_3)_2Fe$ (exocage methyl groups on phosphorus atoms not shown).

the complex is evident as well as the 1,7-nature of the carbon and phosphorus atoms. In one cage the carbon atom appears to occupy two sites in the crystal and no clear assignment could be made. The distortion of the ligand cage due to the phosphorus atom is also apparent.

Initial experiments indicate that the 1,2- and $1,7\text{-}B_9H_9CHPCH_3^-$ ions form neutral bis-ligand complexes with iron and cobalt in which the metal atom is in the $+2$ formal oxidation state. The complex, $(1,7\text{-}B_9H_9CHPCH_3)_2$ Ni has also been prepared but it is somewhat unstable and has properties very different from its iron and cobalt analogs (see Section III,A). Iron(II) and cobalt(III) compounds of the 1,2- and $1,7\text{-}B_9H_9CHP^{2-}$ ions have been fully characterized. Using procedures described in the dicarbollide section it was also possible to form the mixed ligand complexes, $(CO)_3Mn(1,7\text{-}B_9H_9CHPCH_3)$ and $C_5H_5Fe(1,7\text{-}B_9H_9CHPCH_3)$.

E. 1,2- and 1,7-$B_9H_9CHAs^{2-}$ and As-Substituted Ligands

The cage molecules 1,2- and $1,7\text{-}B_{10}H_{10}CHAs$ have been very recently prepared by methods similar to those employed with the phosphorus analogs (*38*). Reflux in piperidine of the icosahedral molecules results in removal of one boron atom from the cage structure to form the 1,2- or $1,7\text{-}B_9H_{10}CHAs^-$ ions (*38*). Reaction of these ions with methyl iodide produces the neutral 1,2- or $1,7\text{-}B_9H_{10}CHAsCH_3$ species. Preliminary studies indicate that reaction of these ligands with base and then transition metal halides does produce metal complexes very similar to their phosphacarba analogs. Attempts to prepare neutral derivatives of the type $(B_9H_9CHAsCH_3)_2M$ by the two methods described in Section II,D for the phosphorus derivatives have been unsuccessful. In both cases partially methylated derivatives are formed [Eq. (7)]. The arsenic atoms in the metal complexes must have rather weak Lewis base properties. The phosphacarba complexes, $(1,7\text{-}B_9H_9CHPCH_3)_2M$ (M = Fe or Co),

$$2\,[1,7\text{-}B_9H_{10}CHAsCH_3] + MCl_2 + \text{Base} \longrightarrow$$

$$(1,7\text{-}B_9H_9CHAsCH_3)M(1,7\text{-}B_9H_9CHAs)^- \qquad (7)$$

$$(1,7\text{-}B_9H_9CHAs)_2M + 2\,CH_3I \longrightarrow$$
$$(M = \text{Fe or Co})$$

can be demethylated to $(1,7\text{-}B_9H_9CHP)M(1,7\text{-}B_9H_9CHPCH_3)^-$ or $(1,7\text{-}B_9H_9CHP)_2M^{2-}$, but this transformation requires sodium hydride in refluxing 1,2-dimethoxyethane (37).

F. $B_{10}H_{10}S^{2-}$ Ligand

The $B_{10}H_{10}S^{2-}$ ligand (thioollide ion) can be generated from decaborane in moderate yield by the following sequence of reactions (21):

$$B_{10}H_{14} + S^{2-} + 4 H_2O \rightarrow B_9H_{12}S^- + B(OH)_4^- + 3 H_2$$

$$CsB_9H_{12}S \xrightarrow{200°C} CsB_{10}H_{11}S^-$$

$$B_{10}H_{11}S^- + n\text{-}C_4H_9Li \rightarrow B_{10}H_{10}S^{2-}$$

(8)

Reaction of $B_{10}H_{10}S^{2-}$ ion generated via n-butyllithium or aqueous alkali with $FeCl_2$ gives the stable maroon complex $(B_{10}H_{10}S)_2Fe^{2-}$ (21). The cobalt(III) compounds $(B_{10}H_{10}S)_2Co^-$ and $C_5H_5Co(B_{10}H_{10}S)$ have been produced from $CoCl_2$ by methods similar to those used with the dicarbollide complexes. The thioollide ion reacts with cis-dichlorobis(triethylphosphine)platinum and dichlorobis(triethylphosphine)cobalt to give $(B_{10}H_{10}S)Pt[P(C_2H_5)_3]_2$ and $(B_{10}H_{10}S)Co[P(C_2H_5)_3]_2$, respectively (21). The structures of these compounds have not yet been elucidated.

Reactivity toward electrophilic reagents of this type of complex was evinced by partial chlorination, bromination, and acid-catalyzed proton exchange of the complex $(B_{10}H_{10}S)_2Fe^{2-}$ (21).

III

PHYSICAL PROPERTIES OF ELEVEN-ATOM LIGAND COMPLEXES

A. Magnetic Susceptibility

The magnetic moments of several sandwich-bonded complexes with d electron configurations d^5 through d^9 are given in Table III. Most of the d^5, Fe(III) complexes have a μ_{eff} of about 2.3 μ_B (Bohr magnetons). This is in good agreement with the experimental value $\mu_{eff} = 2.34$ μ_B obtained for the ferricinium ion. German and Dyatkina (7) have calculated a $\mu_{eff} = 2.38$ μ_B for $Fe(1,2-B_9H_9C_2H_2)_2^-$ using the g values measured earlier in an electron spin resonance study by Maki and Berry (30).

TABLE III

MAGNETIC SUSCEPTIBILITY DATA

d Electron configuration	Compound	Observed μ_{eff} (μ_B)	Reference
d^3	$Cs[Cr(1,2-B_9H_9C_2H_2]_2)]$	3.85	33
d^5	$(CH_3)_4N[Fe(1,7-B_9H_9CHP)_2]$	2.34	37
d^5	$(CH_3)_4N[Fe(1,2-B_9H_9C_2H_2)_2]$	2.10	18
d^5	$(CH_3)_4NFe[1,2-B_9H_9C_2(CH_3)_2]_2$	2.45	37
		1.99	18
d^5	$(CH_3)_4NFe(1,7-B_9H_9CHAs)_2$	2.34	38
d^9	$[(C_2H_5)_4N]_2Cu(1,2-B_9H_9C_2H_2)_2$	1.70	42
d^8	$(C_6H_5)_4AsCu(1,2-B_9H_9C_2H_2)_2$	Diamagnetic	42
d^8	$(C_2H_5)_4NAu(1,2-B_9H_9C_2H_2)_2$	Diamagnetic	42
d^9	$[(C_2H_5)_4N]_2Au(1,2-B_9H_9C_2H_2)_2$	1.79	42
d^8	$[(C_2H_5)_4N]_2Pd(1,2-B_9H_9C_2H_2)_2$	Diamagnetic	42
d^7	$(C_2H_5)_4NPd(1,2-B_9H_9C_2H_2)_2$	1.68	42
d^7	$(CH_3)_4NNi(1,7-B_9H_9C_2H_2)_2$	1.74	18
d^7	$(CH_3)_4NNi(1,2-B_9H_9C_2H_2)_2$	1.74	18
d^8	$[(C_2H_5)_4N]_2Ni(1,2-B_9H_9C_2H_2)_2$	2.90	18
d^7	$Co(1,7-B_9H_9CHPCH_3)_2$ (m.p. 232°–234°)	1.82	37
d^7	$Co(1,7-B_9H_9CHPCH_3)_2$ (m.p. 278°–279°)	1.89	37
d^8	$Ni(1,7-B_9H_9CHPCH_3)_2$ (m.p. 229°–231°)	Diamagnetic	37

All d^6 complexes thus far obtained are diamagnetic. The d^7 and d^9 complexes have a μ_{eff} of 1.8 ± 0.1 μ_B, corresponding to a spin-only formulation of one unpaired electron. All the d^8 cage complexes [Ni(II), Pd(II), Cu(III), and Au(III)] are diamagnetic with the exception of $Ni(1,2-B_9H_9C_2H_2)_2^{2-}$ (2.90 μ_B). This difference may be due to a variance in structure within this set of compounds or to a variance of the magnetic moment–temperature dependence within the set. It should be noted that nickelocene is paramagnetic ($\mu_{eff} = 2.86$ μ_B).

B. Nuclear Magnetic Resonance

The proton NMR spectra of the diamagnetic complexes have been useful in confirming parts of the structure of these compounds. The carborane CH resonance appears as a broad singlet in most proton spectra. The ^{11}B NMR spectra of the diamagnetic complexes have to date been useless for structural purposes because of the complex overlap of the various doublets.

The ^{11}B NMR spectra of paramagnetic species such as $(1,2\text{-}B_9H_9C_2H_2)_2$ Fe$^-$ (Fig. 9) exhibit unusually broad resonances containing no evidence of

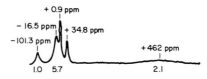

Fig. 9. ^{11}B NMR spectrum of $(CH_3)_4N[(1,2\text{-}B_9C_2H_{11})_2Fe]$ (19.3 MHz) measured relative to $BF_3 \cdot O(C_2H_5)_2$ equal to 0 ppm.

spin–spin coupling of the ^{11}B nuclei with the protons to which they are bonded. In addition, some of the resonance lines exhibit very large chemical shifts (contact and pseudocontact shifts). This type of decoupling phenomenon had been observed earlier in ^{11}B NMR spectra in which an Fe(III) compound was added to a solution of a boron hydride derivative (28). Considering the spectrum in Fig. 9, it appears that three boron atoms on each cage (at -101.3 and $+462$ ppm) are most affected by the unpaired electron on the iron atom. These resonances may well be associated with the three boron atoms in the open face of the 1,2-dicarbollide ligands. Similar decoupling and contact shift phenomena have been observed in the boron NMR spectra of the Fe^{3+} complexes of the $1,7\text{-}B_9H_9CHP^{2-}$ and $1,7\text{-}B_9H_9CHAs^{2-}$ ions.

C. Nuclear Quadrupole Resonance

The nuclear quadrupole resonance spectrum of $Cs(1,2\text{-}B_9C_2H_{11})_2Co$ has been measured and compared with that of $(\pi\text{-}C_5H_5)_2Co(ClO_4)$ (9). The field gradients (along the pseudo-fivefold axis) and the asymmetry parameters are nearly identical in these two compounds. This suggests that the electronic environment about the cobalt atom in these two types of complexes are very similar and that the roles of the boron and carbon atoms in the π-bonding are almost indistinguishable.

D. Oxidation–Reduction Studies

Reduction potentials for a variety of the 11-atom ligand complexes have been measured either by cyclic voltammetry or by polarographic methods (see Table IV). In certain cases the most stable formal oxidation state for

TABLE IV

REDUCTION POTENTIALS FOR ELEVEN-ATOM LIGAND COMPLEXES

Compound	$E_{1/2}$ vs. sce (V)	Method[a]	Reference
$M^{IV} + e \rightarrow M^{III}$			
$Cs[1,2-B_9C_2H_{11})_2Co]$	$+1.57$	c.v.	18
$(1,2-B_9C_2H_{11})_2Ni$	$+0.18$	c.v.	18
$(1,2-B_9C_2H_{11})_2Pd$	-0.14	c.v.	42
$(1,7-B_9C_2H_{11})_2Ni$	$+0.55$	c.v.	18
$Cs_3(B_{10}H_{10}CH)_2Co$	$+0.37$	p	39
$Cs_2(B_{10}H_{10}CH)_2Ni$	-1.12	p	39
$(B_{10}H_{10}CNH_3)_2Ni$	-0.49	p	39
$M^{III} + e^- \rightarrow M^{II}$			
$(CH_3)_4N[(1,2-B_9C_2H_{11})_2Fe]$	-0.424	p	18
$(CH_3)_4N[(1,2-B_9H_9C_2(CH_3)_2)_2Fe]$	-0.538	p	18
$(CH_3)_4N[(1,2-B_9H_9CHC(C_6H_5))_2Fe]$	-0.464	p	18
$(C_5H_5)Fe(1,2-B_9C_2H_{11})$	-0.08	c.v.	18
$Cs[(1,2-B_9C_2H_{11})_2Co]$	-1.42	p	18
	-1.46	c.v.	18
$(CH_3)_4N[1,2-B_9H_6Br_3C_2H_2)_2Co]$	-0.48	c.v.	18
$(CH_3)_4N[(1,2-B_9H_9C_2(CH_3)_2)_2Co]$	-1.16	p	18
	-1.13	c.v.	18
$(CH_3)_4N[(1,2-B_9H_9CHC(C_6H_5))_2Co]$	-1.28	p	18
$(C_5H_5)Co(1,2-B_9C_2H_{11})$	-1.25	c.v.	18
$Cs[(1,7-B_9C_2H_{11})_2Co]$	-1.17	c.v.	18
$(CH_3)_4N(1,2-B_9C_2H_{11})_2Ni^-$	-0.63	p	18
	-0.66	c.v.	18
$(1,7-B_9C_2H_{11})_2Ni^-$	-0.91	c.v.	18
$[(B_{10}H_{10}CH)_2Fe]^{3-}$	-0.733	p	39
$[(B_{10}H_{10}S)_2Fe]^{2-}$	$+0.21$	p	21
$[(B_{10}H_{10}CNH_3)_2Ni]$	-0.60	p	39
$[(C_6H_5)_3PCH_3]Cu(1,2-B_9C_2H_{11})$	-0.35	c.v.	42
$[(C_6H_5)_3PCH_3]Au(1,2-B_9C_2H_{11})$	-0.62	c.v.	42
$(C_2H_5)_4N[(1,2-B_9C_2H_{11})Pd]$	-0.56	c.v.	42
$M^{II} + e^- \rightarrow M^I$			
$(1,2-B_9H_6Br_3C_2H_2)_2Co^{2-}$	-1.58	c.v.	18
$[(C_2H_5)_4N]_2Cu(1,2-B_9C_2H_{11})_2$	-0.99	c.v.	42

[a] c.v. = cyclic voltammetry; p = polarography.

a given metal appears to be markedly dependent upon the charge of the free π-bonding ligand(s). This is illustrated by the reduction potentials

TABLE V

SELECTED REDUCTION POTENTIALS FOR SOME IRON COMPLEXES

d^5	$+e \rightarrow d^6$	$E_{1/2}$ vs. sce (V)
$(C_5H_5)_2Fe^+$	$+e \rightarrow (C_5H_5)_2Fe$	$+0.3$
$(C_5H_5)Fe(1,2-B_9C_2H_{11})$	$+e \rightarrow (C_5H_5)Fe(1,2-B_9C_2H_{11})^-$	-0.08
$Fe(1,2-B_9C_2H_{11})_2^-$	$+e \rightarrow Fe(1,2-B_9C_2H_{11})_2^{2-}$	-0.42
$Fe(B_{10}H_{10}CNH_3)_2^-$	$+e \rightarrow Fe(B_{10}H_{10}CNH_3)_2^{2-}$	-0.49
$Fe(B_{10}H_{10}CH)_2^{3-}$	$+e \rightarrow Fe(B_{10}H_{10}CH)_2^{4-}$	-0.73

listed in Table V. Employing the carbollide and dicarbollide ions, it has been possible to isolate and characterize complexes having the metal in unusual formal oxidation states [i.e., Co(IV), Ni(IV), Cu(III), Pd(III), and Au(II)]. The $Ni(1,2-B_9C_2H_{11})_2^-$ ion ($d^6 + e \rightarrow d^7$; $E_{1/2} = +0.18V$) is oxidized to the Ni(IV) species with ferric chloride, whereas $Ni(B_{10}H_{10}CH)_2^{2-}$ is synthesized directly ($d^6 + e \rightarrow d^7$; $E_{1/2} = -1.12$ V).

When the π-bonding ligand contains a sulfur or phosphorus atom, the redox properties of the corresponding complexes are much more like the corresponding metallocene. Thus the $Fe^{II}(B_{10}H_{10}S)_2^{2-}$ and $Fe^{II}(B_9H_9 CHP)_2^{2-}$ ions are more stable than the Fe(III) species.

E. Electronic Spectral Data

The electronic spectrum is yet another property which illustrates the similarities between the metallocenes and (π-ollyl) metal compounds. In Table VI are listed some data for a series of $Co^{III}(d^6)$ complexes. These particular absorption bands with the small extinction coefficients are probably two of the spin-allowed d–d transitions. Scott (34) has developed an approximate axial ligand field model for the carborane–transition metal complexes and has discussed the optical spectra in relation to this bonding theory. The actual assessment of bonding in the (π-ollyl) metal compound as well as the metallocenes would be greatly aided by accurate assignments of the electronic spectra.

TABLE VI

ELECTRONIC SPECTRAL DATA FOR SOME (π-OLLYL)Co^{III} COMPLEXES

Compound	λ_{max}' m$\mu(\epsilon)$		Reference
$(C_5H_5)_2Co^+$	333 (2500)	410 (220)	3
$(C_5H_5)Co(1,2-B_9C_2H_{11})$	320 (545)	422 (364)	18
$(1,2-B_9C_2H_{11})_2Co^-$	345 (2200)	445 (440)	18
$(1,2-B_9H_6Br_3C_2H_2)_2Co^-$	385 (5730)	470 (600)	18
$[1,2-B_9H_9CHC(C_6H_5)]_2Co^-$	390 (760)	495 (478)	18
$(1,7-B_9C_2H_{11})_2Co^-$	—	433 (272)	18
$(B_{10}H_{10}S)_2Co^-$	370 (268)	445 (445)	21
$(B_{10}CH_{11})_2Co^{3-}$	—	422 (382)	22
$(1,7-B_9H_9CHP)_2Co^-$	—	451 (357)	37
$(1,7-B_9H_9CHP)Co(1,7-B_9H_9CHPCH_3)$	356 (3130)	445 (580)	37

IV

COMPLEXES OF THE $B_7C_2H_9{}^{2-}$ LIGAND

Treatment of $B_7C_2H_{13}$ with two equivalents of sodium hydride in diethyl ether generated the $B_7C_2H_{11}{}^{2-}$ ion (14). Reaction of this ion with cobalt(II) chloride produced hydrogen, cobalt metal, and the stable complex $(B_7C_2H_9)_2Co^-$(I). Similar complexes with other metals have also been characterized. Preliminary X-ray diffraction results obtained with the tetraethylammonium salt of this anion confirmed structure (I), shown in Fig. 10. The structure consists of two bicapped Archimedean antiprisms sharing the cobalt atom in common as an equatorial vertex. The complex in Fig. 10 (I) represents the d,l(meso) isomer. A neutral complex $C_5H_5Co(B_7C_2H_9)$ (A) (see Fig. 11) was prepared by reaction of cobalt(II) chloride with a mixture of the $B_7C_2H_{11}{}^{2-}$ and $C_5H_5{}^-$ ions (14). Available NMR data suggest that the carborane ligand in this complex has the same configuration as shown in Fig. 10 (I).

The cesium salt of $(B_7C_2H_9)_2Co^-$ (I) can be rearranged at 315°C in 24 hours to an isomeric product (III) in almost quantitative yield (6). The cyclopentadienyl derivative (A) can also be rearranged under the same conditions to a new isomer (B). The low-field doublet present in the ^{11}B NMR spectra of (I) and A is not seen in the spectra of the products. The boron NMR spectrum of B consists of four doublets of relative areas 1:2:2:2. This and proton NMR data suggest that the isomerized ligand has both carbon atoms in axial positions [Fig. 10 (III)]. In this formulation the carbon atom in the nonapical position of (I) has rearranged to the new apex position. A similar thermal rearrangement is observed in the transformation (at 350° C) of 1,6-$B_8C_2H_{10}$ to 1,10-$B_8C_2H_{10}$ (35).

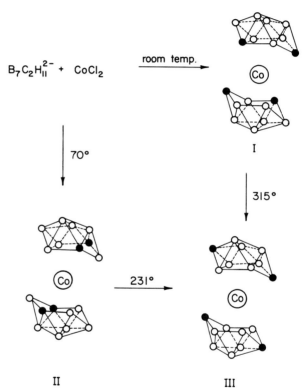

FIG. 10. (I) Structure of the $(B_7C_2H_9)_2Co^-$ ion obtained from room-temperature reaction of $B_7C_2H_{11}^{2-}$ and $CoCl_2$. (II) and (III) Proposed structures of the $(B_7C_2H_9)_2Co^-$ ions after thermal rearrangement.

A third isomeric species (II) has been isolated when $B_7C_2H_{11}^{2-}$ is reacted with cobalt(II) chloride at 70° C. This compound has only one carborane (CH) resonance in the proton NMR, suggesting that the two carbon atoms occupy equivalent positions in the cage. The favored structure of the three possible choices is illustrated in Fig. 10 (II). This choice is based on the observation that one of the symmetrical primary isomerization products via a diamond–square–diamond rearrangement mechanism of 1,6-$B_8C_2H_{10}$ would be 2,3-$B_8C_2H_{10}$.

Two electrophilic substitution reactions of neutral complex A have been reported. These are Friedel-Crafts acetylation and bromination. The proposed positions of substitution are indicated in Fig. 11 (20).

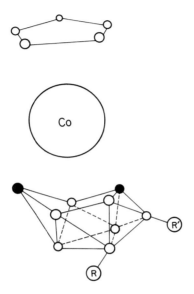

FIG. 11. (i) R = R′ = H, proposed structure of (π-C_5H_5)Co($B_7C_2H_9$) (isomer A) obtained

$$\overset{O}{\underset{\|}{}}$$

from $B_7C_2H_{11}{}^{2-}$, $C_5H_5{}^-$, and $CoCl_2$. (ii) R = CH_3C^-, R′ = H, and R = R′ = Br, proposed structures of substituted complexes.

V

COMPLEX OF THE $B_6C_2H_8{}^{2-}$ LIGAND

When $B_7C_2H_{11}{}^{2-}$ was treated with $BrMn(CO)_5$ at reflux in tetrahydrofuran, the expected $B_7C_2H_9$ complex was not isolated. Instead a 48% yield of a new type of complex, ($B_6C_2H_8$)$Mn(CO)_3{}^-$, was obtained (15). The ^{11}B NMR spectrum contained a 1:2:2:1 pattern of doublets. The proton NMR spectrum suggested that the two carborane CH units are equivalent.

Based on these data, the structure shown in Fig. 12 was proposed. This is a symmetrically tricapped trigonal prism in which the manganese atom occupies a vertex of the trigonal prism.

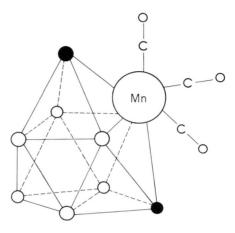

FIG. 12. Proposed structure of the $(B_6C_2H_8)Mn(CO)_3^-$ ion.

VI

MULTIFUNCTIONAL TRANSITION METAL LIGANDS

It was observed in the preparation of the complex $(1,2\text{-}B_9H_9C_2H_2)_2Co^-$ by the aqueous base method that red by-products were produced. A new dianion was isolated from the by-products as a cesium salt with the composition $Cs_2[C_6H_{32}B_{26}Co_2]\cdot H_2O$ (5). A recently completed X-ray diffraction study (49) indicates that the compound has the structure shown in Fig. 13. The bifunctional ligand in this complex resembles a basket and has been given the trivial name (3,6)-1,2-dicarbacanastide ion from the Spanish noun "canasta" meaning basket. This unusual complex may arise by base abstraction of a BH unit from $(1,2\text{-}B_9H_9C_2H_2)_2Co^-$ and subsequent capture of a $(1,2\text{-}B_9H_9C_2H_2)Co$ unit. The mobility of the 1,2-dicarbollide ion was demonstrated by treatment of $(1,2\text{-}B_9H_9C_2H_2)_2Co^-$ with strong aqueous base and nickelous chloride, followed by air oxidation. This produced a low yield of $(1,2\text{-}B_9H_9C_2H_2)_2Ni$ (5). Reaction of $(1,2\text{-}B_9H_9 C_2H_2)_2Co^-$ with 30% aqueous sodium and excess cobaltous chloride gave an improved (15%) yield of the dinuclear complex.

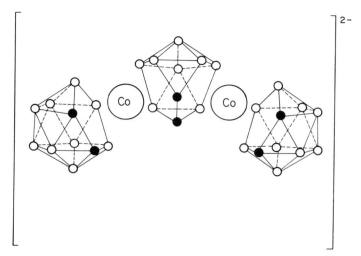

FIG. 13. Structure of the $[(B_9C_2H_{11})Co(B_8C_2H_{10})Co(B_9C_2H_{11})]^{2-}$ ion.

The 1,2- and 1,7-isomers of the $B_9H_{10}CHE^-$ ions (E = P or As) also show multifunctional character. In addition to forming π-complexes as described earlier, these ligands also use the available electron pair on phosphorus or arsenic to form σ-bonded complexes with metals [Eq. (9)] (40).

$$1,2\text{-}B_9H_{10}CHP^- + M(CO)_6 \xrightarrow[\text{THF}]{\text{UV}} (1,2\text{-}B_9H_{10}CHP)M(CO)_5^- \qquad (9)$$
$$+ CO$$

(M = Cr, Mo, W)

ACKNOWLEDGMENTS

I am indebted to Mrs. B. G. Rogainis and Mrs. S. G. Welcker for their capable assistance with the manuscript. Some of the research was supported in part by the National Science Foundation under Grant GP-10148.

REFERENCES

1. Adams, R. M., *Inorg. Chem.* **2**, 1087 (1963).
2. Bird, P. H., and Churchill, M. R., *Chem. Commun.* p. 403 (1967).
3. Cotton, F. A., Whipple, R. O., and Wilkinson, G., *J. Am. Chem. Soc.* **75**, 3586 (1953).
4. DeBoer, B. G., Zalkin, A., and Templeton, D. H., *Inorg. Chem.* **7**, 2288 (1968).

5. Francis, J. N., and Hawthorne, M. F., *J. Am. Chem. Soc.* **90**, 1663 (1968).
6. George, T. A., and Hawthorne, M. F., *J. Am. Chem. Soc.* **90**, 1661 (1968).
7. German, E. D., and Dyatkina, M. E., *Zh. Strukt. Khim.* **7**, 866 (1966).
8. Greenwood, N. N., and Travers, N. F., *Chem. Commun.* p. 216 (1967).
9. Harris, C. B., *Inorg. Chem.* **7**, 1517 (1968).
10. Hawthorne, M. F., Young, D. C., and Wegner, P. A., *J. Am. Chem. Soc.* **87**, 1818 (1965).
11. Hawthorne, M. F., and Andrews, T. D., *J. Am. Chem. Soc.* **87**, 2496 (1965).
12. Hawthorne, M. F., and Andrews, T. D., *Chem. Commun.* p. 443 (1965).
13. Hawthorne, M. F., and Pilling, R. L., *J. Am. Chem. Soc.* **87**, 3987 (1965).
14. Hawthorne, M. F., and George, T. A., *J. Am. Chem. Soc.* **89**, 7114 (1967).
15. Hawthorne, M. F., and Pitts, A. D., *J. Am. Chem. Soc.* **89**, 7115 (1967).
16. Hawthorne, M. F., Andrews, T. D., Garrett, P. M., Olsen, F. P., Reintjes, M., Tebbe, F. N., Warren, L. F., Wegner, P. A., and Young, D. C., *Inorg. Syn.* **10**, 111 (1967).
17. Hawthorne, M. F., *Accounts Chem. Res.* **1**, 281 (1968).
18. Hawthorne, M. F., Young, D. C., Andrews, T. D., Howe, D. V., Pilling, R. L., Pitts, A. D., Reintjes, M., Warren, L. F., and Wegner, P. A., *J. Am. Chem. Soc.* **90**, 879 (1968).
19. Hawthorne, M. F., and Wegner, P. A., *J. Am. Chem. Soc.* **90**, 896 (1968).
20. Hawthorne, M. F., George, T. A., and Pitts, A. D., *Abstr. 156th Meeting Am. Chem. Soc.*, *Atlantic City, 1968* INOR-150.
21. Hertler, W. R., Klanberg, F., and Muetterties, E. L., *Inorg. Chem.* **6**, 1696 (1967).
22. Hyatt, D. E., Little, J. L., Moran, J. T., Scholer, F. R., and Todd, L. J., *J. Am. Chem. Soc.* **89**, 3342 (1967).
23. Kaesz, H. D., Fellmann, W., Wilkes, G. R., and Dahl, L. F., *J. Am. Chem. Soc.* **87**, 2753 (1965).
24. Klanberg, F., Muetterties, E. L., and Gugenberger, L. J., *Inorg. Chem.* **7**, 2272 (1968).
25. Klanberg, F., Wegner, P. A., Parshall, G. W., and Muetterties, E. L., *Inorg. Chem.* **7**, 2072 (1968).
26. Knoth, W. H., *J. Am. Chem. Soc.* **89**, 3343 (1967).
27. Knoth, W. H., Little, J. L., and Todd, L. J., *Inorg. Syn.* **11**, 41 (1968).
28. Lipscomb, W. N., and Kaezmarczyk, A., *Proc. Natl. Acad. Sci. U.S.* **47**, 1796 (1961).
29. Little, J. L., Moran, J. T., and Todd, L. J., *J. Am. Chem. Soc.* **89**, 5495 (1967).
30. Maki, A. H., and Berry, T. E., *J. Am. Chem. Soc.* **87**, 4437 (1965).
31. Noeth H., and Schmid, G., *Z. Anorg. Allgem. Chem.* **345**, 69 (1966).
32. Parshall, G. W., *J. Am. Chem. Soc.* **86**, 361 (1964).
33. Ruhle, H. W., and Hawthorne, M. F., *Inorg. Chem.* **7**, 2279 (1968).
34. Scott, D. R., *J. Organometal. Chem. (Amsterdam)* **6**, 429 (1966).
35. Tebbe, F. N., Garrett, P. M., and Hawthorne, M. F., *J. Am. Chem. Soc.* **90**, 869 (1968).
36. Todd, L. J., Paul, I. C., Little, J. L., Welcker, P. S., and Peterson, C. R., *J. Am. Chem. Soc.* **90**, 4489 (1968).
37. Todd, L. J., and Little, J. L., to be published.
38. Todd, L. J., and Burke, A., to be published.
39. Todd, L. J., and Welcker, P. S., to be published.
40. Todd, L. J., and Silverstein, H. T., to be published.
41. Warren, L. F., and Hawthorne, M. F., *J. Am. Chem. Soc.* **89**, 470 (1967).
42. Warren, L. F., and Hawthorne, M. F., *J. Am. Chem. Soc.* **90**, 4823 (1968).
43. Wegner, P. A., and Hawthorne, M. F., *Chem. Commun.* p. 861 (1966).
44. Wing, R. M., *J. Am. Chem. Soc.* **89**, 5599 (1967).

45. Wing, R. M., *J. Am. Chem. Soc.* **90**, 4828 (1968).
46. Zalkin, A., Templeton, D. H., and Hopkins, T. E., *J. Am. Chem. Soc.* **87**, 3988 (1965).
47. Zalkin, A., Hopkins, T. E., and Templeton, D. H., *Inorg. Chem.* **5**, 1189 (1966).
48. Zalkin, A., Hopkins, T. E., and Templeton, D. H., *Inorg. Chem.* **6**, 1911 (1967).
49. Zalkin, A., St. Clair, D. J., and Templeton, D. H., *Abstr. Am. Cryst. Assoc.* p. 58 (1968).

Metal Carbonyl Cations

E. W. ABEL and S. P. TYFIELD

Department of Inorganic Chemistry,
The University
Bristol, England

I

INTRODUCTION

The phenomenal growth in the chemistry of metal carbonyl compounds has been heavily concentrated upon the neutral and anionic complexes. Only very recently have the occurrence and usefulness of the cationic metal carbonyls come into real prominence.

The first recognized member of this species $\{Co(CO)_3[(C_6H_5)_3P]_2\}^+$ $[Co(CO)_4]^-$ was reported (219) as recently as 1958; though it seems likely that cationic metal carbonyls had been made, but unrecognized, at a much earlier date. Thus, for example, Schützenburger (222) reported in 1870 the interaction of ammonia and dichloroplatinum(II) dicarbonyl.

$$2 NH_3 + Pt(CO)_2Cl_2 \rightarrow Pt(CO)_2Cl_2 \cdot 2 NH_3$$

The product of this reaction was reformulated (70) as the cationic complex $[Pt(CO)_2(NH_3)_2]^{2+}Cl_2^{2-}$, and even more recent work (149, 243) suggests the cationic species present is actually $[Pt(CO)(NH_3)_2Cl]^+$, according to the equation

$$4\,NH_3 + 2\,Pt(CO)_2Cl_2 \rightarrow [Pt(CO)(NH_3)_2Cl]^+[Pt(CO)Cl_3]^- + (NH_2CHO)_2$$

This, however, is a moot point, and the reaction merits further investigation.

The cationic metal carbonyls are, by definition, electrolytes in polar solvents, and are in general diamagnetic and conform to the "rare gas" formalism of metal carbonyls (1, 57).

The nature of bonding in the cationic metal carbonyls has been investigated by both vibrational and electronic spectroscopy, and molecular orbital calculations have been carried out; these are consistent with a bonding scheme for carbon monoxide coordinated to a metal, consisting of a dative σ-bond from carbon to metal, augmented by a synergic metal-to-carbonyl dative π-bond (1, 57).

The characteristic high carbonyl frequencies of cationic metal carbonyls are rationalized in terms of a reduced π-back-donation from the metal (1, 56). Such a proposal implicitly assumes that the carbon–metal σ-bond is independent of charge on the metal, and has been supported by semi-empirical molecular orbital calculations. This assumption of invariance of carbon–metal and carbon–oxygen σ-bonds with change in the metal's oxidation state has been invalidated by a Raman and infrared investigation of $M(CO)_6^+$, $M(CO)_6$, and $M(CO)_6^-$ species (3, 4). It appears that the carbon–metal σ-bond increases with rise of positive charge on the metal, with concomitant decrease of metal–carbon π-bonding.

The relative reactivity of cationic metal carbonyls has been predicted in the case of $Mn(CO)_6^+$ relative to $Cr(CO)_6$ and $V(CO)_6^-$ (35). The substitution behavior of $Re(CO)_6^+$ relative to $W(CO)_6$ indicates that the cation is not more readily substituted by neutral ligands (3, 6), which correlates with the spectroscopic investigations (3, 4) of the metal–carbonyl bond.

The high carbon–oxygen bond order in cationic metal carbonyls due to the reduced π-back-donation is further supported by the intensities of the infrared carbonyl stretching bands (25), estimates of the degree of back-donation from electronic spectra (4), and from calculations of π-overlap populations (35).

II

SYNTHESES OF THE CATIONIC METAL CARBONYLS

The methods that have been utilized to prepare the cationic metal carbonyls cover an exceedingly wide range of chemical reactions. An effort has been made to classify these in Sections A to H below.

A. Halide Displacement

1. Halide Displacement by Neutral Ligand

The substitution of a halide anion by a neutral ligand may occur in several ways. The direct displacement of halogen bonded to metal by a neutral ligand occurs with a variety of bases, some of which are cited below (Cp = cyclopentadienyl):

$$NH_3 + Mn(CO)_5Cl \rightarrow [Mn(CO)_3(NH_3)_3]^+Cl^- \tag{22}$$

$$PEt_3 + CpMo(CO)_3Cl \rightarrow [CpMo(CO)_3(PEt_3)]^+Cl^- \tag{195}$$

$$PPh_3 + CpFe(CO)_2I \rightarrow [CpFe(CO)_2(PPh_3)]^+I^- \tag{234}$$

$$p\text{-}MeC_6H_4NC + Ir(CO)_2(p\text{-}MeC_6H_4NH_2)Cl \rightarrow [Ir(CO)(p\text{-}MeC_6H_4NC)_3]^+Cl^- \tag{192}$$

A variation on this reaction is the expulsion of the halide from the incipient ligand, thereby raising the formal oxidation state of the metal.

$$CpCo(CO)_2 + C_3H_5Br \rightarrow [CpCo(CO)C_3H_5]^+Br^- \tag{82}$$

2. Assisted Halide Displacement Using Halide Acceptor

The use of a Lewis acid to promote the abstraction of a halide anion from a metal halide complex, in the presence of a neutral ligand, has been widely employed to produce both substituted and totally unsubstituted metal carbonyl cations. The halide acceptor most generally used is aluminum trichloride, but other Lewis acids have been employed, as indicated below. The requirements for the halide acceptor include that the anion product should be sufficiently large to stabilize the salt formed (15).

$$Mn(CO)_5Cl + AlCl_3 + CO \xrightarrow[300 \text{ atm}]{100°C} [Mn(CO)_6]^+[AlCl_4]^- \tag{80, 93}$$

$$Re(CO)_3(PPh_3)_2Cl + FeCl_3 + CO \xrightarrow[400 \text{ atm}]{20°C} [Re(CO)_4(PPh_3)_2]^+[FeCl_4]^- \tag{173}$$

$$Co(PPh_3)_2I_2 + Cu + CO \xrightarrow[200\ atm]{70°C} [Co(CO)_3(PPh_3)_2]^+[CuI_2]^- \qquad (219)$$

$$CpCo(CO)(C_2F_5)I + AgClO_4 + CO \xrightarrow[80\ atm]{25°C} [CpCo(CO)_2C_2F_5]^+[ClO_4]^- \ (235)$$

Analogous reactions have been effected by expulsion of other anions by base attack, for instance the displacement of nitrate.

$$(bipy)Mn(CO)_3(NO_3) + PPh_3 \rightarrow [(bipy)(PPh_3)Mn(CO)_3]^+NO_3^- \qquad (9)$$

B. Disproportionation of Metal Carbonyls

The reaction of binuclear metal carbonyls with bases has been recognized (137, 229, 244) as producing cationic metal carbonyl species, although the earlier derivatives defied isolation. This has been classed as a "base reaction" (122).

$$Mn_2(CO)_{10} + BuNH_2 \xrightarrow{20°C} [Mn(CO)_5(BuNH_2)]^+[Mn(CO)_5]^- \qquad (122)$$

$$Co_2(CO)_8 + C_5H_{10}NH \xrightarrow{20°C} [Co(CO)_4(C_5H_{10}NH)]^+[Co(CO)_4]^- \qquad (244)$$

Recently, stable cationic carbonyl derivatives have been isolated and fully characterized.

$$Co_2(CO)_8 + PEt_3 \xrightarrow[Dioxane]{20°C} [Co(CO)_3(PEt_3)_2]^+[Co(CO)_4]^- \qquad (130, 219)$$

$$Mn_2(CO)_{10} + Diphos \xrightarrow[dioxane]{100°C} [Mn(CO)_2(diphos)_2]^+[Mn(CO)_5]^- \qquad (217)$$

$$[CpMo(CO)_3]_2 + PPh_3 \xrightarrow[50\ hr]{UV} [CpMo(CO)_2(PPh_3)_2]^+[CpMo(CO)_3]^- \ (115)$$

$$Hg[Co(CO)_4]_2 + Tdp \xrightarrow[benzene]{20°C} [Co(CO)_3(Tdp)_2]^+[Co(CO)_4]^- \qquad (158)$$

The alternative reaction for these systems is direct substitution without disproportionation.

$$Co_2(CO)_8 + PEt_3 \xrightarrow[benzene]{60°C} [Co(CO)_3(PEt_3)]_2 \qquad (130)$$

The reaction favoring ionic rather than neutral products is controlled by a fine balance of factors, such as temperature, solvent, nature of ligand, and character of the metal–metal bonded system. As yet, the relative importance of these factors remains undefined.

C. *Protonation and Alkylation*

1. *Protonation*

The addition of a proton to a metal carbonyl compound may occur in either of two modes: the formation of metal–hydrogen bond, or protonation of a ligand attached to the central metal atom. If the ligand protonated is an organic radical, a carbonium ion is produced, which may be stabilized by suitable delocalization of charge over the complex, including the central metal atom. Consequently, such protonated species may be legitimately considered as examples of cationic metal carbonyl compounds.

Because of the legion of unsaturated organic systems bonded to metal carbonyl residues, this method provides an abundance of cationic carbonyls. Our classification of the method employed is according to the nature of the protonation site.

a. *σ-Allyl or σ-Pentadienyl Ligands.* This reaction has been studied particularly by Green and co-workers for the π-cyclopentadienyliron dicarbonyl system. They were the first to recognize that the facile protonation of σ-allyliron compounds yields π-ethylenic cations (*105, 108, 109*).

(*105, 108*)

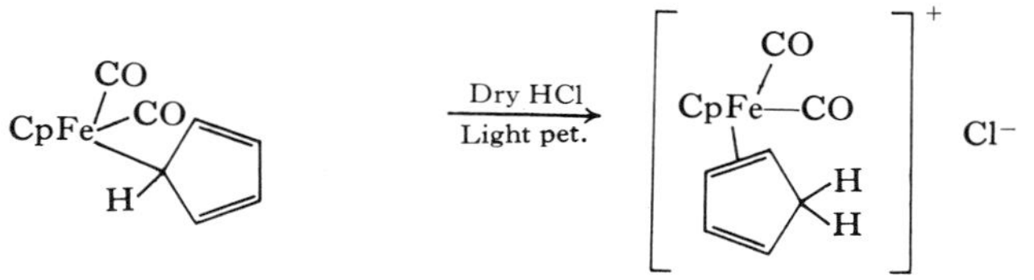

(*109*)

The proposed mechanism proceeds via a nonclassical carbonium intermediate.

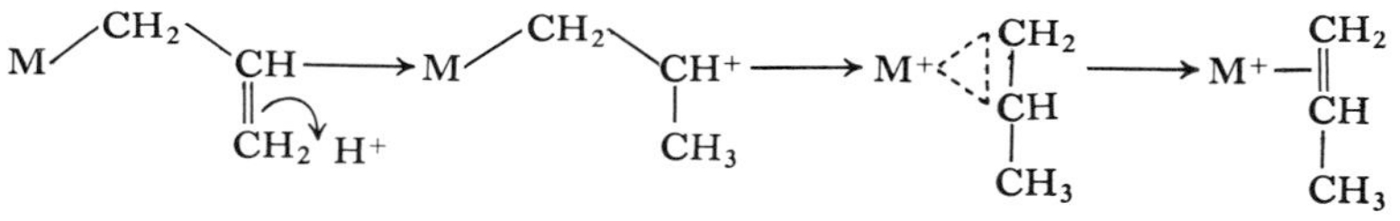

b. *Protonation of π-Olefinic Ligands.* The protonation of dienes complexed to a metal carbonyl may yield a π-allyl cation.

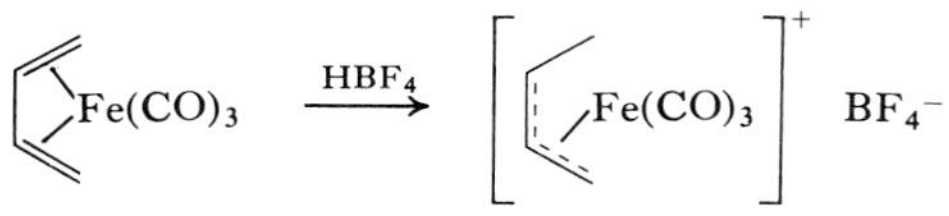

$$(74)$$

However, in cases when π-delocalization is restricted, the alternative of protonation on the metal may occur.

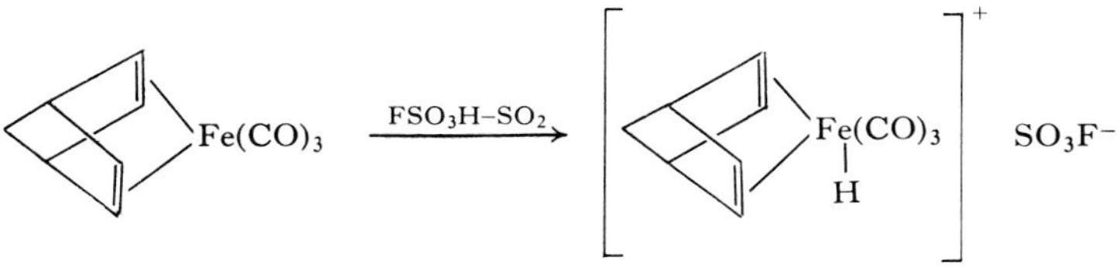

$$(75)$$

This reaction with cyclic olefins has produced some interesting carbonium species; a topical example is the C_8H_9 ligand. This has been complexed to iron as bicyclic cyclooctatrienyl (*64, 144, 221*), and to molybdenum as a monocyclic homocarbonium ion (*247*). The latter is an example of a non-classical carbonium ion.

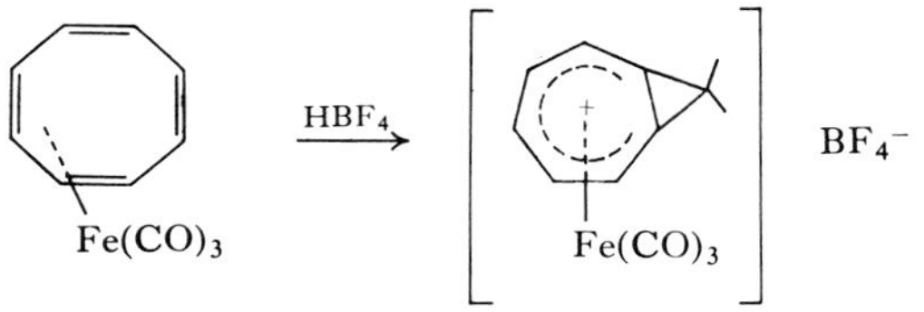

$$(64)$$

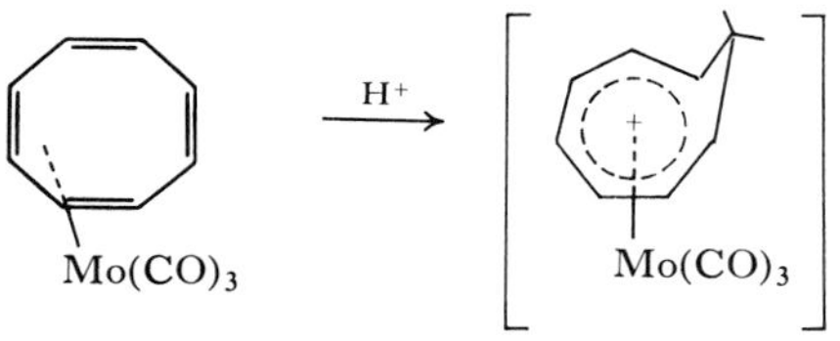

$$(247)$$

The protonation of fulvenes produces π-cyclopentadienyls.

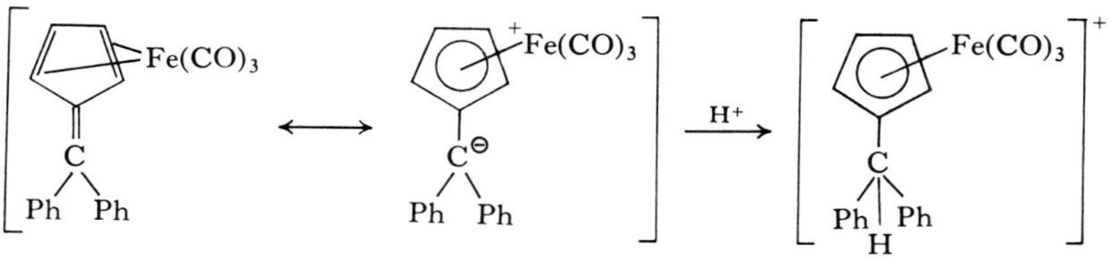

(241, 242)

c. *π-Allyl Alcohols or Ethers.* The reaction of allylic iron tricarbonyl systems on acidification has been examined in detail, particularly by Pettit and co-workers. The yields of cations are often high and even quantitative.

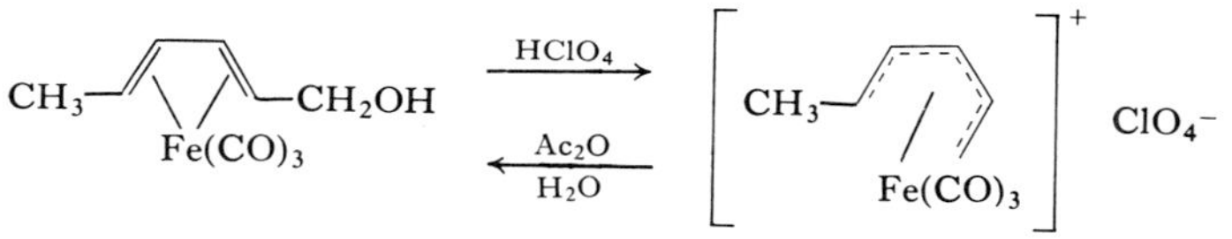

(187)

The reaction of methyl tropyl ether–iron tricarbonyl with acid yields a tropylium complex, in which the ring formally contains an uncoordinated double bond (*188*), but is actually an example of a fluxional molecule.

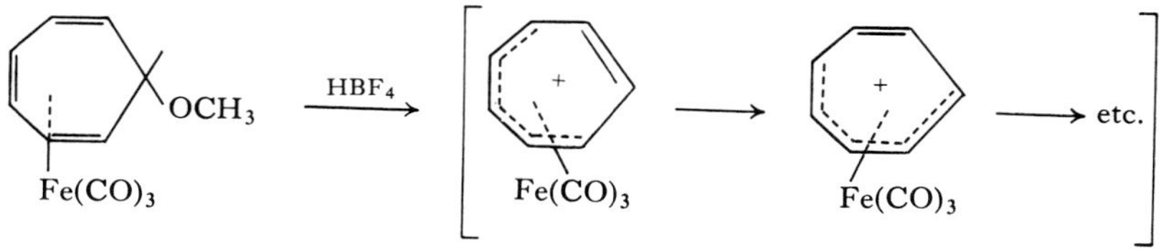

(188)

d. *Alkoxycarbonyls.* The alkoxycarbonyl metal carbonyls may yield a cationic metal carbonyl on acidification, as well as an alcohol.

$$Mn(CO)_5CO_2Et \xrightarrow{\text{Dry HCl}} [Mn(CO)_6]^+[HCl_2]^- + EtOH \qquad (175)$$

Since, in many cases the metal "esters" may be prepared from the metal carbonylate anion and a chloroformate, this reaction scheme enables cationic metal carbonyl compounds to be prepared from a neutral carbonyl compound, without recourse to high-pressure carbonylation, as required in Section A,2. Such a synthetic route is indicated as follows:

$$[Co(CO)_3(PPh_3)]_2 \xrightarrow{\text{2 Na}} Na^+[Co(CO)_3(PPh_3)]^-$$

$$\xrightarrow{\text{ClCO}_2\text{Et}} Co(CO)_3(PPh_3)CO_2Et \xrightarrow[-50°C]{\text{Dry HCl}} [Co(CO)_4(PPh_3)]^+[HCl_2]^- \quad (123)$$

A competing reaction to the formation of a cationic carbonyl is the subsequent reaction of the reactive cation with its associated anion, so that the overall product may not be a carbonyl salt.

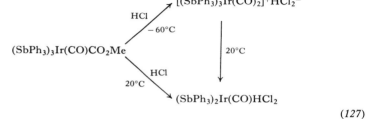

$$(127)$$

e. *Basic Carbonyls.* The metal carbonyl compounds which incorporate a proton on acidification to form a metal–hydrogen bond are considered in this section. This has already been mentioned as a possibility (Section C,1,b) when treating an olefinic metal carbonyl with an acid. There are several other compounds which can react in this manner with strong acids, and although the cationic species formed in this way are labile they have been identified spectroscopically.

$$Fe(CO)_3(PPh_3)_2 \xrightarrow{\text{Conc. H}_2\text{SO}_4} [HFe(CO)_3(PPh_3)_2]^+ \quad (63, 65)$$

$$Os(CO)_3(PPh_3)_2 \xrightarrow{\text{HCl}} [HOs(CO)_3(PPh_3)_2]^+[HCl_2]^-$$

$$\downarrow$$

$$Os(CO)_2(PPh_3)_2Cl_2 \quad (52)$$

$$[CpFe(CO)_2]_2 \xrightarrow{\text{CF}_3\text{CO}_2\text{H}} [(CpFe(CO)_2)_2H]^+CF_3CO_2^- \quad (63)$$

$$IrH(CO)(PPh_3)_3 \xrightarrow{\text{HCl}} [IrH_2(CO)(PPh_3)_3]^+Cl^- \quad (238)$$

Because of the increase in the coordination number of the metal, often with the formation of an octahedral complex, this reaction is restricted to what may be termed "coordinately unsaturated" metal carbonyls, as illustrated in the examples above.

f. *Metal Hydrides.* Metal hydrides react on protonation to produce molecular hydrogen. If this is effected in the presence of a neutral ligand, such as carbon monoxide, cationic metal complexes of the ligand may be obtained by incorporation.

$$\textit{trans-PtHCl(PEt}_3)_2 \xrightarrow[\text{Aq. HBF}_4]{\text{CO}} \textit{trans-}[Pt(CO)Cl(PEt_3)_2]^+BF_4^- + H_2 \quad (40, 42)$$

This reaction may be considered analogous to those of Section A,2, in which the acid is the anion acceptor, this time for the hydride rather than the halide anion.

2. *Alkylation*

The alkylation of an unsaturated atom or moiety bonded to a metal carbonyl produces an onium salt. This complex may be considered to be a further example of a cationic metal carbonyl, provided that the formal charge on the alkylated center can interact with the atomic orbitals of the metal. This may be justified on the basis of electronegativity differences between the metal atom and the alkylated center. It seems reasonable to argue that the location of the formal positive charge is on the electropositive metal atom, although the metal is probably only fractionally charged.

$$CpW(CN)(CO)_3 + MeI \rightarrow [CpW(CO)_3(CNMe)]^+I^- \qquad (46)$$

$$CpFe(CO)_2SMe + MeI \rightarrow [CpFe(CO)_2(SMe_2)]^+I^- \qquad (164)$$

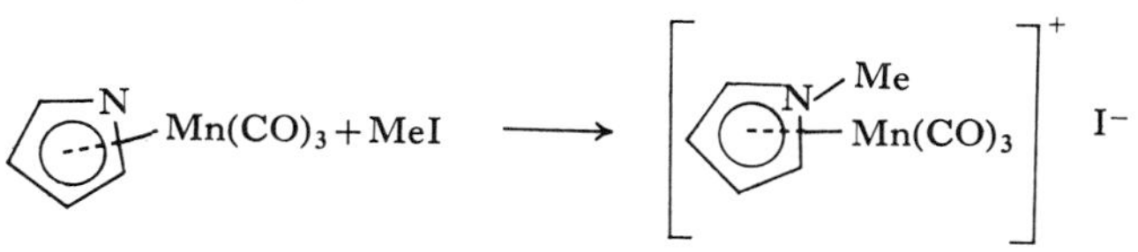

$$(153)$$

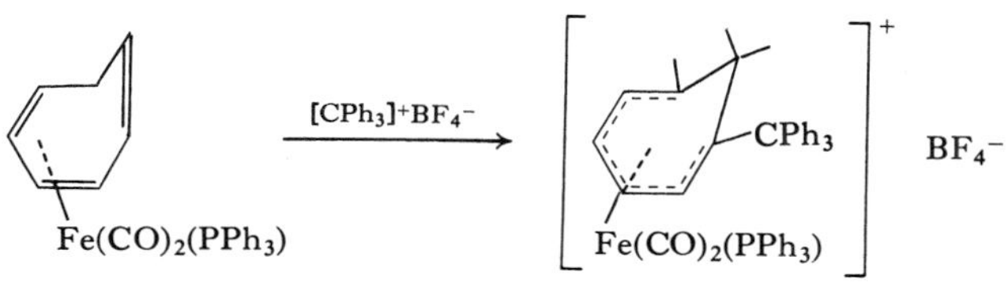

$$(36)$$

D. Hydride Abstraction

1. *Hydride Abstraction from Organic Ligands*

The removal of a hydride ion from an organic radical is an important method of generating carbonium ions stabilized by metal carbonyl systems. Dauben and Honnen (*61*) in 1958 were the first to exploit this method by use of the powerful hydride abstractor, triphenyl methyl (or trityl) carbonium ion, which is converted thereby into triphenylmethane.

$$C_7H_8Mo(CO)_3 + [Ph_3C]^+BF_4^- \xrightarrow{\text{Instantly}} [C_7H_7Mo(CO)_3]^+BF_4^- + Ph_3CH \quad (61)$$

When this hydride abstraction was tried with the iron-cycloheptatriene analog, a carbonium ion complex was formed, but by trityl alkylation.

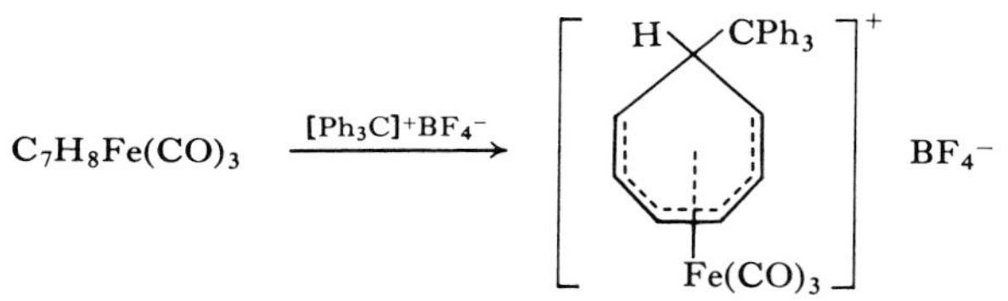

$$(60)$$

This point has already been mentioned in Section C,2 (*36*).

The range of cations formed by hydride removal is illustrated by a few examples, as follows:

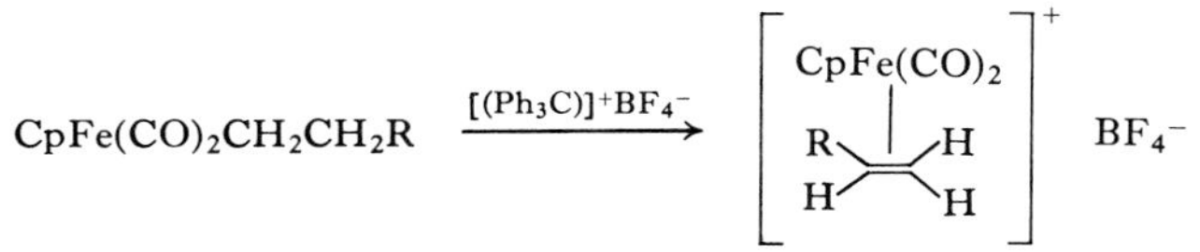

$$(110)$$

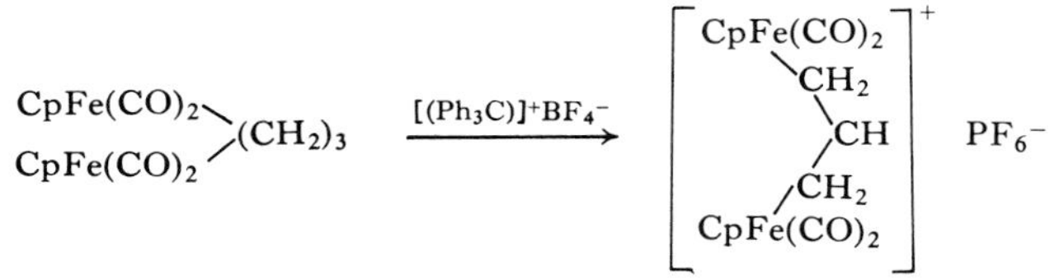

$$(166)$$

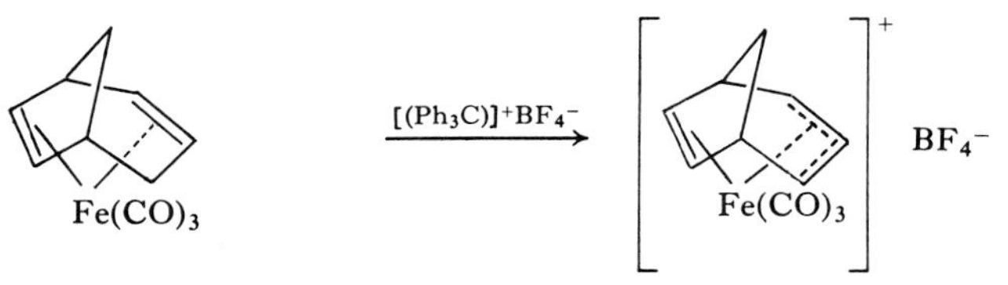

$$(196)$$

$$(81)$$

2. Hydride Abstraction from Metal Hydrides

Hydride abstraction of a hydrogen directly bonded to a metal atom has been used to synthesize cationic metal carbonyls. This may be accomplished by protonation, as outlined in Section C,1,f, or with a Lewis acid, such as boron trifluoride, in the presence of carbon monoxide (98).

$$CpCr(CO)_3H + BF_3 \cdot Me_2O + CO \rightarrow [CpCr(CO)_4]^+BF_4^- \qquad (98)$$

The relationship between the hydride abstraction and the protonation reactions is illustrated by an example in which the hydride removal is reversible.

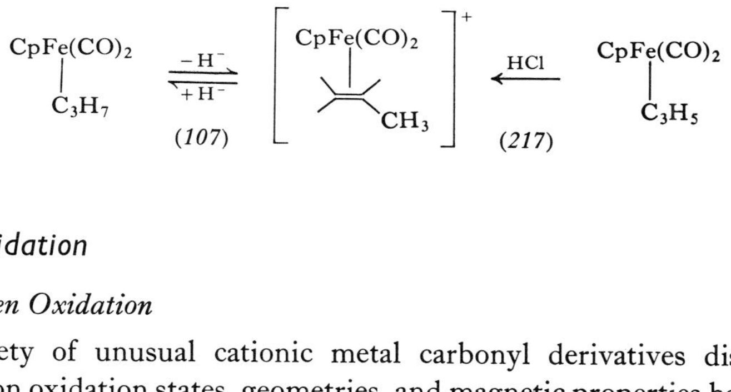

E. Oxidation

1. Halogen Oxidation

A variety of unusual cationic metal carbonyl derivatives displaying uncommon oxidation states, geometries, and magnetic properties have been obtained by the controlled oxidation of some metal carbonyl compounds.

Thus the addition of two equivalents of iodine to diarsinetungsten tetracarbonyl yielded a seven-coordinate tungsten(II) cation, while with excess bromine the first seven-coordinate tungsten(III) derivative was obtained (181).

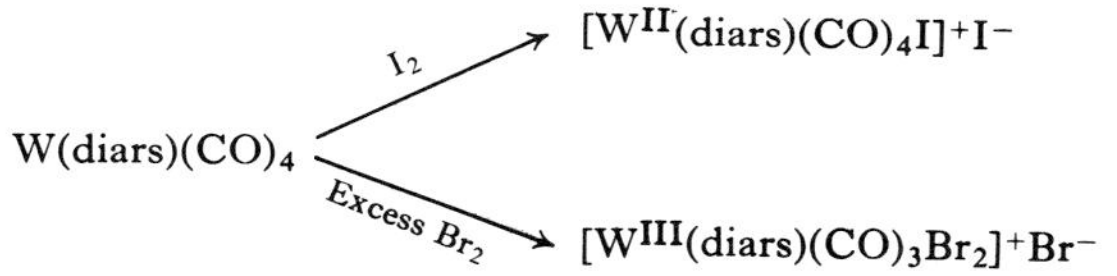

$$W(diars)(CO)_4 \xrightarrow{I_2} [W^{II}(diars)(CO)_4I]^+I^- \qquad (181)$$

$$W(diars)(CO)_4 \xrightarrow{Excess\ Br_2} [W^{III}(diars)(CO)_3Br_2]^+Br^- \qquad (181)$$

Similarly,

$$W(triars)(CO)_3 \xrightarrow{I_2} [W^{II}(triars)(CO)_3I]^+I^- \qquad (197)$$

Analogous cations may be formed using antimony pentachloride as a mild chlorinating agent.

$$W(C_6Me_6)(CO)_3 \xrightarrow{SbCl_5} [W^{II}(C_6Me_6)(CO)_3Cl]^+SbCl_6^- \qquad (226)$$

The mode of action of the antimony pentachloride may be formulated in general terms.

$$M(CO)_x + 2\,SbCl_5 \rightarrow [M(CO)_xCl]^+SbCl_6^- + SbCl_3 \qquad (161)$$

Pseudo-halogens may be used in a similar manner. The pseudo-inter-halogens ClN_3 or BrN_3 react with iron pentacarbonyl to produce a remarkable bridged, binuclear, doubly charged cation, which is paramagnetic.

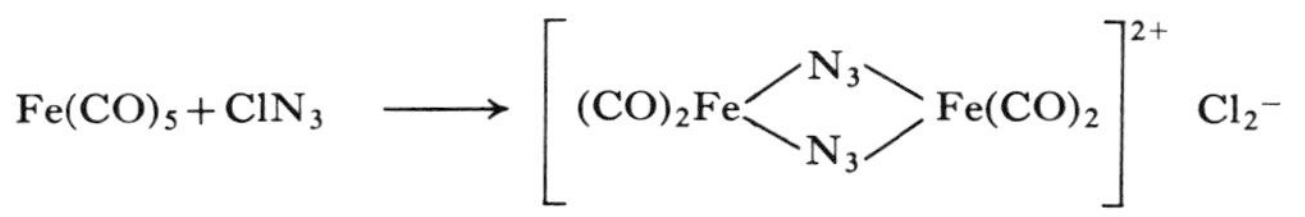

$$Fe(CO)_5 + ClN_3 \longrightarrow \left[(CO)_2Fe \underset{N_3}{\overset{N_3}{<>}} Fe(CO)_2 \right]^{2+} Cl_2^- \qquad (180)$$

An extension of this category should include the mixed "pseudo-halogen" X—HgX. Thus in the iron triad, cationic products have been prepared with mercuric chloride.

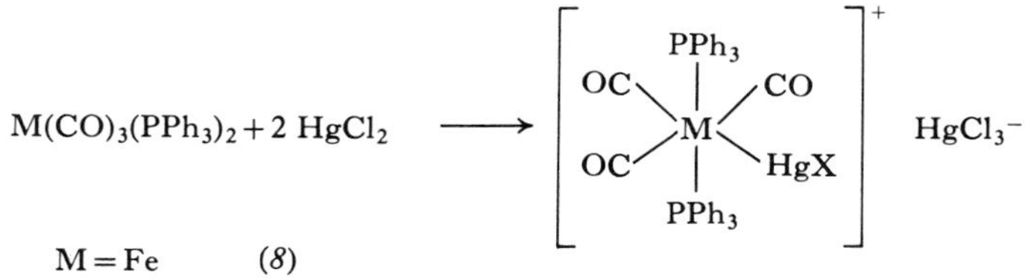

$$M(CO)_3(PPh_3)_2 + 2\,HgCl_2 \longrightarrow \left[\begin{array}{c} PPh_3 \\ OC \diagdown \mid \diagup CO \\ >M< \\ OC \diagup \mid \diagdown HgX \\ PPh_3 \end{array} \right]^+ HgCl_3^-$$

$$M = Fe \qquad (8)$$
$$= Ru,\,Os \quad (51)$$

This is closely related to the reaction

$$Fe(CO)_3(PPh_3)_2 + 2\,I_2 \rightarrow [Fe(CO)_2(PPh_3)_2I_2]^+I_3^- \qquad (44)$$

2. Oxidation in Acid Media

The oxidation of metal carbonyl compounds by atmospheric oxygen in acid media, or alternatively with an oxidizing acid, have only been reported in a few instances.

$$C_6H_6V(CO)_4 + O_2 + HCl \rightarrow [C_6H_6V(CO)_4]^+Cl^- \qquad (47)$$

$$[CpFe(CO)_2]_2 + O_2 + HClO_4 \rightarrow [CpFe(CO)_2(H_2O)]^+ClO_4^- \qquad (215)$$

$$[Mn(CO)_2(diphos)_2]^+ \xrightarrow{HNO_3} [Mn(CO)_2(diphos)_2]^{2+} \qquad (227)$$

The doubly charged Mn(II) cation is paramagnetic and isoelectronic with vanadium hexacarbonyl and neutral chromium pentacarbonyl iodide. These are examples of some of the few carbonyl compounds which contravene the rare gas formalism (1, 57).

3. Oxidation with Vanadium Hexacarbonyl

Vanadium hexacarbonyl readily reverts to its anionic form, thereby filling all of its bonding orbitals, and is consequently a useful, albeit expensive, one-electron oxidizing agent.

$$V(CO)_6 + Cp_2V \xrightarrow[15°C]{\substack{CO \\ atm.\ pressure}} [Cp_2V(CO)_2]^+[V(CO)_6]^- \qquad (32)$$

$$V(CO)_6 + Arene \xrightarrow{35°C} [V(CO)_4(Arene)]^+[V(CO)_6]^- \qquad (28, 29)$$

In the second example, the intermediate is probably $V(CO)_3Arene$, which is oxidatively carbonylated to give the ionic product.

4. Electrochemical Oxidation

The electrochemical oxidation of metal carbonyl compounds has not been developed as a synthetic route to cationic complexes, although the electrochemistry of this oxidation process has been investigated (67, 198).

F. Carbonylation of Cations

Several cationic metal carbonyls have been prepared by carbonylation of cationic metal complexes. The metal remains in the same oxidation state in this process, but can either change or retain its coordination number.

$$[Ir(diphos)_2]^+ + CO \longrightarrow [Ir(diphos)_2(CO)]^+$$

(131, 150, 220)

$$\left[(PEt_3)_2Pt \overset{Cl}{\underset{Cl}{\diagup\diagdown}} Pt(PEt_3)_2 \right]^{2+} \left[BF_4 \right]_2^- + CO \longrightarrow [(PEt_3)_2PtCl(CO)]^+[BF_4]^-$$

$$(41)$$

G. Ligand Substitution on Cationic Metal Carbonyls

The replacement of ligands of a metal carbonyl cation, either carbon monoxide or another ligand, serves as a valuable route to substituted cationic metal carbonyls.

1. CO Substitution

The displacement of carbon monoxide from a cationic metal carbonyl has been reported in only a few instances [v-triars = tris-1,1,1-(dimethyl-arsinomethyl)ethane].

$$[Re(CO)_6]^+ClO_4^- + v\text{-triars} \rightarrow [(v\text{-triars})Re(CO)_3]^+ClO_4^- \qquad (208)$$

$$[Ir(CO)_3(PPh_3)_2]^+ClO_4^- + PPh_3 \rightarrow [Ir(CO)_2(PPh_3)_3]^+ClO_4^- \qquad (193)$$

2. Arene or Olefin Substitution

Just as with neutral arene or olefin metal carbonyls, the cationic analogs are useful intermediates for the preparation of other substituted cations. This is because the organic moiety is often readily displaced by another ligand and, further, allows carbonyl derivatives to be prepared, which may not be readily accessible by direct reaction (dien = diethylene triamine).

$$[C_6H_6Mn(CO)_3]^+I^- + dien \rightarrow [(dien)Mn(CO)_3]^+I^- \qquad (2)$$

$$[(C_2H_4)Mn(CO)_5]^+BF_4^- + PPh_3 \rightarrow [(PPh_3)_2Mn(CO)_4]^+BF_4^- \qquad (104)$$

H. Miscellaneous

The general classification outlined in Sections A to G covers the vast majority of the types of reaction whereby cationic metal carbonyl compounds are prepared. There remain, nevertheless, a few further reactions which, for the sake of completeness, are included in this survey, but which do not belong to a specific section.

The substitution of a neutral ligand, including carbon monoxide, attached to a metal carbonyl complex by the nitrosyl cation, produces a cationic metal carbonyl nitrosyl.

$$CpMn(CO)_3 \xrightarrow[\text{(NaNO}_2\text{/HCl)}]{NO^+} [CpMn(CO)_2(NO)]^+ \qquad (215)$$

A stabilized diazonium salt, prepared in a similar manner, is of interest.

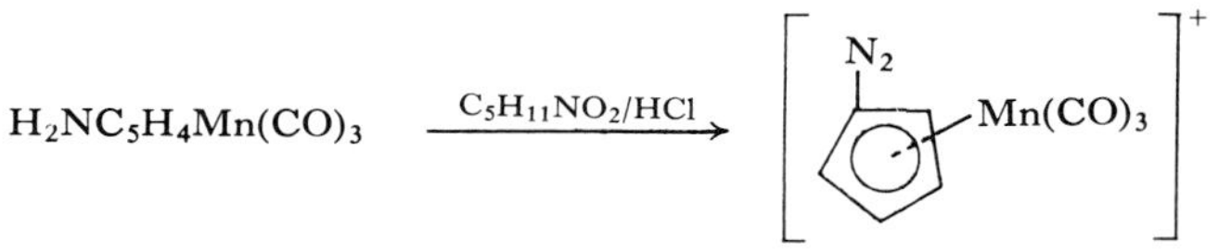

$$H_2NC_5H_4Mn(CO)_3 \xrightarrow{C_5H_{11}NO_2\text{/HCl}} \left[\begin{array}{c} N_2 \\ \overset{\displaystyle}{\bigodot} \!\!-\! Mn(CO)_3 \end{array} \right]^+$$

$$(27)$$

III

SURVEY OF THE CATIONIC METAL CARBONYLS AND THEIR PROPERTIES BY GROUPS

A. Titanium Group

There is a patent claim for the preparation of $[C_6H_6Ti(CO)_4]^+Br^-$ and $[C_6H_6Ti(CO)_3(NO)]^+Br^-$ (47). Apart from this, there appear to be no examples of Group IV metal cationic carbonyls.

B. Vanadium Group

The known vanadium carbonyl cations are of two types, namely a tetracarbonyl, $[\text{Arene}V(CO)_4]^+$, and a dicarbonyl, $[Cp_2V(CO)_2]^+$. The tetracarbonyl derivatives are readily prepared under mild conditions by the reaction of an arene with vanadium hexacarbonyl. The arenes used include benzene, its methyl derivatives (28, 29), naphthalene, and anisole (29). The cation is probably formed by oxidation of the intermediary arene vanadium tricarbonyl.

$$V(CO)_6 \xrightarrow[35°C]{\text{Arene}} V(CO)_3\text{Arene} \xrightarrow{V(CO)_6} [V(CO)_4\text{Arene}]^+[V(CO)_6]^-$$

The red cations are isolated with bulky anions such as fluorophosphate or phenylborate (28, 29).

An alternative preparation is reported in the patent literature, in which benzenevanadium tetracarbonyl is oxidized in acid media to $[C_6H_6V(CO)_4]^+$ (47).

The dicarbonyl cation $[Cp_2V(CO)_2]^+$ is prepared by facile oxidation in nonpolar solvents (32).

$$V(CO)_6 + Cp_2V \xrightarrow[15°C]{\overset{CO}{\text{atm. pressure}}} [Cp_2V(CO)_2]^+[V(CO)_6]^- \qquad (32)$$

The yield of the stable salt is almost quantitative. The dicarbonyl di-π-cyclopentadienylvanadium cation is particularly stable; thus it may be recovered after treatment with excess iodine, as the triiodide.

$$[Cp_2V(CO)_2]^+[V(CO)_6]^- + 3 I_2 \rightarrow [Cp_2V(CO)_2]^+I_3^- + V^{3+} + 3 I^- + 6 CO \quad (32)$$

The infrared and proton NMR spectra of the vanadium carbonyl cations have been reported (28, 29, 32). The investigation of the reactivity of these cations with nucleophiles is limited to a study of the hydridic reduction of the tetracarbonyl species (31). A mechanism is proposed which involves primary coordination of hydride to vanadium; the hydrogen then transfers to an arene carbon atom to form a methylene group, with the hydridic hydrogen in the endo position, viz., H_α. The product is π-cyclohexadienyl-vanadium tetracarbonyl.

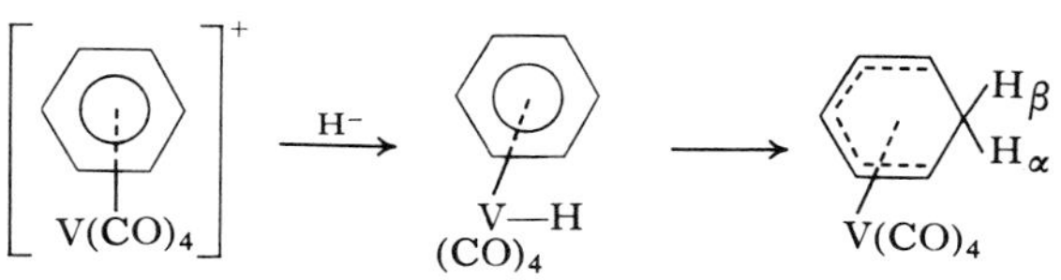

$$(31)$$

No carbonyl cations of vanadium with σ-donor ligands have been reported. When Lewis bases react with vanadium hexacarbonyl, either a substituted derivative or a hexacarbonylvanadate salt is obtained. The latter is formed by disproportionation, which also produces a carbonyl-free cation.

$$V(CO)_6 + NH_3 \xrightarrow{-30°- +20°C} [V(NH_3)_6]^{2+}[V(CO)_6]_2^- \qquad (19)$$

$$V(CO)_6 + \text{diphos} \xrightarrow{20°C} [V(\text{diphos})_3]^{2+}[V(CO)_6]_2^- \qquad (20)$$

$$V(CO)_6 + \text{diphos} \xrightarrow{120°C} V(CO)_4(\text{diphos}) \text{ or}$$
$$V(CO)_2(\text{diphos})_2 \tag{20}$$

No cationic carbonyl complexes of niobium and tantalum have been reported.

C. Chromium Group

1. *Lewis Base Derivatives and Related Compounds*

The Group VI metals would not be expected to form binary metal carbonyl cations, but they do form some substituted cations with nitrogen and phosphorus ligands. The paramagnetic monomeric and dimeric chromium pentacarbonyl iodides react in liquid ammonia with iodide expulsion.

$$Cr(CO)_5I + NH_3 \rightarrow [Cr(CO)_2(NH_3)_4]^+I^- + CO(NH_2)_2 + NH_4I \tag{18}$$
$$Cr_2(CO)_{10}I + NH_3 \rightarrow [Cr(CO)_2(NH_3)_4]^+[Cr(CO)_3(NH_3)_2I]^- \tag{18}$$

The carbonyls are disposed trans to each other, which is deduced from the single carbonyl in the infrared spectrum. A similar cation may be formed by mild oxidation of the chromium dicarbonyl phosphine, $Cr(CO)_2QP$ [QP = tris(o-diphenylphosphinophenyl)phosphine].

$$Cr(CO)_2QP \xrightarrow{SbCl_5} [Cr(CO)_2QP]^+ \tag{145}$$

This cation decomposes fairly rapidly in air; treatment with excess halogen causes complete expulsion of carbon monoxide (145).

A bis(diphos) cation has been reported to be formed for molybdenum and tungsten by oxidation. Both cations are paramagnetic with a *trans*-carbonyl configuration, rather than the expected cis arrangement. This is probably caused by steric effects (184).

$$M(CO)_2(\text{diphos})_2 \xrightarrow{3\,I} [M(CO)_2(\text{diphos})_2]^+I_3^- \tag{183, 184}$$
$$M = Mo, W$$

The more usual product of halogen oxidation of a substituted Group VI hexacarbonyl is a seven-coordinate halogen complex, which may be cationic or neutral. Thus the mono(diphos) tungsten tetracarbonyl yields, on controlled oxidation, a transient heptacoordinate cation, which reverts on carbonyl expulsion to an isolatable neutral diiodide (184). The overall reaction is an example of an oxidative elimination reaction (185).

$$W(CO)_4(diphos) + I_2 \longrightarrow [W(CO)_4(diphos)I]^+I^-$$

$$\downarrow$$

$$[W(CO)_3(diphos)I_2] \qquad\qquad (184)$$

An analogous cation has been isolated for the diarsine derivative, namely $[W(CO)_4(diars)I]^+$ (181). The reaction of stannic bromide with $Mo(CO)_4$ (diars) [and $W(CO)_4(diars)$] appears to proceed in a similar way, so that an unstable mixed metal cation $[Mo(CO)_4(diars)SnBr_3]^+Br_3^-$ is produced, which decomposes to $(diars)Mo(CO)_2Br_3$ (179).

The substituted tricarbonyl complexes react in a similar fashion to the tetracarbonyl complexes with halogens, whereas electrochemical oxidation appears to cause the formation of a doubly charged cation, $[M(CO)_3$ $(triars)]^{2+}$, where $M = Cr$, Mo, W (67). Allyl halides react with the tripyridine-substituted carbonyls of the type $Mo(CO)_3(bipy)py$ [or $Mo(CO)_3$ (phen)py; phen = phenanthroline] to form a π-allylic carbonyl cation, such as $[Mo(CO)_2(bipy)py(\pi\text{-allyl})]^+$ (146). The cation reacts with sulfur dioxide to form $Mo(CO)_2(bipy)(SO_2)_2$ (147).

A list of the reported ionic heptacoordinate halide carbonyl cations and related complexes is presented in Table I.

2. π-Cyclopentadienyl Carbonyl Cations

The Group VI metals form a series of π-cyclopentadienyl carbonyl cations. $[CpM(CO)_4]^+$ is the parent cation and is comparable to the Group VII hexacarbonyls $[M'(CO)_6]^+$, where $M' = Mn$ $(80, 93)$, Tc (135), Re $(173, 177)$.

The $[CpM(CO)_4]^+$ cations have been prepared by high-pressure carbonylation in the presence of the Lewis acids $AlCl_3$ (80) or BF_3 (98).

$$CpCr(CO)_3H + BF_3 \cdot OMe_2 + CO \xrightarrow[70°C]{240\ atm} [CpCr(CO)_4]^+BF_4^- \qquad (98)$$

$$CpM(CO)_3Cl + AlCl_3 + CO \xrightarrow[70°C]{240\ atm} [CpM(CO)_4]^+AlCl_4^- \qquad (80, 93)$$

$$M = Mo, W$$

TABLE I

ISOLATED CATIONIC HEPTACOORDINATE CARBONYL CATIONS[a]
OF GROUP VI METALS

$[Cr(CO)_2(diars)_2Br]^+$ Br^- and Br_3^-	(182)
$[Cr(CO)_2(diars)_2I]^+$ I^- and I_3^-	(182)
$[Mo(CO)_2(diars)_2Br]^+$ Br^- and Br_3^-	(205)
$[Mo(CO)_2(diars)_2I]^+$ I^-	(205)
$[W(CO)_4(diars)I]^+$ I^- and I_3^-	(181)
$[W(CO)_3(diars)Br_2]^+$ Br^-, 1.54^b	(181)
$[W(CO)_2(diars)_2Br]^+$ Br^- and Br_3^-	(181)
$[W(CO)_2(diars)_2I]^+$ I^- and I_3^-	(181)
$[W(CO)_3(l\text{-triars})Br]^+$ Br^-	(197)
$[W(CO)_3(l\text{-triars})I]^+$ I^-	(197)
$[W(CO)_3(C_6Me_6)Cl]^+$ $SbCl_6^-$ and BPh_4^-	(226)
$[Cr(CO)_3(ttas)I]^+$ I^-	(54)
$[Mo(CO)_3(ttas)Br]^+$ Br^-	(54)
$[Mo(CO)_3(ttas)I]^+$ I^-	(54)
$[W(CO)_3(ttas)Br]^+$ Br^-	(54)
$[W(CO)_3(ttas)I]^+$ I^-	(54)
$[Mo(CO)_2(bipy)_2I]^+$ I^- and I_3^-	(21)
$[Mo(CO)_2(phen)_2I]^+$ I^- and I_3^-	(21)
$[W(CO)_2(phen)_2I]^+$ I^- and I_3^-	(21)
$[Mo(CO)_3(bipy)(MeNO_2)I]^+$ $I^{-\,c}$	(230)
$[W(CO)_3(bipy)(MeNO_2)Br]^+$ $Br^{-\,c}$	(230)
$[W(CO)_4(diphos)I]^+$ $I_3^{-\,c}$	(184)
$[W(CO)_3(v\text{-triars})I]^+$ I^- and ClO_4^-, 0.5^b	(209)
$[W(CO)_3(v\text{-triars})Br]^+$ Br^-, ClO_4^- and BPh_4^-, 0.5^b	(209)
$[Mo(CO)_3(v\text{-triars})I]^+$ I^-, 0.9^b	(209)
$[Mo(CO)_3(v\text{-triars})Br]^+$ Br^- and BPh_4^-, 0.8^b	(209)
$[Mo(CO)_3(l\text{-triars})I]^+$ I^-	(209)
$[Mo(CO)_3(l\text{-triars})Br]^+$ BPh_4^-	(209)
$[Cr(CO)_2(v\text{-triars})I]^+$ BPh_4^-, 2.83^b	(209)
$[Cr(CO)_3(v\text{-triars})I]^+$ BPh_4^-, 1.38^b	(209)
$[Mo(CO)_2(bipy)py(\pi\text{-allyl})]^+$ BPh_4^- and BF_4^-	(146)
$[Mo(CO)_2(bipy)py(2\text{-Me-}\pi\text{-allyl})]^+$ BF_4^-	(146)
$[Mo(CO)_2(phen)py(\pi\text{-allyl})]^+$ BF_4^-	(146)
$[W(CO)_2(bipy)py(\pi\text{-allyl})]^+$ BPh_4^- and BF_4^-	(146)
$[W(CO)_2(phen)py(\pi\text{-allyl})]^+$ BF_4^-	(146)
$[Mo(CO)_4(diars)SnBr_3]^+$ $Br_3^{-\,d}$	(179)
$[W(CO)_4(diars)SnBr_3]^+$ $Br_3^{-\,d}$	(179)

[a] Diamagnetic unless stated to contrary. l-triars = methylbis
(-dimethylarsino-3-propyl) arsine; ttas = bis-o (o-dimethylar-
sinophenyl) methylarsine; v-triars=tris-1,1,1-(dimethylarsino)
ethane.

[b] Paramagnetism in μ_B.

[c] Not isolated.

[d] Ill-defined.

Several monosubstituted derivatives have been made in a similar manner, except that a unidentate ligand is used in place of the carbon monoxide, namely NH_3, N_2H_4 (88); C_2H_4 (78); PPh_3, $AsPh_3$, CH_3CN (231). Alternatively, substituted monoolefin derivatives have been synthesized by protonation of the metal σ-allyl π-cyclopentadienyl tricarbonyl (58, 112).

$$CpMo(CO)_3(C_3H_5) \xrightarrow{HCl} [CpMo(CO)_3(CH_3CH\!=\!CH_2)]^+Cl^- \qquad (58)$$

Dicarbonyl cationic derivatives, $[CpM(CO)_2L_2]^+$, containing Lewis bases, L, have been obtained in a variety of ways. Direct halide displacement from the π-cyclopentadienyl tricarbonyl halides has been reported for

$L = PEt_3$	and	$M = Mo$	(114, 195, 231)
		$M = W$	(91)
$L = PPh_3$	and	$M = Mo$	(114, 195)
$L_2 = $ diphos	and	$M = Mo$	(114)
$L_2 = $ diars	and	$M = Mo$	(114)
$L_2 = $ dipy	and	$M = Mo, W$	(231)
$L_2 = $ phen	and	$M = Mo, W$	(231)

while $AlCl_3$-promoted halide expulsion is reported for $L = PPh_3$, $AsPh_3$, CH_3CN, and $M = Mo$, W (231).

The same substituted cations are obtained by the photochemical disproportionation of dimeric π-cyclopentadienylmolybdenum tricarbonyl, $[CpMo(CO)_3]_2$, with the following ligands; PEt_3, PPh_3, $PBu_3{}^n$, and diphos (115).

The disubstituted methyl isocyanide cations $[CpM(CO)_2(CNMe)_2]^+$, $M = Mo$, W, have been obtained by methylation of the corresponding dicyanide anions, $[CpM(CO)_2(CN)_2]^-$ (46). Direct interaction between phenyl isocyanide and $CpMo(CO)_3Cl$ causes total carbonyl displacement (154).

The monocarbonyl cationic derivatives, $[CpM(CO)L_3]^+$, are known for the "tridentate" ligands benzene and mesitylene, as well as for σ-donors such as $L_3 = (CH_3CN)(PPh_3)_2$ (231).

$$CpMo(CO)_3Cl + AlCl_3 + C_6H_6 \xrightarrow[\text{reflux}]{2\ hr} [CpMo(CO)C_6H_6]^+AlCl_4^- \quad (47, 83)$$

On treatment with hydridic reagents, the molybdenum monocarbonyl cation yields the red paramagnetic $CpMoC_6H_6$ (84, 85), whereas the tungsten analog is converted into a cyclohexadien-1,3-monocarbonyl hydride, $CpW(CO)(C_6H_8)H$ (85).

Other π-cyclopentadienyl cations reported are $[CpCr(CO)(NO)_2]^+PF_6^-$ (86), $[(CpMo(CO)_3)_2H]_2^+SO_4^{2-}$, $[(CpW(CO)_3)_2H]^+PF_6^-$, $[(Cp_2MoW (CO)_6)H]_2^+SO_4^{2-}$ (63).

3. Carbonium Carbonyl Cations

The first successful attempt to prepare a tropylium cationic metal carbonyl was by hydride abstraction from cycloheptatrienylmolybdenum tricarbonyl using triphenylmethyl tetrafluoroborate to produce $[C_7H_7 Mo(CO)_3]^+$ (61).

The reaction of the tropylium chromium cation, $[C_7H_7Cr(CO)_3]^+$, with a variety of anionic nucleophiles has been studied by Munro and Pauson (204, 213). Anionic attack at the ring neutralizes the positive charge, yielding a heptatrienyl derivative (204), or resulting in ring contraction to form $Cr(CO)_3C_6H_6$ as with $[C_5H_5]^-$ (204), or causing the "abnormal" reaction of ring coupling (204) to produce a bi(cycloheptatrienyl)chromium tricarbonyl complex.

The stabilization of the tropylium residue by chromium tricarbonyl has been investigated by Pettit and co-workers (143). They failed to synthesize a π-benzyl analog.

The monohomotropylium molybdenum cation $[C_8H_9Mo(CO)_3]^+$ is formed on protonation of the cyclooctatetraene carbonyl $C_8H_8Mo(CO)_3$ (247). An N-methylisoquinolinium chromium tricarbonyl cation is described by Öfele (210).

D. Manganese Group

The Group VII metals form stable, diamagnetic binary metal carbonyl cations from their pentacarbonyl halides.

$$M(CO)_5X + AlCl_3 + CO \rightarrow [M(CO)_6]^+[AlCl_3X]^-$$

M = Mn   (80, 93, 173, 177); 90° C, 300 atm CO, 20 hr

= Tc   (135); 90° C, 300 atm CO, 15 hr

= Re   (133, 134, 173, 177); 90° C, 320 atm CO, 20 hr

A nonpressure synthesis may be used; the treatment of metal ethoxycarbonyl pentacarbonyl with dry hydrogen chloride releases the metal hexacarbonyl cation.

$$M(CO)_5CO_2Et \xrightarrow{\text{HCl}} [M(CO)_6]^+[HCl_2^-]$$

$$M = Mn~(178);~Re~(176)$$

Rhenium pentacarbonyl chloride is converted into its hexacarbonyl cation by refluxing with mesitylene (208) or thiophene (225), and aluminum chloride.

The manganese hexacarbonyl cation is stable in a dry atmosphere, but moisture causes hydrolysis to the hydride and subsequently to manganese carbonyl.

$$[Mn(CO)_6]^+ \xrightarrow{\text{H}_2\text{O}} Mn(CO)_5H \longrightarrow Mn_2(CO)_{10} + H_2~(80, 93, 176)$$

Similarly, ammonolysis causes the same reaction (22). Both these solvolysis reactions are examples of base attack upon the carbonyl carbon and not on the central metal. The reaction with alkoxide results in the formation of an alkoxycarbonyl (176, 178).

$$[Mn(CO)_4(PPh_3)_2]^+ + MeO^- \to Mn(CO)_3(PPh_3)_2CO_2Me \qquad (178)$$

The equivalent solvolysis product rearranges, as is exemplified by the ammonolysis reaction scheme:

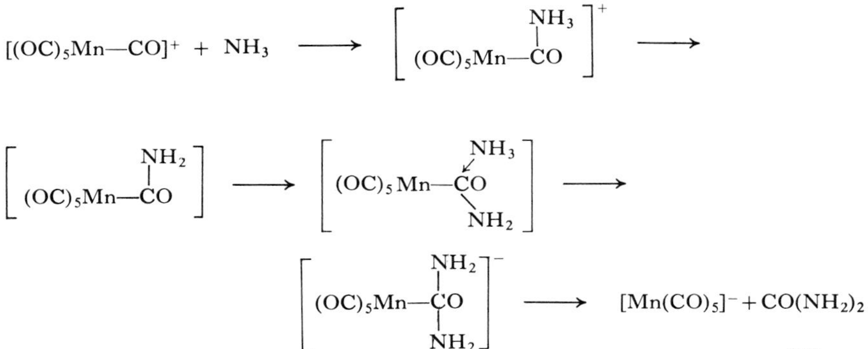

$$[(OC)_5Mn-CO]^+ + NH_3 \longrightarrow \left[(OC)_5Mn-\overset{\overset{\displaystyle NH_3}{|}}{CO} \right]^+ \longrightarrow$$

$$\left[(OC)_5Mn-\overset{\overset{\displaystyle NH_2}{|}}{CO} \right] \longrightarrow \left[(OC)_5Mn-\overset{\nearrow NH_3}{\underset{\searrow NH_2}{CO}} \right] \longrightarrow$$

$$\left[(OC)_5Mn-\overset{\overset{\displaystyle NH_2}{|}}{\underset{\underset{\displaystyle NH_2}{|}}{CO}} \right]^- \longrightarrow [Mn(CO)_5]^- + CO(NH_2)_2$$

$$(22)$$

In contrast, technetium and rhenium hexacarbonyl cations are much more stable and may be recovered from aqueous solution (133–135).

The intensities of the infrared-active carbonyl stretching mode of the hexacarbonyl cations $[Mn(CO)_6]^+$ and $[Re(CO)_6]^+$ have been measured (17). The absolute intensities of the hexacarbonyl cations are in the order $[M(CO)_6]^+ < M(CO)_6 < [M(CO)_6]^-$, which corresponds to the reduction of π-back-bonding from metal to carbonyl as the positive charge on the metal is increased. (34)

A full vibrational analysis of $[Re(CO)_6]^+$ (3, 4) has indicated that σ-bonding between metal and carbon increases as the charge on the metal is increased, while the π-bonding concomitantly decreases.

The chemical reactivity of these cations has been investigated. Carbon monoxide exchange studies (142) show that the manganese and rhenium hexacarbonyl cations are inert to carbon monoxide substitution, while, interestingly, the rhenium cation exchanges oxygen in acid media with $H_2^{18}O$ (203). The substitution reactions of $[Re(CO)_6]^+ClO_4^-$ have been investigated with a variety of ligands; thus the cations $[Re(CO)_4(diphos)]^+$ (5, 6) and $[Re(CO)_3(v\text{-trias})]^+$ (208) are obtained directly. Monodentate phosphorus(III) ligands do not yield substituted cations, but instead a redox reaction occurs to yield neutral carbonyl chloride complexes. Triphenylphosphine reacts to form $trans$-$Re(CO)_3(PPh_3)_2Cl$ (5, 6).

The substituted hexacarbonyl cations of Group VII metals are generally prepared from the corresponding pentacarbonyl halide derivative by carbonylation (6, 80, 93, 135, 173, 177).

The known substituted hexacarbonyl cations of Group VII metals are presented in Table II, together with their method of preparation. There are several general observations concerning the preparation which are noteworthy. The ease of halide expulsion from the pentacarbonyl derivatives appears to be facilitated by decrease in carbonyl substitution, change of halogen in the order of Cl > Br > I, change of metal Mn > Tc and Re, and by use of polar solvents. Thus diphos reacts with $Mn(CO)_5Br$ to produce $Mn(CO)_3(diphos)Br$ (211), while excess ligand reacts in boiling methanol to form $[Mn(CO)_2(diphos)_2]^+Br^-$ (212). The latter reaction also applies to $Mn(CO)_5Cl$, but not to $Mn(CO)_5I$. The same cationic carbonyl was previously reported to be formed in the reaction of diphos with manganese carbonyl (217). The analogous rhenium cation is not formed from rhenium carbonyl (101), presumably because of the more robust nature of the Re—Re bond relative to the Mn—Mn bond, which prevents disproportionation of rhenium carbonyl. A recent report (113), however, suggests that the lower reactivity of rhenium carbonyl is governed by its less

TABLE II

ISOLATED SUBSTITUTED CARBONYL CATIONS OF GROUP VII METALS

Cation	Method of preparation	References
Manganese		
$[Mn(CO)_5NH_2Me]^+$	$Mn(CO)_4(CONH_2Me)(NH_2Me)/HCl$	10
$[Mn(CO)_5PPh_3]^+$	$Mn(CO)_4(PPh_3)Cl/AlCl_3/CO$	173, 177
$[Mn(CO)_5P(C_6H_{11})_3]^+$	$Mn(CO)_4(P(C_6H_{11})_3)Cl/AlCl_3/CO$	173, 177
$[Mn(CO)_4(PPh_3)_2]^+$	$Mn(CO)_3(PPh_3)_2Cl/AlCl_3/CO$	173, 177
	$[Mn(CO)_5C_2H_4]^+/PPh_3$	110
	$[Mn(CO)_5C_2H_3Me]^+/PPh_3$	104
$\{Mn(CO)_4[P(C_6H_{11})_3]_2\}^+$	$Mn(CO)_3(P(C_6H_{11})_3)_2Cl/AlCl_3/CO$	173, 177
$[Mn(CO)_4(TePh_2)_2]^+$	$Mn(CO)_3(TePh_2)_2Cl/AlCl_3/CO$	173, 177
$[Mn(CO)_4phen]^+$	$Mn(CO)_3phenCl/AlCl_3/CO$	173, 177
$\{Mn(CO)_4[P(OPh)_3]_2\}^+$	$Mn(CO)_3[P(OPh)_3]_2Cl/AlCl_3/CO$	173, 177
$[Mn(CO)_4(NH_3)_2]^+$	$Mn(CO)_5Cl/NH_3$	138
$[Mn(CO)_4(DMSO)_2]^+$	$Mn(CO)_5Cl/DMSO$	138
$[Mn(CO)_4(en)]^+$	$Mn(CO)_5Cl/en$	138
$[Mn(CO)_3(NH_3)_3]^+$	$Mn(CO)_5Cl/NH_3$ 60°C	22
	$Mn_2(CO)_{10}/NH_3$ 20°C	22
$[Mn(CO)_5C_2H_4]^+$	$Mn(CO)_5Cl/AlCl_3/C_2H_4$	93
	$EtMn(CO)_5/(Ph_3C)^+$	110
$[Mn(CO)_5C_2H_3Me]^+$	$C_3H_5Mn(CO)_5/HCl$	104
$[Mn(CO)_3Arene]^+$	$Mn(CO)_5X/Arene/AlCl_3$	
	Arene $= C_6H_6, C_6H_5Me, C_6H_5Et,$	
	$C_6H_3Me_3$, *p*-hexylbiphenyl	47
	$= C_6H_6, C_6H_5Me, C_6H_3Me_3$	48
	$= C_6H_6, C_6H_5Me, C_6H_3Me_3,$	
	$C_6Me_6, C_{10}H_8$	245
	$= C_4Me_4S$	224, 225
	$= C_4Me_{4-x}H_xS$ $x = 1, 2, 3, 4$	225
	$= C_4H_4NMe$	153
$[Mn(CO)_3dien]^+$	$[Mn(CO)_3C_6H_3Me_3]^+/dien$	2
$[Mn(CO)_3QP]^+$	$Mn(CO)_5Cl/QP$	38
$[Mn(CO)_3TP]^+$	$Mn(CO)_5Cl/TP$	38
$[Mn(CO)_2QP]^+$	$[Mn(CO)_3QP]^+/UV$	38
$[Mn(CO)_2(diphos)_2]^+$	$Mn(CO)_5Cl/diphos$	212
	$Mn(CO)_5Br/diphos$	212
	$Mn_2(CO)_{10}/diphos$	217
$[Mn(CO)_3v-triars]^+$	$[Mn(CO)_3C_6H_3Me_3]^+/v-triars$	208
$[MnCp(CO)_2NO]^+$	$CpMn(CO)_3/NaNO_2/HCl$	162, 163
$[Mn(CO)_3C_5H_4N_2]^+$	$Mn(CO)_3C_5H_4NH_2/C_5H_{11}NO_2/HCl$	27
$[Mn(CO)_5—H—Fe(CO)_2Cp]^+$	$CpFeMn(CO)_7/H_2SO_4$	63
$[Mn(CO)_2(diphos)_2]^{2+}$	$[Mn(CO)_2(diphos)_2]^+/HNO_3$, etc.	227

TABLE II—*continued*

Cation	Method of preparation	References
Technetium		
$[Tc(CO)_4(PPh_3)_2]^+$	$Tc(CO)_3(PPh_3)_2Cl/AlCl_3/CO$	*135*
Rhenium		
$[Re(CO)_4(PPh_3)_2]^+$	$Re(CO)_3(PPh_3)_2Cl/AlCl_3/CO$	*173*
	$Re(CO)_5Cl/PPh_3$ [a]	*139*
$[Re(CO)_4phen]^+$	$Re(CO)_3phenCl/AlCl_3/CO$	*173*
$[Re(CO)_4(C_7H_7NC)_2]^+$	$Re(CO)_5Cl/C_7H_7NC$	*139*
$[Re(CO)_4(NH_3)_2]^+$	$Re(CO)_5Cl/NH_3$	*139*
$[Re(CO)_4(C_2H_4)_2]^+$	$Re(CO)_5Cl/AlCl_3/C_2H_4$	*94*
$[Re(CO)_4diphos]^+$	$[Re(CO)_6]^+/diphos$	*5, 6*
	$Re(CO)_3(diphos)Cl/AlCl_3/CO$	*6*
$[Re(CO)_3(v\text{-triars})]^+$	$[Re(CO)_6]^+/v\text{-triars}$	*208*
$[Re(CO)_3C_{12}H_{18}]^+$	$Re(CO)_5Cl/C_{12}H_{18}/AlCl_3$	*246*
$[Re(CO)_2(diphos)_2]^+$	$Re(CO)_5I/diphos$	*102*
$[Re(CO)(diars)_2X_2]^+$	$Re(CO)(diars)_2X/X_2$	*171*
$\quad X = Br, I$		
$[ReCp(CO)_2NO]^+$	$CpRe(CO)_3/NO^+$	*97*
$[ReCp(CO)_3Cl]^+$	$CpRe(CO)_3/SbCl_5$	*161*
$[Re(CO)_3(HNC_4H_8)_3]^+$	$[Re(CO)_3SePh_2I]_2/HNC_4H_8$	*136*

[a] Subsequently shown to be $Re(CO)_3(PPh_3)_2Cl$ (7).

favorable activation entropy. Instead the $[Re(CO)_2(diphos)_2]^+$ cation is made by iodide displacement from $Re(CO)_5I$ by diphos under forcing conditions (*102*).

$$Re(CO)_5I + 2 \text{ diphos} \xrightarrow[\substack{in \ vacuo \\ 2 \ hr}]{240°C} [Re(CO)_2(diphos)_2]^+I^- \qquad (102)$$

The disproportionation of manganese carbonyl with nitrogen ligands is reviewed by Hieber *et al.* (*122*). Calderazzo has investigated the catalytic carbonylation of amines by manganese carbonyl (*30*). The corresponding chemistry of rhenium carbonyl remains to be investigated.

The reaction of $Mn(CO)_5Br$ and probably $Re(CO)_5Br$ with primary aliphatic amines forms neutral carboxamido complexes with bromide expulsion. The carboxamido complex is converted to a pentacarbonyl amine cation upon protonation (10).

$$Mn(CO)_5Br \xrightarrow{RNH_2} Mn(CO)_4(RCONH_2)(NH_2R) \underset{RNH_2}{\overset{HCl}{\rightleftharpoons}} [Mn(CO)_5(NH_2R)]^+Cl^- \quad (10)$$

The $[Mn(CO)_2(diphos)_2]^+$ cation forms a paramagnetic doubly charged cation $[Mn(CO)_2(diphos)_2]^{2+}$ on oxidation (227). The reaction of aromatic ligands with the pentacarbonyl halides generally results in the formation of a substituted metal tricarbonyl cation.

$$C_6H_6 + Mn(CO)_5I \xrightarrow{80°C} [C_6H_6Mn(CO)_3]^+I^- \quad (47, 48, 245)$$

$$C_4Me_4S + Mn(CO)_5Cl + AlCl_3 \xrightarrow{100°C} [C_4Me_4SMn(CO)_3]^+AlCl_4^- \quad (224, 225)$$

This does not apply to the reaction of mesitylene (208) and thiophenes (225) with rhenium pentacarbonyl chloride, but hexamethylbenzene (246) and ethylene (94) react to form carbonyl cations.

The manganese mesitylene tricarbonyl cation has been employed as an intermediate in the synthesis of Lewis base, L, tricarbonyl cations for L = dien (2) and v-triars (208). The ammonolysis of $Mn(CO)_5Cl$ directly yields $[Mn(CO)_3(NH_3)_3]^+Cl^-$ (22) or $[Mn(CO)_4(NH_3)_2]^+Cl^-$ (138), depending on the reaction conditions. The ammonolysis products of $Re(CO)_5Cl$ are reported to be $[Re(CO)_4(NH_3)_2]^+Cl^-$ and $Re(CO)_4(NH_3)Cl$ (139), but this is questionable (6).

The hexacarbonyl and substituted carbonyl cations react with a variety of anionic nucleophiles to produce neutral species. The metathetical reaction of metal carbonylate salts with those of the cationic carbonyls yield primarily a mixed metal carbonyl salt; however, on warming, anionic attack may cause expulsion of carbon monoxide and the formation of a metal–metal bond (174–176).

$$[Mn(CO)_6]^+X^- + Na^+[Co(CO)_4]^- \xrightarrow[THF]{-20°C} [Mn(CO)_6]^+[Co(CO)_4]^-$$

$$20°C \Big| THF$$

$$(OC)_5Mn-Co(CO)_4 \quad (177)$$

The anion $[V(CO)_6]^-$ formed no Mn—V or Re—V metal–metal bond, probably because of steric effects, although heptacoordinate vanadium carbonyl derivatives exist, thus $HV(CO)_5(PPh_3)$ (141).

An important reaction of the carbonyl cations of Group VII metals is their base attack by alkoxides (*135, 167, 168, 175*), which has already been mentioned. The product is an alkoxycarbonyl complex.

$$[Mn(CO)_4(PPh_3)_2]^+ \xrightleftharpoons[HCl]{OMe^-} Mn(CO)_3(PPh_3)_2CO \cdot OMe \qquad (175)$$

The alkoxycarbonyl reverts to the cationic carbonyl on protonation, providing the acid anion is inert.

The reaction of halide anions with substituted carbonyl cations does not appear to have been studied, but the reactions with cyanide (*47*) and hydride (*245, 246*) have been reported. Thus $[Mn(CO)_3C_6H_6]^+$ gives the cyclohexadienyl complex in which the hydride is in the endo position of the methylene group. (viz., H_α).

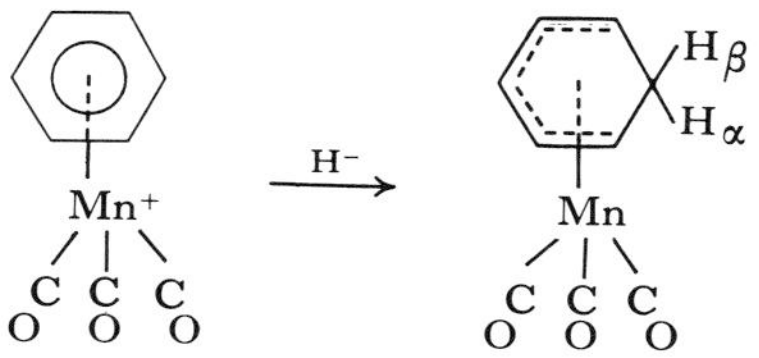

$$(245)$$

This hydrogen is reactive and may be readily abstracted.

E. Iron Group

By analogy with Group VII metals, the hexacarbonyl cations $[M(CO)_6]^{2+}$ (M = Fe, Ru, Os) could be anticipated to exist. Hieber and Kruck reported infrared evidence for the formation of the iron and osmium cations (*133*). Subsequently their existence has been disclaimed (*128*). Before the attempted synthesis of the hexacarbonyl cations from the tetracarbonyl dihalides, Sternberg and associates suggested that iron pentacarbonyl dissociates in certain amines, such as piperidine, into $[Fe(CO)_6]^{2+}$ and $[Fe(CO)_4]^{2-}$ (*228*). The amine reaction with iron pentacarbonyl proved to be highly complex (*68, 69, 132, 140*), but the reaction does not involve any cationic iron carbonyl. Despite the absence of the hexacarbonyl cations of this group, several related substituted cations are known. The bis(triphenylphosphine)osmium tetracarbonyl cation $[Os(CO)_4(PPh_3)_2]^{2+}$ is reported (*126, 128*) but it defies isolation; instead the stable univalent chloro cation is obtained $[Os(CO)_3(PPh_3)_2Cl]^+$ (*128*).

The analogous bromo and iodo cations have been prepared by oxidative addition of halogen to bis(triphenylphosphine)osmium tricarbonyl (52).

$$Os(CO)_3(PPh_3)_2 \xrightarrow{X_2} [Os(CO)_3(PPh_3)_2X]^+X^- \qquad (52)$$
$$X = Br, I$$

The iron analog reacts in a slightly different manner.

$$Fe(CO)_3(PPh_3)_2 \xrightarrow{I_2} [Fe(CO)_2(PPh_3)_2I_2]^+I_3^- \qquad (44)$$

It is deduced from infrared data (206) that iron pentacarbonyl reacts with bromine at $-80°$ C to form the cationic halogeno carbonyl $[Fe(CO)_5Br]^+Br^-$ (206), and the covalent Br—CO—Fe(CO)$_4$Br. From similar infrared data, however, it is concluded that the product is a seven-coordinate non-ionic complex $Fe(CO)_5Br_2$ (76)! The reaction of $Os(CO)_3(PPh_3)_2$ with hydrogen chloride produces a cationic hydrido carbonyl, which may be isolated, although it readily reverts to a neutral dichloride by attack of the associated chloride anion (52).

$$Os(CO)_3(PPh_3)_2 \xrightarrow{HCl} [Os(CO)_3(PPh_3)_2H]^+Cl^-$$
$$\downarrow HCl \qquad (52)$$
$$[Os(CO)_2(PPh_3)_2Cl_2] + CO + H_2$$

The basic nature of $Os(CO)_3(PPh_3)_2$, as well as the iron and ruthenium analogs, is further shown by the reaction with mercuric halides.

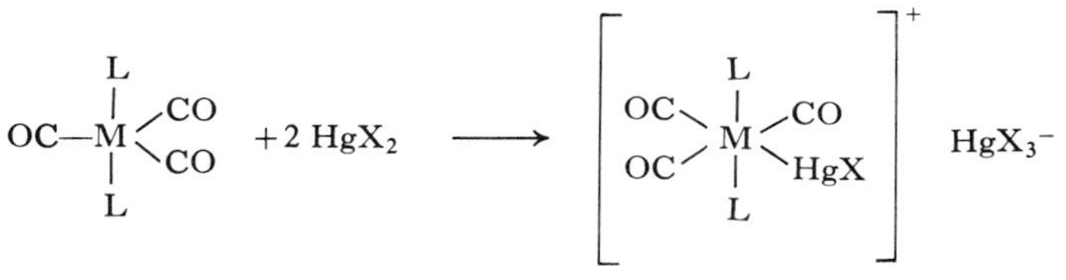

$$M = Os, Ru; \ L = PPh_3; \ X = Cl, Br, I \qquad (51)$$
$$M = Fe; \ L = PPh_3, P(OPh)_3; \ X = Cl \qquad (8)$$

Protonation of $Fe(CO)_5$, $Fe(CO)_4(PPh_3)$, $Fe(CO)_3(PPh_3)_2$, $Fe(CO)_4$ (AsPh$_3$), and $Fe(CO)_3(AsPh_3)_2$ in strong acids is recognized to form octahedral iron cations [comparable to $HMn(CO)_5$ derivatives] by the high-field proton NMR signal (65). Norbornadieneiron tricarbonyl behaves in a similar manner (75). Such octahedral hydrido-carbonyl cations have been proposed to explain the accelerated carbonyl exchange rates of iron carbonyl complexes in strong acids (16).

A remarkable cationic iron carbonyl is formed by the reaction of iron pentacarbonyl with chloro or bromo azide, $[Fe_2(CO)_4(N_3)_2]^{2+}$ (180). This cation has the highest recorded paramagnetism for a metal carbonyl of 5.29 μ_B. A planar (D_{2h}) structure is proposed; the azido groups are bridging and the carbonyls are terminal.

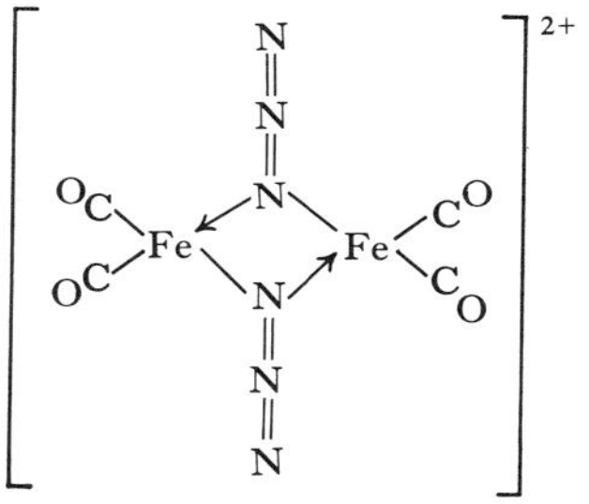

An iron nitrosyl carbonyl cation is prepared by disproportionation of mercurybis(iron nitrosyl tricarbonyl) in the presence of tris(dimethylamino)phosphine (159).

$$Hg[Fe(CO)_3NO]_2 + Tdp \rightarrow [Fe(CO)_2(NO)(Tdp)_2]^+[Fe(CO)_3NO]^- \quad (159)$$

The analogous triphenylphosphine cation is obtained from $Fe(CO)_3(PPh_3)_2$ with nitrosyl halides, NOX, and with dinitrogen tetroxide (59).

$$Fe(CO)_3(PPh_3)_2 + NOX \rightarrow [Fe(CO)_2(NO)(PPh_3)_2]^+X^- \quad (59)$$
$$X = Cl, Br, NO_3$$

An aged carbonylated ruthenium chloride solution reacts with tetrabenzyl-dithiaramyl disulfide to produce a dithiocarbamate cation (170).

$$RuCl_3(alc.) + CO + \{(C_6H_5CH_2)_2NCS_2\}_2 \rightarrow [Ru(CO)_2\{S_2CN(CH_2C_6H_5)_2\}_2]^+Cl^- \quad (170)$$

The product of stannous chloride and the carbonylated ruthenium chloride solution gives a cationic complex with diethyl sulfide (169).

$$RuCl_3(alc.) + CO + SnCl_2 + Et_2S \rightarrow [Ru(CO)_2(SEt_2)_3SnCl_3]^+Cl^- \quad (169)$$

1. Carbonium Iron Carbonyl Complexes

Iron tricarbonyl allyls and dienyls have been reviewed (214).

a. *π-Allyl Cations.* A variety of π-allyliron tricarbonyl cations have been obtained by treatment of diene-$Fe(CO)_3$ complexes with strong acids (71, 74). Alternatively, the π-allyl cations are obtained by reaction of $AgBF_4$ or $AgClO_4$ with covalent allyl halides of type (I).

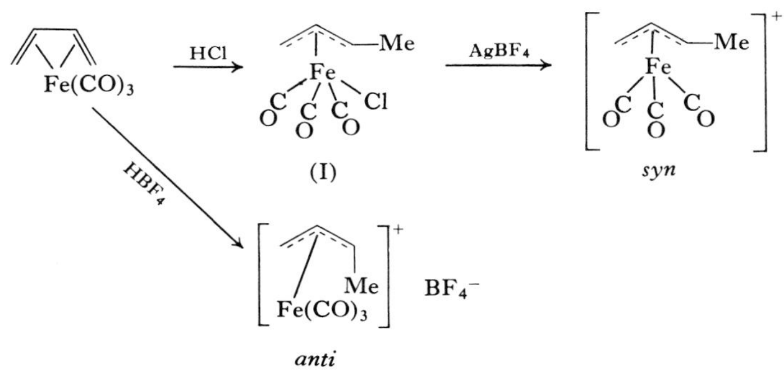

$$(72)$$

$$(74)$$

The π-allyliron tricarbonyl cations are unusual because the iron atom is two electrons short of the next rare gas configuration.

Several π-allyliron tetracarbonyl cations have been prepared and their Mössbauer spectra investigated (50).

b. *Cyclic Dienyliron Carbonyl Cations.* Fischer and Fischer obtained cyclohexadienyliron tricarbonyl fluoroborate by reaction of triphenylmethyl fluoroborate with cyclohexadieneiron tricarbonyl.

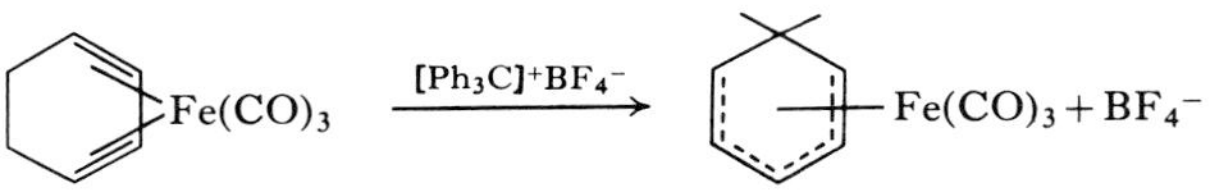

$$(81)$$

The methoxycyclohexadiene derivatives behave similarly (23). On hydride treatment the cations revert to a diene, thus

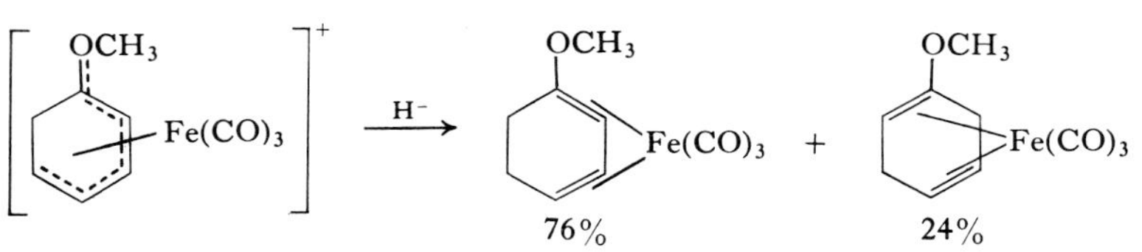

$$(23)$$

The trityl reaction with cycloheptatrieneiron tricarbonyl yields a cyclo-heptadienyl complex, by trityl addition to the uncoordinated double bond, rather than hydride abstraction.

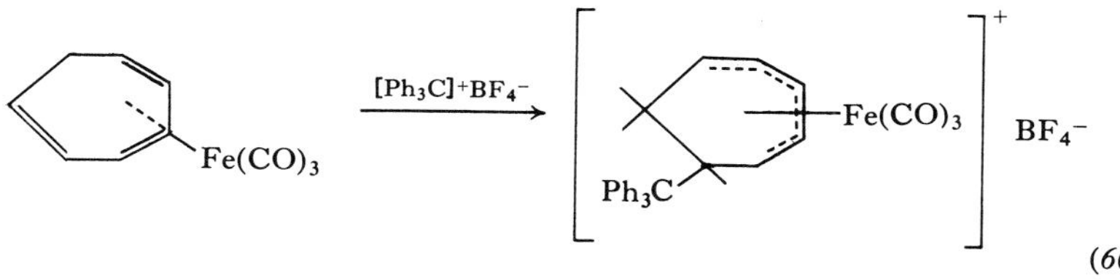

$$(60)$$
also (36)

The unsubstituted cycloheptadienyliron tricarbonyl cation is similarly prepared by protonation of the free double bond $(26, 60, 152)$. The Möss-bauer spectra of some pentadienyl and hexadienyl cations have been examined $(49, 121)$.

Protonation of cyclooctatetreneiron tricarbonyl by strong acids produces a bicyclic dienyl cation (II) (64), rather than the direct proton adduct (III) as initially proposed (221).

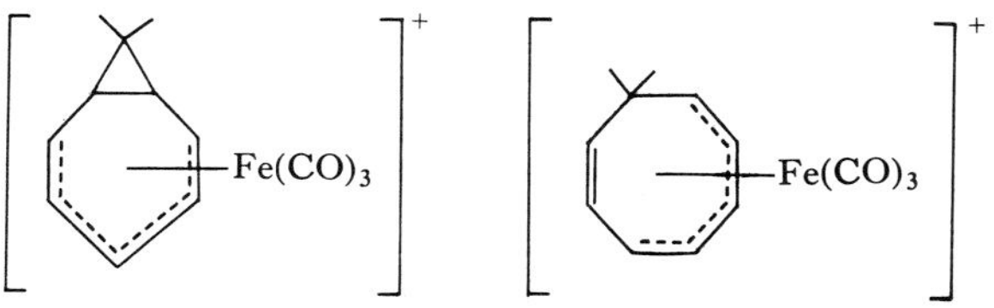

(II)   (III)

This formulation was deduced from the proton NMR spectra (64), but has subsequently been confirmed by the reaction of the cation (II) with hydroxide (144).

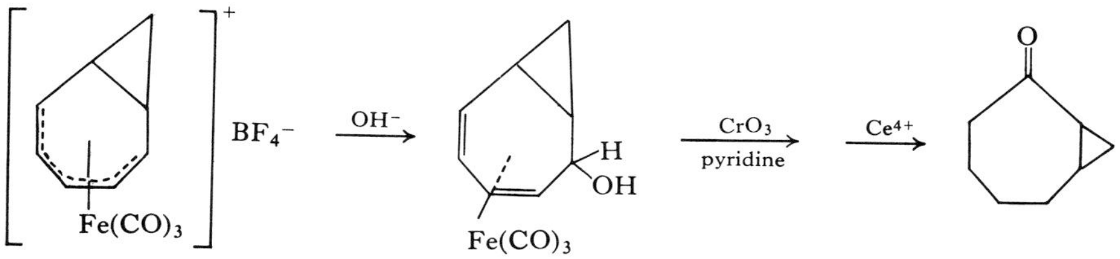

$$(144)$$

However, protonation of the cyclooctatrieneiron tricarbonyl by strong acids yields a stable monocyclic complex (200).

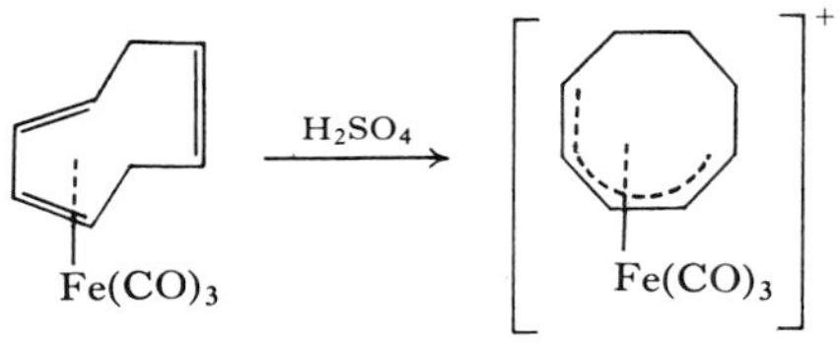

(200)

A bimetallic iron tricarbonyl cation $\{C_8H_9Fe_2(CO)_6\}^+$ (IV) is derived from cyclooctatriene (157). The Mössbauer resonance of spectrum of (IV) and the related cycloheptatrienyl-$Fe_2(CO)_6$ cation (V) (73) suggest that these cations may be considered as containing a bis(π-allyl) and an allyl-diene system, respectively. In the second case (V), rapid valence tautomerism is invoked to account for the unique proton NMR signal in solution.

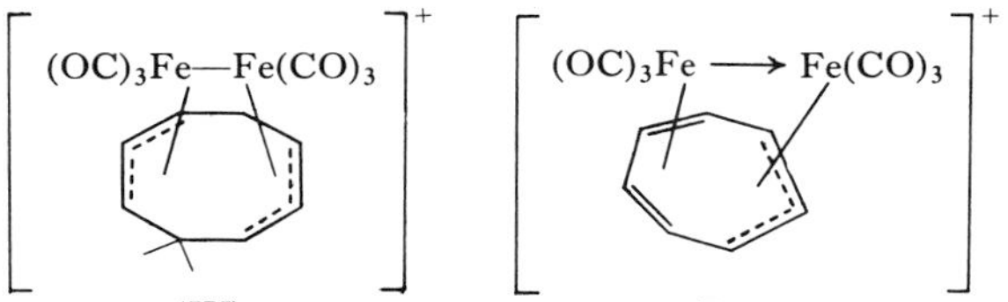

Moreover, to account for the diamagnetism of (V) a Fe → Fe dative bond is postulated (73).

The tropylium iron tricarbonyl was eventually synthesized in 1964 by protonation of methyl tropyl ether–iron tricarbonyl.

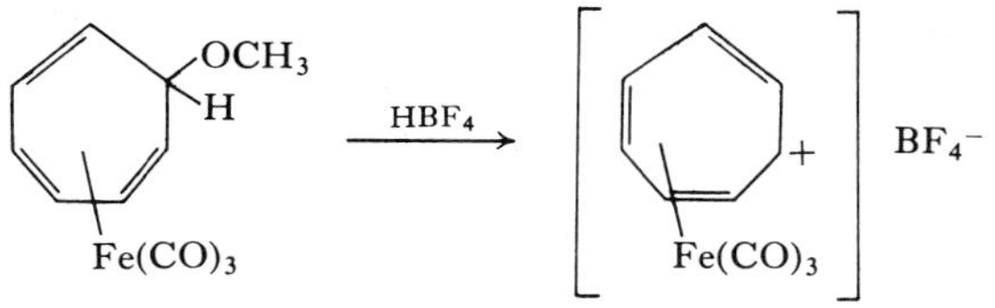

(188)

Because the tropylium cation contains one uncoordinated olefinic bond, yet displays a single proton NMR signal in solution, rapid valence tautomerism or "ring whizzing" is proposed (188).

The iron tricarbonyl cation of the smallest cyclic dienyl, cyclobutadienyl cation, has recently been prepared (100). It may be considered to exist in one of two possible forms (VI or VII), of which the allylic formulation (VII) is favored.

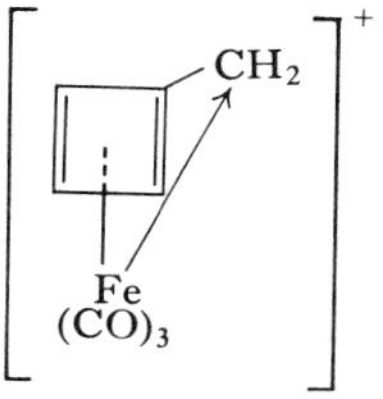

(VI) 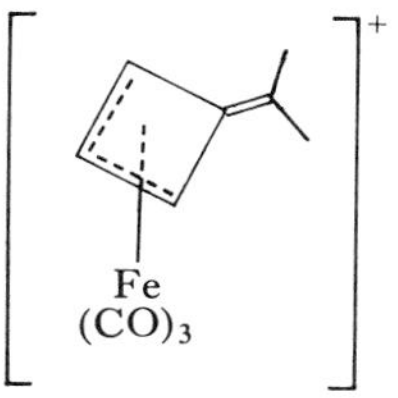   (VII)

A π-allylic together with an olefin system are considered to be involved in bonding to the iron tricarbonyl residue of bicyclo[3.2.1]octadienyliron tricarbonyl cation (VIII) (*196*). This is apparent from the X-ray structural determination, although a delocalized form cannot be discounted (IX).

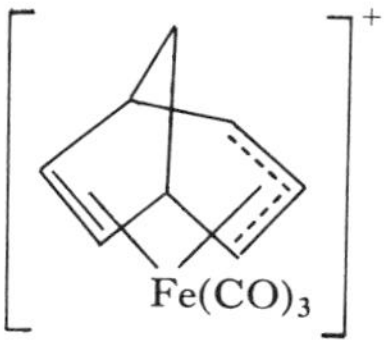

(VIII) 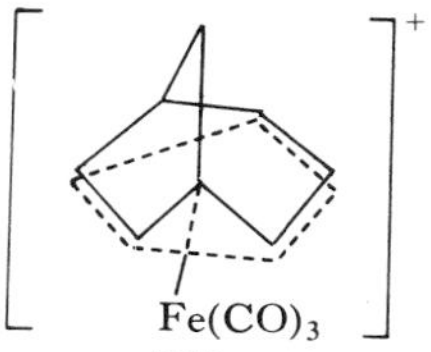   (IX)   (*196*)

c. *Noncyclic Pentadienyl Systems.* The acyclic π-pentadienyliron tricarbonyl cations are prepared by protonation of a dienol (*187, 189, 190*) or by hydride abstraction from a *cis*-1,3 pentadieneiron tricarbonyl (*190*) to form *cis*-π-pentadienyl-Fe(CO)$_3$ cation.

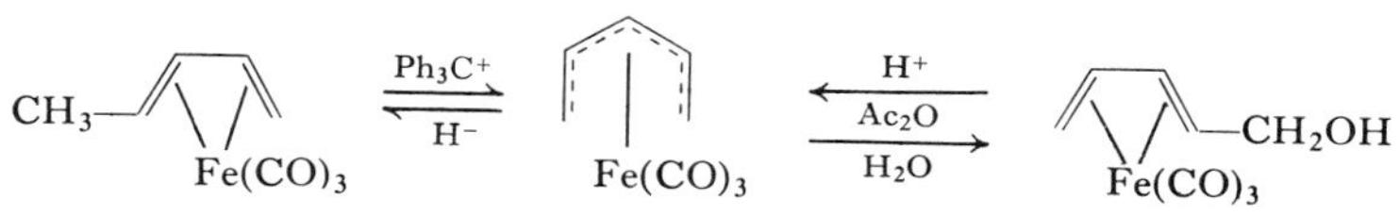

The mechanism of the dienol protonation is considered (*187*) to be

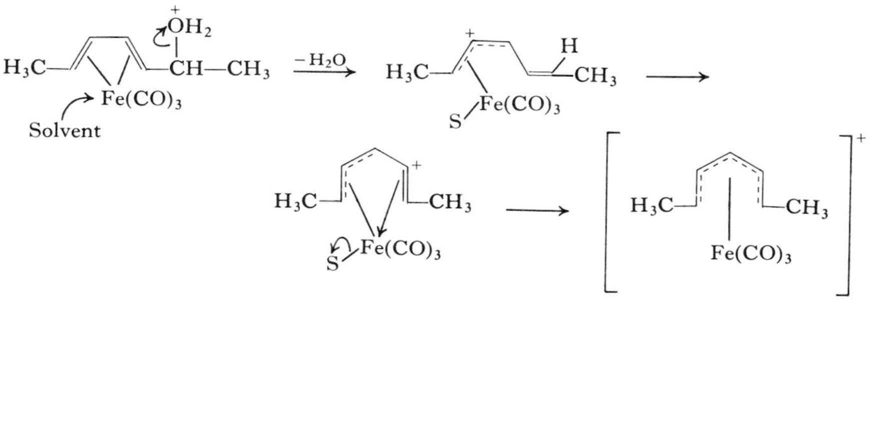

There is recent kinetic evidence from the solvolysis rates of iron tricarbonyl dienyls that *trans*-π-pentadienyl species are reaction intermediates (*45*). There are two structures proposed, as shown:

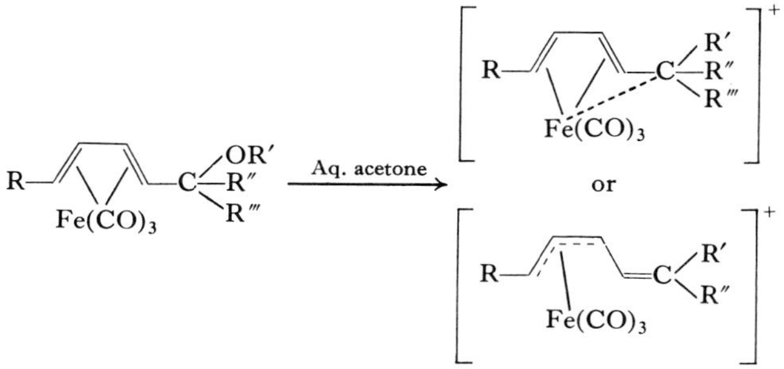

d. *π-Cyclopentadienyliron Carbonyl Cations.* The π-cyclopentadienyliron tricarbonyl cation was initially prepared by high-pressure carbonylation of π-cyclopentadienyliron dicarbonyl chloride (*62, 78*). An improved carbonylation synthesis using π-cyclopentadienyliron dicarbonyl iodide is reported (*156*). A nonpressure method is available in which methoxycarbonyl-π-cyclopentadienyliron dicarbonyl is treated with dry hydrogen chloride (*167, 168*).

$$CpFe(CO)_2CO \cdot OMe \xrightarrow{HCl} [CpFe(CO)_3]^+[HCl_2^-] \qquad (167, 168)$$

Similarly protonation of diphenylfulveneiron tricarbonyl generates a substituted π-cyclopentadienyl cation (*241, 242*). Hydride abstraction from cyclopentadieneiron tricarbonyl releases the π-cyclopentadienyl cation complex (*172*). The Mössbauer spectra of the $[CpFe(CO)_3]^+$ cation and related iron carbonyl cations have been determined (*121*).

There are many substituted π-cyclopentadienyliron carbonyl cations, which are often prepared from a π-cyclopentadienyliron dicarbonyl halide. The halide may be expelled by a variety of ligands to form a cation in preference to carbonyl displacement. Which actually occurs in practice appears to be a function of the reaction parameters, such as reaction temperature, basicity of the ligand, and halide used. The cations of this type are listed in Table III.

Whereas the monoolefin derivatives of π-cyclopentadienylruthenium tricarbonyl cation are known (*99*), the unsubstituted carbonyl has only recently been prepared (*24*). The osmium analog has not been reported.

Monoolefin-substituted π-cyclopentadienyliron dicarbonyl cations may be prepared by protonation of σ-allyl and related carbonyls (*12–14, 105, 108, 109*) or else by hydride abstraction from some σ-alkyl derivatives (*106, 107, 110*).

Methylation to form substituted cations with $L = SMe_2$ (*164*) and $L_2 = (CNMe)_2$ (*46*) are reported.

TABLE III

Isolated Substituted $[CpFe(CO)_{3-n}L_n]^+$ ($n = 1, 2$) Cations

$n = 1$	
$L = NH_3, N_2H_4, PEt_3$	(*91*)
$= PPh_3$	(*62, 108, 154*)
$= AsPh_3, SbPh_3$	(*62*)
$= Tdp$	(*158*)
$= py, CH_3CN, \frac{1}{2} diphos$	(*234*)
$= H_2O$	(*106, 116, 215*)
$= C_2H_4$	(*78, 79, 106, 107, 110*)
$= C_3H_6, C_4H_8$	(*79, 105, 106, 108, 110*)
$= C_{18}H_{36}, C_6H_{10}, C_8H_{14}, C_4H_6, C_6H_8$	(*79*)
$= C_5H_6$	(*109*)
$= C_3H_4$	(*12*)
$= CH_2CNH, CHMeCNH$	(*13*)
$= CH_2CHOH, CH_2CMeOH$	(*14*)
$= CH_3CN, CH_2CHCN$	(*92*)
$= py, PhCN, PhNH_2$	(*89*)
$= py, CH_3CN$	(*90*)
$= SMe_2$	(*164*)
$= XFe(CO)_2Cp$ $X = Cl, Br, I$	(*89, 90, 92*)
$X = PMe_2$	(*118*)
$X = C_3H_5$	(*166*)
$X = C_2H_4NMe_2$	(*165*)
$= H[Mn(CO)_5]$	(*63*)
$= H[Fe(CO)_2Cp]$	(*63*)
$n = 2$	
$L_2 = (PPh_3)_2, (PPh_3)(CH_3CN), diphos, bipy$	(*234*)
$= (CNMe)_2$	(*46*)

The reaction of some π-cyclopentadienyliron carbonyl cations with NaBH$_4$ (*62*), phenyl- and perfluorophenyllithium (*232, 233*) have been studied.

A number of related bridged cations of the type $[CpFe(CO)_2—X—Fe(CO)_2Cp]^+$ have been prepared (see Table III). The bridged halides are very sensitive to nucleophilic attack, yielding cations of the type $[CpFe(CO)_2L]^+$ (see Table III).

A stable polynuclear paramagnetic π-cyclopentadienyliron carbonyl bromide $[CpFe(CO)]_4^+Br_3^-$ is reported to be formed by oxidation of the tetrameric carbonyl $[CpFe(CO)]_4$ (160).

The occurrence of stable carbene complexes of Groups VI and VII is well documented (87, 96, 202), but the π-cyclopentadienyliron carbene carbonyl cations (103, 151) appear unstable.

The π-cyclopentadienyliron dicarbonyl dimer and the mixed metal carbonyl π-cyclopentadienyliron-manganese heptacarbonyl are protonated by strong acids to form isolatable salts, namely $\{[CpFe(CO)_2]_2H\}^+PF_6^-$ and $[CpFeMn(CO)_7H]^+PF_6^-$. The ruthenium dimer $[CpRu(CO)_2]_2$ behaves similarly, although no salt was isolated (63).

F. Cobalt Group

1. Cobalt

In 1952 Wender and co-workers proposed (244) that dicobalt octacarbonyl disproportionates in the presence of a Lewis base (B) to yield a cobalt(II) carbonylcobaltate(− I) salt.

$$3 \ Co_2(CO)_8 + 12 \ B \ \rightarrow \ 2 \ [CoB_6]^{2+}[Co(CO)_4]_2^- + 8 \ CO \qquad (244)$$

In the case of the strong base piperidine the intermediary cation was studied.

$$C_5H_{10}NH + Co_2(CO)_8 \ \rightarrow \ [C_5H_{10}NHCo(CO)_4]^+[Co(CO)_4]^- \qquad (244)$$

This mode of reaction was substantiated by further investigations of cobalt carbonyl with nitrogen (137, 229) and oxygen (229, 237) bases. With high-pressure carbon monoxide (201) cobalt carbonyl appeared to incorporate an extra molecule of carbon monoxide to form, most probably, the pentacarbonylcobalt carbonylcobaltate, in line with base reactions.

$$CO + Co_2(CO)_8 \ \rightleftharpoons \ [Co(CO)_5]^+[Co(CO)_4]^- \qquad (201)$$

The pentacarbonylcobalt cation is the parent carbonyl cation of cobalt, but it is only the phosphine and arsine derivatives which are stable enough to be isolated. A nonpressure synthesis of the pentacarbonylcobalt cation by acidification of the ethoxycarbonylcobalt tetracarbonyl failed, although the monosubstituted phosphine cation succeeded (*123*).

$$(C_6H_5)_3PCo(CO)_3CO \cdot OEt \xrightarrow[\text{ether, } -50°C]{\text{HCl}} [Co(CO)_4(PPh_3)]^+[HCl_2]^- + EtOH \quad (123)$$

The first stable cobalt carbonyl cation to be prepared was *trans*-bis(triphenylphosphine)cobalt tricarbonyl cation (*219*), obtained by disproportionation of cobalt carbonyl and carbonylation of $Co(PPh_3)_2I_2$. Hieber and Freyer examined the reaction of triphenylphosphine with cobalt carbonyl (*129, 130*). The product from the reaction with triphenylarsine and triphenylstibine is a salt at low temperatures, but, on warming, a redox reaction occurs, producing a substituted cobalt carbonyl (*129*). Thus the reaction scheme is

$$Co_2(CO)_8 + L \xrightarrow[\text{polar solvent}]{\text{Low temp.}} [Co(CO)_3L_2]^+[Co(CO)_4]^-$$

$$\downarrow \text{High temp.}$$

$$[Co(CO)_3L]_2 + CO$$

$$L = AsPh_3, SbPh_3, PEt_3$$

The cobalt cations may be prepared by carbonylation (*124*).

$$Co(CO)_2(PPh_3)_2Cl \xrightarrow[\text{AlCl}_3]{\text{CO 1 atm}} [Co(CO)_3(PPh_3)_2]^+AlCl_4^- \quad (124)$$

Several analogous derivatives have been reported; these are compiled in Table IV.

The kinetics of the formation of $[Co(CO)_3(PPh_3)_2]^+[Co(CO)_4]^-$ from cobalt carbonyl (*119*) suggest the formation of a 1:1 complex as an intermediate.

The geometry of the tricarbonyl cations $[Co(CO)_3L_2]^+$ has been established, from the infrared-active carbonyl stretching modes [L = PPh_3 (*240*)] and the proton NMR [L = Tdp (*158*)], to be trigonal bipyramidal with the phosphines disposed trans in the axial positions.

The reaction of $[Co(CO)_3(PPh_3)_2]^+[Co(CO)_4]^-$ with anionic nucleophiles has been studied. Hydridic reduction with $NaBH_4$ (*199*) proceeds via the formation of the hydride $Co(CO)_3(PPh_3)H$ to $[Co(CO)_3(PPh_3)]_2$. Hydrazine reacts similarly (*199*), while with NaI in refluxing acetone $Co(CO)_2(PPh_3)_2I$ is produced (*216*). LiCl, LiBr, and KCNO react with the cation to form a polymeric carbonyl $[Co(CO)_2(PPh_3)]_n$ (*216*).

TABLE IV

The Isolated Carbonyl Cations of Cobalt, Rhodium, and Iridium of the Type
$[M(CO)_{5-n}L_n]^+$ ($n = 1, 2, 3, 4$)

$n = 1$

M = Co	L = PPh_3[a]	(*123*)
	= Me_2NH[a]	(*229*)
	= $C_5H_{10}NH$[a]	(*244*)
	= MeOH, $C_8H_{17}OH$, H_2O[a]	(*237*)

$n = 2$

M = Co	L = PPh_3	(*124, 129, 130, 219*)
	= $P(OPh)_3$	(*124*)
	= PEt_3	(*130, 218*)
	= $P(C_6H_{11})_3$	(*130*)
	= $AsPh_3$, $SbPh_3$	(*130*)
	= Tdp	(*158*)
M = Ir	L = PPh_3, PPh_2Me	(*39, 193*)
	= PPh_2Me	(*39*)

$n = 3$

M = Co	$L_3 = \frac{3}{2}$ diphos	(*217*)
M = Ir	L = PPh_3	(*193*)
M = Ir, Rh	L = $SbPh_3$	(*127*)

$n = 4$

M = Co, Rh	$L_4 = QP$, QAs	(*117*)
M = Co	$L_2 =$ diphos	(*220*)
M = Ir	$L_2 =$ diphos	(*125, 131, 133, 150, 220, 239*)
	$L_2 = MDP$[b]	(*131*)
	$L_4 = (PPh_3)_2[P(OMe)_3]_2$	(*39*)
M = Rh	$L_2 =$ v-diars	(*186*)

[a] Detected only.
[b] MPD = Methylenebis (diphenylphosphine).

Cobalt forms an unusual triangular cluster cation $[Co_3(CO)_2(C_6H_6)_3]^+$, which was discovered independently by Fischer and Beckert (*77*) and Chini and Ercoli (*37*). The structure proposed is analogous to $Ni_3(CO)_2Cp_3$ (*95*).

Several cationic π-cyclopentadienylcobalt derivatives have been prepared. Treichel and Werber (235, 236) have produced a number of perfluoroalkyl complexes $[CpCo(CO)LR_F]^+ClO_4^-$, where $L = PPh_3$, CH_3CN, py and $R_F = C_2F_5$, C_3F_7. The stability of these cations approximately parallels the

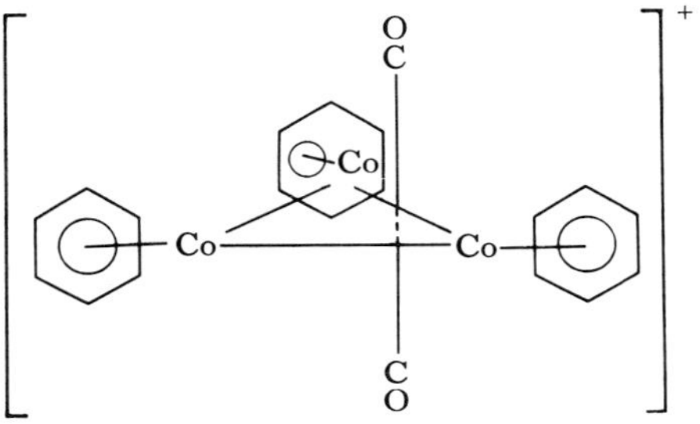

$$(95)$$

ligand's basicity. The reaction of mercuric chloride with π-cyclopentadienylcobalt dicarbonyl was suggested (55) to produce a salt, but an X-ray examination showed that the product was an adduct (207). The π-cyclopentadienyl-π-allylcobalt monocarbonyl cation is known (82, 120).

$$CpCo(CO)_2 + C_3H_5Br \rightarrow [CpCo(CO)C_3H_5]^+Br^- \qquad (82, 120)$$

2. Rhodium and Iridium

There are marked differences between the carbonyl cations of cobalt and its congeners, rhodium and iridium. For instance, the heavier elements form square-planar carbonyl cations as well as higher coordinate complexes. This is paralleled by the isocyanide cations; thus cobalt forms $[Co(CNR)_5]^+$ cations (191), whereas rhodium and iridium form $[M(CNR)_4]^+$ cations (191, 192, 194).

The tetracarbonyl cations of rhodium and iridium are not known, but the isocyanide derivatives of the type $[Ir(CO)(RNC)_3]^+$ are reported (192).

$$Ir(CO)_2(C_7H_7NH_2)Cl + RNC \rightarrow [Ir(CO)(RNC)_3]^+Cl^- \qquad (192)$$

The dicarbonyl iridium and rhodium phosphine $[M(CO)_2L_2]^+$ cations are known $[L = PPh_3, P(C_6H_{11})_3]$ (127), obtained by mild carbonylation.

$$M(CO)L_2Cl + AlCl_3 + CO \rightarrow [M(CO)_2L_2]^+AlCl_4^- \qquad (127)$$
$$M = Rh, Ir$$

Derivatives of the hypothetical pentacarbonyl rhodium and iridium cations have recently attracted much interest. The iridium bis(diphos) derivative has been prepared by several methods.

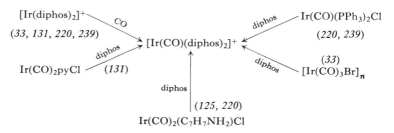

The carbonyl is reversibly coordinated.

$$[Ir(CO)(diphos)_2]^+ \underset{+CO}{\overset{-CO}{\rightleftharpoons}} [Ir(diphos)_2]^+ \qquad (239)$$

The crystal structure of $[Ir(CO)(diphos)_2]^+$ has been determined. The iridium atom is in a distorted trigonal bipyramidal environment (150).

The rhodium quadridentate phosphine, QP, and arsine, QAs, monocarbonyl cations (117) are more labile than the cobalt analog (117), since the rhodium's carbonyl is readily lost by halide displacement.

A number of pentacoordinated dicarbonyl cations of rhodium and iridium have been prepared. The triphenylstibine derivatives are obtained by mild carbonylation.

$$M(CO)L_3Cl + CO + AlCl_3 \rightarrow [M(CO)_2L_3]^+AlCl_4^- \qquad (127)$$
$$M = Rh, Ir; L = SbPh_3$$

The tris(triphenylphosphine)iridium dicarbonyl cation has been prepared from the tricarbonyl cation as follows:

$$[Ir(CO)_3(PPh_3)_2]^+ \xrightarrow{PPh_3} [Ir(CO)_2(PPh_3)_3]^+ \qquad (193)$$

The tricarbonyl cation is obtained by carbonylation of the cationic hydrido carbonyl or by protonation of the alkoxycarbonyl.

$$Ir(CO)H_n(PPh_3)_2 \xrightarrow{HClO_4} [IrH_2(CO)(PPh_3)_2]^+ClO_4^-$$
$$n = 1 \text{ or } 3 \qquad\qquad\qquad \downarrow CO$$
$$Ir(CO)_2(PPh_3)_2CO \cdot OMe \xrightarrow{HClO_4} [Ir(CO)_3(PPh_3)_2]^+ClO_4^- \qquad (193)$$

Alternatively, mild carbonylation of an iridium halide may be used (*39*).

$$Ir(CO)(PPh_3)_2Cl + 2\ CO \xrightarrow{NaClO_4} [Ir(CO)_3(PPh_3)_2]^+ClO_4^- \qquad (39)$$

Alkali halides react with $[Ir(CO)_3(PPh_3)_2]^+$ to form iridium carbonyl halides $Ir(CO)(PPh_3)_2Cl$ or $Ir(CO)_2(PPh_3)_2I$ (*193*).

The dihydrido-monocarbonyl cations have been prepared by Angoletta and Caglio (*11*), both by protonation of $Ir(CO)H_{1,3}(PPh_3)_2$ and by carbonylation of $[IrH_3(PPh_3)_3]^+$. There are two forms of the cation. A closely related hydrido-carbonyl cation, $[Ir(CO)H_2(PPh_3)_3]^+$, is also prepared from $[IrH_3(PPh_3)_3]^+$ (*11*), or by addition of hydrogen halide to $Ir(CO)$ $(PPh_3)_2H$ (*238*).

$$[IrH_3(PPh_3)_3]^+ \underset{\underset{CO}{\nearrow}}{\overset{\overset{CO}{\searrow}}{}} \begin{matrix} [IrH_2(CO)(PPh_3)_2]^+ & \qquad (11) \\[2ex] [IrH_2(CO)(PPh_3)_3]^+ & \qquad (11) \end{matrix}$$

$$IrH(CO)(PPh_3)_3 \xrightarrow{HX} [IrH_2(CO)(PPh_3)_3]^+X^- \qquad (238)$$

A further example of the hydrido-carbonyl cations of the type $[Ir(CO)_x$ $(PPh_3)_y(H)_z]^+$ is the dicarbonyl $[Ir(CO)_2(PPh_3)_2H_2]^+$ formed by additive protonation (*53*).

$$Ir(CO)_2(PPh_3)_2H + HPF_6 \rightarrow [Ir(CO)_2(PPh_3)_2H_2]^+PF_6^- \qquad (53)$$

The addition products of allyl halides and *trans*-$Ir(CO)Cl(PPh_2Me)_2$ lose their halide to form a monocarbonyl cation (*66*), in methanolic sodium tetraphenylborate solution.

$$Ir(CO)Cl(PPh_2Me)_2(C_3H_5)X \xrightarrow[\text{MeOH}]{NaBPh_4}$$

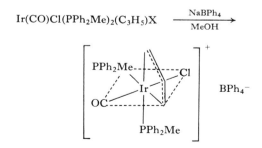

G. Nickel Group

There are few cationic carbonyl complexes reported for the nickel group metals. Although the π-allylnickel tricarbonyl cation is predicted (111) to exist, there is no report of an isolated nickel carbonyl cation. However, palladium and platinum form several stable cationic carbonyl complexes. Platinum-bridged halide carbonyls react with 2,2′-bipyridyl to form the salt $[Pt(CO)bipy\ X]^+[Pt(CO)X_3]^-$ (148, 149). Schützenburger in 1870 (222) reported the formation of an ammonia adduct with platinum dicarbonyl dichloride, $Pt(CO)_2Cl_2 \cdot 2\ NH_3$. This has subsequently been formulated as $[Pt(CO)_2(NH_3)_2]^{2+}Cl_2^{2-}$ (70), but this may well be a disproportionation product by analogy with the bipyridyl reaction, that is, of the type $[Pt(CO)(NH_3)_2Cl]^+[Pt(CO)Cl_3]^-$. This deserves reinvestigation.

The phosphine cation, trans-$[Pt(CO)(PEt_3)_2Cl]^+$, has recently been prepared (39–41). It was first obtained (40) as a low yield product from the reaction of trans-$PtHCl(PEt_3)_2$ with C_2F_4 in Pyrex glass tubes, and was isolated as the $[SiF_5]^-$ or $[BF_4]^-$ salt. The fluoroborate salt structure has been determined (40). The cations trans-$[Pt(CO)L_2Cl]^+$ (L = phosphine) are obtained by pressure (40, 42) or by atmospheric (11, 41) carbonylations. Their carbonyl absorptions are particularly high; thus, that of trans-$[Pd(PEt_3)_2(CO)Cl]^+$ is 2135 cm^{-1} (41).

$$Pt(PEt_3)_2HCl + aq.\ HBF_4 \xrightarrow[\text{benzene, }120°C]{\text{CO 5 atm}} [Pt(CO)(PEt_3)_2Cl]^+BF_4^- \quad (40, 42)$$

$$\left[(PR_3)_2M \overset{X}{\underset{X}{\diagdown}} M(PR_3)_2 \right]^{2+} \left[BF_4 \right]_2^- \xrightarrow{CO} \left[M(CO)(PR_3)_2Cl \right]^+ BF_4^-$$

$$(41)$$

$$M = Pt, \quad R = Et\ or\ Ph, \quad X = Cl,\ Br\ or\ I$$
$$M = Pd, \quad R = Et\ or\ Ph, \quad X = Cl$$

$$Pt(PEt_3)_2Cl_2 + CO \xrightarrow{NaClO_4} [Pt(CO)(PEt_3)_2Cl]^+ClO_4^- \quad (39)$$

Similarly,

$$Pt(PEt_3)_2HCl + CO \xrightarrow{NaClO_4} [Pt(CO)(PEt_3)_2H]^+ClO_4^- \quad (39)$$

The platinum trans-$[Pt(CO)Cl(PR_3)_2]^+$ cations are less reactive than the isoelectronic neutral Ir(I) and Rh(I) carbonyls (42).

The carbonyl may be displaced by chloride to form cis-$[PtCl_2(PR_3)_2]$, or by reaction with water to form the $trans$-hydride (43).

$$trans\text{-}[Pt(CO)Cl(PEt_3)_2]^+ \xrightarrow{\ H_2O\ } trans\text{-}[PtHCl(PEt_3)_2] \qquad (43)$$

The triphenylphosphine cation reacts with alcohols to form an alkoxy-carbonyl which yields the hydride on reaction with water.

$$trans\text{-}[Pt(CO)Cl(PPh_3)_2]^+ \underset{HBF_4}{\overset{\substack{ROH,\ 25^\circ C \\ (R\,=\,Me,\ Et)}}{\rightleftharpoons}} PtCl(PPh_3)_2CO\cdot OR$$

$$\Big\downarrow H_2O,\ 120^\circ C$$

$$trans\text{-}[PtClH(PPh_3)_2] \qquad (43)$$

Although it has been tempting to formulate the reported complex $Pt(CO)_2F_8$ (223) as a cationic carbonyl complex, the existence of this material is now questionable (155).

REFERENCES

 1. Abel, E. W., *Quart. Rev.* (*London*) **17**, 133 (1963).
 2. Abel, E. W., Bennett, M. A., and Wilkinson, G., *J. Chem. Soc.* p. 2323 (1959).
 3. Abel, E. W., McLean, R. A. N., Norton, M. G., and Tyfield, S. P., *Chem. Commun.* p. 900 (1968).
 4. Abel, E. W., McLean, R. A. N., Tyfield, S. P., Braterman, P. S., Walker, A. P., and Hendra, P. J., *J. Mol. Spectry.* **30**, 29 (1969).
 5. Abel, E. W., and Tyfield, S. P., *Chem. Commun.* p. 465 (1968).
 6. Abel, E. W., and Tyfield, S. P., unpublished work (1968).
 7. Abel, E. W., and Wilkinson, G., *J. Chem. Soc.* p. 1501 (1959).
 8. Adams, D. M., Cook, D. J., and Kemmitt, R. D. W., *Chem. Commun.* p. 103 (1966).
 9. Addison, C. C., and Kilner, M., *J. Chem. Soc., A* p. 1249 (1966).
10. Angelici, R. J., and Denton, D. L., *Inorg. Chim. Acta* **2**, 3 (1968).
11. Angoletta, M., and Caglio, G., *Rend. Ist. Lombardo Sci. Lettere* **A97**, 823 (1963).
12. Ariyaratne, J. K. P., and Green, M. L. H., *J. Organometal. Chem.* (*Amsterdam*) **1**, 90 (1963).
13. Ariyaratne, J. K. P., and Green, M. L. H., *J. Chem. Soc.* p. 2976 (1963).
14. Ariyaratne, J. K. P., and Green, M. L. H., *J. Chem. Soc.* p. 1 (1964).
15. Basolo, F., *Coord. Chem. Rev.* **3**, 213 (1968).
16. Basolo, F., Brault, A. T., and Poë, A. J., *J. Chem. Soc.* p. 676 (1964).
17. Beck, W., and Nitzschmann, R. E., *Z. Naturforsch.* **17b**, 577 (1962).
18. Behrens, H., and Herrman, D., *Z. Anorg. Allgem. Chem.* **351**, 225 (1967).
19. Behrens, H., and Lutz, K., *Z. Anorg. Allgem. Chem.* **354**, 184 (1967).
20. Behrens, H., and Lutz, K., *Z. Anorg. Allgem. Chem.* **356**, 225 (1967).
21. Behrens, H., and Rosenfelder, J., *Z. Anorg. Allgem. Chem.* **352**, 61 (1967).

22. Behrens, H., Ruyter, E., and Wakamatsu, H., *Z. Anorg. Allgem. Chem.* **349**, 241 (1967).
23. Birch, A. J., Cross, P. E., Lewis, J., and White, D. A., *Chem. & Ind.* (*London*) p. 838 (1964).
24. Blackmore, T., and Bruce, M. I., private communication (1968).
25. Brown, T. L., and Darensburg, D. J., *Inorg. Chem.* **6**, 971 (1967).
26. Burton, R., Pratt, L., and Wilkinson, G., *J. Chem. Soc.* p. 594 (1961).
27. Cais, M., and Narkis, N., *J. Organometal. Chem.* (*Amsterdam*) **3**, 269 (1965).
28. Calderazzo, F., *Inorg. Chem.* **3**, 1207 (1964).
29. Calderazzo, F., *Inorg. Chem.* **4**, 223 (1965).
30. Calderazzo, F., *Inorg. Chem.* **4**, 293 (1965).
31. Calderazzo, F., *Inorg. Chem.* **5**, 429 (1966).
32. Calderazzo, F., and Bacciarelli, S., *Inorg. Chem.* **2**, 721 (1963).
33. Cansiani, F., Sartorelli, V., and Zingales, F., *Rend. Ist. Lombardo Sci. Lettere* **A101**, 227 (1967).
34. Cattrall, R. W., and Clark, R. H. J., *J. Organometal. Chem.* (*Amsterdam*) **6**, 167 (1966).
35. Caulton, K. G., and Fenske, R. F., *Inorg. Chem.* **7**, 1273 (1968).
36. Chaudhari, F. M., and Pauson, P. L., *J. Organometal. Chem.* (*Amsterdam*) **5**, 73 (1966).
37. Chini, P., and Ercoli, R., *Gazz. Chim. Ital.* **88**, 1170 (1958).
38. Chiswell, B., and Venanzi, L., *J. Chem. Soc.*, *A* p. 417 (1966).
39. Church, M. J., and Mays, M. J., *Chem. Commun.* p. 435 (1968).
40. Clark, H. C., Corfield, P. W. R., Dixon, K. R., and Ibers, J. A., *J. Am. Chem. Soc.* **89**, 3360 (1967).
41. Clark, H. C., Dixon, K. R., and Jacobs, W. J., *Chem. Commun.* p. 93 (1968).
42. Clark, H. C., Dixon, K. R., and Jacobs, W. J., *J. Am. Chem. Soc.* **90**, 2259 (1968).
43. Clark, H. C., Dixon, K. R., and Jacobs, W. J., *Chem. Commun.* p. 548 (1968).
44. Clifford, A. F., and Mukherjee, A. K., *Inorg. Chem.* **2**, 151 (1963).
45. Clinton, N. A., and Lillya, C. P., *Chem. Commun.* p. 597 (1968).
46. Coffey, C. E., *J. Inorg. & Nucl. Chem.* **25**, 179 (1963).
47. Coffield, T. H., and Closson, R. D., U.S. Patent 3,130,214; *Chem. Abstr.* **61**, 4397 (1964).
48. Coffield, T. H., Sandel, V., and Closson, R. D., *J. Am. Chem. Soc.* **79**, 5826 (1957).
49. Collins, R. L., and Pettit, R., *J. Am. Chem. Soc.* **85**, 2332 (1963).
50. Collins, R. L., and Pettit, R., *J. Chem. Phys.* **39**, 3433 (1963).
51. Collman, J. P., and Roper, W. R., *Chem. Commun.* p. 244 (1966).
52. Collman, J. P., and Roper, W. R., *J. Am. Chem. Soc.* **88**, 3504 (1966).
53. Collman, J. P., Vastine, F. D., and Roper, W. R., *J. Am. Chem. Soc.* **90**, 2282 (1968).
54. Cook, C. D., Nyholm, R. S., and Tobe, M. L., *J. Chem. Soc.* p. 4194 (1965).
55. Cook, D. J., and Kemmitt, R. D. W., *Chem. & Ind.* (*London*) p. 946 (1966).
56. Cotton, F. A., *Inorg. Chem.* **3**, 702 (1964).
57. Cotton, F. A., and Wilkinson, G., "Advanced Inorganic Chemistry," 2nd ed., Chapter 27. Wiley (Interscience), New York, 1966.
58. Cousins, M., and Green, M. L. H., *J. Chem. Soc.* p. 889 (1963).
59. Crooks, G. R., and Johnson, B. F. G., *J. Chem. Soc.*, *A* p. 1238 (1968).
60. Dauben, H. P., and Bertelli, D. J., *J. Am. Chem. Soc.* **83**, 497 (1961).
61. Dauben, H. P., and Honnen, L. R., *J. Am. Chem. Soc.* **80**, 5570 (1958).
62. Davison, A., Green, M. L. H., and Wilkinson, G., *J. Chem. Soc.* p. 3172 (1961).
63. Davison, A., McFarlane, W., Pratt, L., and Wilkinson, G., *J. Chem. Soc.* p. 3653 (1962).

64. Davison, A., McFarlane, W., Pratt, L., and Wilkinson, G., *J. Chem. Soc.* p. 4821 (1962).
65. Davison, A., and Wilkinson, G., *Proc. Chem. Soc.* p. 356 (1960).
66. Deeming, A. J., and Shaw, B. L., *Chem. Commun.* p. 751 (1968).
67. Dessy, R. E., Stary, F. E., King, R. B., and Waldrop, M., *J. Am. Chem. Soc.* **88**, 471 (1966).
68. Edgell, W., and Bulkin, B., *J. Am. Chem. Soc.* **88**, 4839 (1966).
69. Edgell, W., Yang, M., Bulkin, B., Bayer, R., and Koizumi, N., *J. Am. Chem. Soc.* **87**, 3080 (1965).
70. Emeleus, H. J., and Anderson, J. S., "Modern Aspects of Inorganic Chemistry," Chapter 8. Routledge & Kegan Paul, London, 1960.
71. Emerson, G. F., Ehrlich, K., Giering, W. P., and Lauterbur, P. C., *J. Am. Chem. Soc.* **88**, 3172 (1966).
72. Emerson, G. F., Mahler, J. E., and Pettit, R., *Chem. & Ind. (London)* p. 836 (1964).
73. Emerson, G. F., Mahler, J. E., Pettit, R., and Collins, R., *J. Am. Chem. Soc.* **85**, 3590 (1964).
74. Emerson, G. F., and Pettit, R., *J. Am. Chem. Soc.* **84**, 4591 (1962).
75. Falkowski, D. R., Hunt, D. F., Lillya, C. P., and Rausch, M. D., *J. Am. Chem. Soc.* **89**, 6387 (1967).
76. Farona, M. F., and Camp, G. R., *Proc. 1st Intern. Symp. Metal Carbonyls, Venice, 1968* Paper A4, published by *Inorg. Chim. Acta* (1968).
77. Fischer, E. O., and Beckert, O., *Angew. Chem.* **70**, 744 (1958).
78. Fischer, E. O., and Fichtel, K., *Chem. Ber.* **94**, 1200 (1961).
79. Fischer, E. O., and Fichtel, K., *Chem. Ber.* **95**, 2063 (1962).
80. Fischer, E. O., Fichtel, K., and Öfele, K., *Chem. Ber.* **95**, 249 (1962).
81. Fischer, E. O., and Fischer, R. D., *Angew. Chem.* **72**, 919 (1960).
82. Fischer, E. O., and Fischer, R. D., *Z. Naturforsch.* **16b**, 475 (1961).
83. Fischer, E. O., and Kohl, F. J., *Z. Naturforsch.* **18b**, 504 (1963).
84. Fischer, E. O., and Kohl, F. J., *Angew. Chem. Intern. Ed. English* **3**, 134 (1964).
85. Fischer, E. O., and Kohl, F. J., *Chem. Ber.* **98**, 2134 (1965).
86. Fischer, E. O., and Kuzel, P., *Z. Anorg. Allgem. Chem.* **317**, 226 (1962).
87. Fischer, E. O., and Maasböl, A., *Angew. Chem.* **76**, 645 (1964).
88. Fischer, E. O., and Moser, E., *J. Organometal. Chem. (Amsterdam)* **2**, 230 (1964).
89. Fischer, E. O., and Moser, E., *J. Organometal. Chem. (Amsterdam)* **3**, 16 (1965).
90. Fischer, E. O., and Moser, E., *Z. Naturforsch.* **20b**, 184 (1965).
91. Fischer, E. O., and Moser, E., *J. Organometal. Chem. (Amsterdam)* **5**, 63 (1966).
92. Fischer, E. O., and Moser, E., *Z. Anorg. Allgem. Chem.* **342**, 156 (1966).
93. Fischer, E. O., and Öfele, K., *Angew. Chem.* **73**, 581 (1961).
94. Fischer, E. O., and Öfele, K., *Angew. Chem. Intern. Ed. English* **1**, 52 (1962).
95. Fischer, E. O., and Palm, C., *Chem. Ber.* **91**, 1725 (1958).
96. Fischer, E. O., and Riedel, A., *Chem. Ber.* **101**, 156 (1968).
97. Fischer, E. O., and Strametz, H., *Z. Naturforsch.* **23b**, 278 (1968).
98. Fischer, E. O., and Ulm, K., *Z. Naturforsch.* **16b**, 757 (1961).
99. Fischer, E. O., and Vogler, A., *Z. Naturforsch.* **17b**, 421 (1962).
100. Fitzpatrick, J. D., Watts, L., and Pettit, R., *Tetrahedron Letters* p. 1299 (1966).
101. Freni, M., Biusto, D., and Romiti, P., *J. Inorg. & Nucl. Chem.* **29**, 761 (1967).
102. Freni, M., Valenti, V., and Giusto, D., *J. Inorg. & Nucl. Chem.* **27**, 2635 (1965).
103. Green, M. L. H., and Hurley, C. R., *J. Organometal. Chem. (Amsterdam)* **10**, 188 (1967).

104. Green, M. L. H., Massey, A. G., Moelwyn-Hughes, J. T., and Nagy, P. L. I., *J. Organometal. Chem. (Amsterdam)* **8**, 511 (1967).
105. Green, M. L. H., and Nagy, P. L. I., *Proc. Chem. Soc.* p. 378 (1961).
106. Green, M. L. H., and Nagy, P. L. I., *Proc. Chem. Soc.* p. 74 (1962).
107. Green, M. L. H., and Nagy, P. L. I., *J. Am. Chem. Soc.* **84**, 1310 (1962).
108. Green, M. L. H., and Nagy, P. L. I., *J. Chem. Soc.* p. 189 (1963).
109. Green, M. L. H., and Nagy, P. L. I., *Z. Naturforsch.* **18b**, 162 (1963).
110. Green, M. L. H., and Nagy, P. L. I., *J. Organometal. Chem. (Amsterdam)* **1**, 58 (1963).
111. Green, M. L. H., and Nagy, P. L. I., *Advan. Organometal. Chem.* **2**, 325 (1964).
112. Green, M. L. H., and Stear, A. N., *J. Organometal. Chem. (Amsterdam)* **1**, 230 (1964).
113. Haines, L. I. B., and Poë, A. J., *Chem. Commun.* p. 964 (1968).
114. Haines, R. J., Nyholm, R. S., and Stiddard, M. H. B., *J. Chem. Soc., A* p. 94 (1967).
115. Haines, R. J., Nyholm, R. S., and Stiddard, M. H. B., *J. Chem. Soc., A* p. 43 (1968).
116. Hallam, B. F., and Pauson, P. L., *J. Chem. Soc.* p. 3030 (1956).
117. Hartley, J. G., Kerfoot, D. G. E., and Venanzi, L. M., *Inorg. Chim. Acta* **1**, 145 (1967).
118. Hayter, R. G., and Williams, L. P., *Inorg. Chem.* **3**, 613 (1964).
119. Heck, R. F., *J. Am. Chem. Soc.* **85**, 657 (1963).
120. Heck, R. F., *J. Org. Chem.* **28**, 604 (1963).
121. Herber, R. H., King, R. B., and Wertheim, G. K., *Inorg. Chem.* **3**, 101 (1964).
122. Hieber, W., Beck, W., and Zeitler, G., *Angew. Chem.* **73**, 364 (1961).
123. Hieber, W., and Duchatsch, H., *Chem. Ber.* **98**, 1744 (1965).
124. Hieber, W., and Duchatsch, H., *Chem. Ber.* **98**, 2350 (1965).
125. Hieber, W., and Frey, V., *Chem. Ber.* **99**, 2607 (1966).
126. Hieber, W., and Frey, V., *Z. Naturforsch.* **21b**, 704 (1966).
127. Hieber, W., and Frey, V., *Chem. Ber.* **99**, 2614 (1966).
128. Hieber, W., Frey, V., and John, P., *Chem. Ber.* **100**, 1961 (1967).
129. Hieber, W., and Freyer, W., *Chem. Ber.* **91**, 1230 (1958).
130. Hieber, W., and Freyer, W., *Chem. Ber.* **93**, 462 (1960).
131. Hieber, W., and Kummer, R., *Chem. Ber.* **100**, 148 (1967).
132. Hieber, W., and Kahlen, N., *Chem. Ber.* **91**, 2223 (1958).
133. Hieber, W., and Kruck, T., *Angew. Chem.* **73**, 580 (1961).
134. Hieber, W., and Kruck, T., *Z. Naturforsch.* **16b**, 709 (1961).
135. Hieber, W., Lux, F., and Herget, C., *Z. Naturforsch.* **20b**, 1159 (1965).
136. Hieber, W., Opavsky, W., and Rohm, W., *Chem. Ber.* **101**, 2244 (1968).
137. Hieber, W., Sedlmeier, J., and Abeck, W., *Chem. Ber.* **86**, 700 (1953).
138. Hieber, W., and Schropp, W., *Z. Naturforsch.* **14b**, 460 (1959).
139. Hieber, W., and Schuster, L., *Z. Anorg. Allgem. Chem.* **287**, 214 (1956).
140. Hieber, W., and Werner, R., *Chem. Ber.* **90**, 286 (1957).
141. Hieber, W., Winter, E., and Schubert, E., *Chem. Ber.* **95**, 3070 (1962).
142. Hieber, W., and Wollman, K., *Chem. Ber.* **95**, 1552 (1962).
143. Holmes, J. D., Jones, D. A. K., and Pettit, R., *J. Organometal. Chem. (Amsterdam)* **4**, 324 (1965).
144. Holmes, J. D., and Pettit, R., *J. Am. Chem. Soc.* **85**, 2531 (1963).
145. Howell, I. V., and Venanzi, L. M., *J. Chem. Soc., A* p. 1007 (1967).
146. Hull, C. G., and Stiddard, M. H. B., *J. Organometal. Chem. (Amsterdam)* **9**, 519 (1967).
147. Hull, C. G., and Stiddard, M. H. B., *J. Chem. Soc., A.* p. 710 (1968).
148. Irving, R. J., and Magnusson, E. A., *J. Chem. Soc.* p. 1860 (1956).
149. Irving, R. J., and Magnusson, E. A., *J. Chem. Soc.* p. 2018 (1957).

150. Jarvis, J. A. J., Mais, R. H. B., Owston, P. G., and Taylor, K. A., *Chem. Commun.*
     p. 906 (1966).
151. Jolly, P. W., and Pettit, R., *J. Am. Chem. Soc.* **88**, 5044 (1966).
152. Jones, D., Pratt, L., and Wilkinson, G., *J. Chem. Soc.* p. 4458 (1962).
153. Joshi, K. K., Pauson, P. L., Qazi, A. R., and Stubbs, W. H., *J. Organometal. Chem.*
     (*Amsterdam*) **1**, 471 (1964).
154. Joshi, K. K., Pauson, P. L., and Stubbs, W. H., *J. Organometal. Chem.* (*Amsterdam*)
     1, 51 (1963).
155. Kenmitt, R. D. W., Peacock, R. D., and Wilson, I. L., *Chem. Commun.* p. 772 (1968).
156. King, R. B., *Inorg. Chem.* **1**, 964 (1962).
157. King, R. B., *Inorg. Chem.* **2**, 807 (1963).
158. King, R. B., *Inorg. Chem.* **2**, 936 (1963).
159. King, R. B., *Inorg. Chem.* **2**, 1275 (1963).
160. King, R. B., *Inorg. Chem.* **5**, 2227 (1966).
161. King, R. B., *J. Inorg. & Nucl. Chem.* **29**, 2119 (1967).
162. King, R. B., and Bisnette, M. B., *J. Am. Chem. Soc.* **85**, 2527 (1963).
163. King, R. B., and Bisnette, M. B., *Inorg. Chem.* **3**, 791 (1964).
164. King, R. B., and Bisnette, M. B., *J. Am. Chem. Soc.* **86**, 1267 (1964); *Inorg. Chem.*
     4, 482 (1965).
165. King, R. B., and Bisnette, M. B., *Inorg. Chem.* **5**, 293 (1966).
166. King, R. B., and Bisnette, M. B., *J. Organometal. Chem.* (*Amsterdam*) **7**, 311 (1967).
167. King, R. B., Bisnette, M. B., and Fronzaglia, A., *J. Organometal. Chem.* (*Amsterdam*)
     4, 256 (1965).
168. King, R. B., Bisnette, M. B., and Fronzaglia, A., *J. Organometal. Chem.* (*Amsterdam*)
     5, 341 (1966).
169. Kingston, J. V., Jamieson, J. W. S., and Wilkinson, G., *J. Inorg. & Nucl. Chem.* **29**,
     133 (1967).
170. Kingston, J. V., and Wilkinson, G., *J. Inorg. & Nucl. Chem.* **28**, 2709 (1966).
171. Kirkham, W. J., Osborne, A. G., Nyholm, R. S., and Stiddard, M. H. B., *J. Chem.
     Soc.* p. 550 (1965).
172. Kochhar, R. K., and Pettit, R., *J. Organometal. Chem.* (*Amsterdam*) **6**, 272 (1966).
173. Kruck, T., and Höfler, M., *Chem. Ber.* **96**, 3035 (1963).
174. Kruck, T., and Höfler, M., *Angew. Chem. Intern. Ed. English* **3**, 701 (1964).
175. Kruck, T., and Höfler, M., *Chem. Ber.* **97**, 2289 (1964).
176. Kruck, T., Höfler, M., and Noack, M., *Chem. Ber.* **99**, 1153 (1966).
177. Kruck, T., and Noack, M., *Chem. Ber.* **96**, 3028 (1963).
178. Kruck, T., and Noack, M., *Chem. Ber.* **97**, 1693 (1964).
179. Kummer, R., and Graham, W. A. G., *Inorg. Chem.* **7**, 310 (1968).
180. Lange, G., and Dehnicke, K., *Z. Anorg. Allgem. Chem.* **344**, 167 (1966).
181. Lewis, J., Nyholm, R. S., Pande, C. S., and Stiddard, M. H. B., *J. Chem. Soc.* p. 3600
     (1963).
182. Lewis, J., Nyholm, R. S., Pande, C. S., Sandhu, S. S., and Stiddard, M. H. B., *J.
     Chem. Soc.* p. 3009 (1964).
183. Lewis, J., and Whyman, R., *Chem. Commun.* p. 159 (1965).
184. Lewis, J., and Whyman, R., *J. Chem. Soc.* p. 5486 (1965).
185. Lewis, J., and Wild, S. B., *J. Chem. Soc., A* p. 69 (1966).
186. Mague, J. T., and Mitchener, J. P., *Chem. Commun.* p. 911 (1968).
187. Mahler, J. E., Gibson, D. H., and Pettit, R., *J. Am. Chem. Soc.* **85**, 3955 and 3959
     (1963).

188. Mahler, J. E., Jones, D. A. K., and Pettit, R., *J. Am. Chem. Soc.* **86**, 3589 (1964).
189. Mahler, J. E., and Pettit, R., *J. Am. Chem. Soc.* **84**, 1511 (1962).
190. Mahler, J. E., and Pettit, R., *J. Am. Chem. Soc.* **85**, 3955 (1963).
191. Malatesta, L., *Progr. Inorg. Chem.* **1**, 283 (1959).
192. Malatesta, L., *U.S. Dept. Comm., Office Tech. Serv.*, AD **262,065**, 10–17 (1961); *Chem. Abstr.* **58**, 4596 (1963).
193. Malatesta, L., Caglio, G., and Angoletta, M., *J. Chem. Soc.* p. 6974 (1965).
194. Malatesta, L., and Valarino, L., *J. Chem. Soc.* p. 1867 (1956).
195. Manning, A. R., *J. Chem. Soc.*, *A* p. 1984 (1967).
196. Margulis, T. N., Schiff, L., and Rosenblum, M., *J. Am. Chem. Soc.* **87**, 3269 (1965).
197. Masek, J., Nyholm, R. S., and Stiddard, M. H. B., *Collection Czech. Chem. Commun.* **29**, 1714 (1964).
198. Masek, J., *Collection Czech. Chem. Commun.* **30**, 4117 (1965).
199. McCleverty, J. A., Davison, A., and Wilkinson, G., *J. Chem. Soc.* p. 3890 (1965).
200. McFarlane, W., Pratt, L., and Wilkinson, G., *J. Chem. Soc.* p. 2162 (1963).
201. Metlin, S., Wender, I., and Sternberg, H. W., *Nature* **183**, 457 (1959).
202. Mills, O. S., and Redhouse, A. D., *Angew. Chem.* **76**, 645 (1964).
203. Muetterties, E. L., *Inorg. Chem.* **4**, 1841 (1965).
204. Munro, J. D., and Pauson, P. L., *J. Chem. Soc.* pp. 3475, 3479, and 3486 (1961).
205. Nigam, H. L., Nyholm, R. S., and Stiddard, M. H. B., *J. Chem. Soc.* p. 1806 (1960).
206. Noack, K., *J. Organometal. Chem. (Amsterdam)* **13**, 411 (1968).
207. Nowell, I. N., and Russell, D. R., *Chem. Commun.* p. 817 (1967).
208. Nyholm, R. S., Snow, M. R., and Stiddard, M. H. B., *J. Chem. Soc.* p. 6564 (1965).
209. Nyholm, R. S., Snow, M. R., and Stiddard, M. H. B., *J. Chem. Soc.* p. 6570 (1965).
210. Öfele, K., *Angew. Chem. Intern. Ed. English.* **6**, 988 (1967).
211. Osborne, A. G., and Stiddard, M. H. B., *J. Chem. Soc.* p. 4715 (1962).
212. Osborne, A. G., and Stiddard, M. H. B., *J. Chem. Soc.* p. 700 (1965).
213. Pauson, P. L., and Munro, J. D., *Proc. Chem. Soc.* p. 267 (1959).
214. Pettit, R., and Emerson, G. F., *Advan. Organometal. Chem.* **1**, 1 (1964).
215. Piper, T. S., Cotton, F. A., and Wilkinson, G., *J. Inorg. & Nucl. Chem.* **1**, 165 (1955).
216. Sacco, A., *Gazz. Chim. Ital.* **93**, 542 (1963).
217. Sacco, A., *Gazz. Chim. Ital.* **93**, 698 (1963).
218. Sacco, A., and Freni, M., *Ann. Chim. (Rome)* **48**, 218 (1958); *Chem. Abstr.* **52**, 19656 (1958).
219. Sacco, A., and Freni, M., *J. Inorg. & Nucl. Chem.* **8**, 566 (1958).
220. Sacco, A., Rossi, M., and Nobile, C. F., *Chem. Commun.* p. 589 (1966).
221. Schrauzer, G. N., *J. Am. Chem. Soc.* **83**, 2966 (1961).
222. Schützenburger, M., *Bull. Soc. Chim. France* **14**, 97 (1870).
223. Sharp, D. W. A., *Proc. Chem. Soc.* p. 317 (1960).
224. Singer, H., *Z. Naturforsch.* **21b**, 810 (1966).
225. Singer, H., *J. Organometal. Chem. (Amsterdam)* **9**, 135 (1967).
226. Snow, M. R., and Stiddard, M. H. B., *Chem. Commun.* p. 580 (1965).
227. Snow, M. R., and Stiddard, M. H. B., *J. Chem. Soc.*, *A* p. 777 (1966).
228. Sternberg, H. W., Friedel, R. A., Shufler, S. L., and Wender, I., *J. Am. Chem. Soc.* **77**, 2675 (1955).
229. Sternberg, H. W., Wender, I., Friedel, R. A., and Orchin, M. J., *J. Am. Chem. Soc.* **75**, 3148 (1953).
230. Stiddard, M. H. B., *J. Chem. Soc.* p. 4712 (1962).

231. Treichel, P. M., Barnett, K. W., and Shubkin, R. L., *J. Organometal. Chem. (Amsterdam)* **7**, 449 (1967).
232. Treichel, P. M., and Shubkin, R. L., *J. Organometal. Chem. (Amsterdam)* **5**, 488 (1966).
233. Treichel, P. M., and Shubkin, R. L., *Inorg. Chem.* **6**, 1328 (1967).
234. Treichel, P. M., Shubkin, R. L., Barnett, K. W., and Reichard, D., *Inorg. Chem.* **5**, 1177 (1966).
235. Treichel, P. M., and Werber, G., *Inorg. Chem.* **4**, 1098 (1965).
236. Treichel, P. M., and Werber, G. P., *J. Organometal. Chem. (Amsterdam)* **1**, 157 (1967).
237. Tucci, E. R., and Gwynn, B. H., *J. Am. Chem. Soc.* **86**, 4838 (1964).
238. Vaska, L., *Chem. Commun.* p. 614 (1966).
239. Vaska, L., and Catone, D. L., *J. Am. Chem. Soc.* **88**, 5324 (1966).
240. Vohler, O., *Chem. Ber.* **91**, 1235 (1958).
241. Weiss, E., and Hübel, W., *Angew. Chem.* **73**, 298 (1961).
242. Weiss, E., and Hübel, W., *Chem. Ber.* **95**, 1186 (1962).
243. Wender, I., and Pino, P., ed., "Organic Syntheses via Metal Carbonyls," Vol. 1. Wiley (Interscience), New York, 1968.
244. Wender, I., Sternberg, H. W., and Orchin, M., *J. Am. Chem. Soc.* **74**, 1216 (1952).
245. Winkhaus, G., Pratt, L., and Wilkinson, G., *J. Chem. Soc.* p. 3807 (1961).
246. Winkhaus, G., and Singer, H., *Z. Naturforsch.* **18b**, 418 (1963).
247. Winstein, S., Kaesz, H. D., Kreiter, C. G., and Friedlich, E. C., *J. Am. Chem. Soc.* **87**, 3267 (1965).

Fast Exchange Reactions of Group I, II, and III Organometallic Compounds

JOHN P. OLIVER

Department of Chemistry
Wayne State University
Detroit, Michigan

I

INTRODUCTION

The development of nuclear magnetic resonance spectroscopy for the measurement of the rates of fast reactions (preexchange lifetimes 1–0.001 second) has made it possible to study many alkyl–metal exchange processes which heretofore were experimentally inaccessible. A substantial number of papers dealing with the exchange reactions of Group I, II, and III organo derivatives has appeared, and it is now possible to begin to correlate the rates and mechanisms of these exchange processes with the structure and properties of the metal derivatives.

This review will remain for the most part within the limit of "fast" reactions of σ-bonded alkyl derivatives of these metals, where the rate of reaction may be determined by PMR relaxation methods, and we will deviate from this only to point out possible explanations for differences in exchange rates or mechanisms. No attempts will be made to dwell on the chemistry of these compounds except as it pertains to the exchange reactions discussed.

The chemistry and earlier work on exchange reactions of these compounds have been reviewed adequately in the past few years [for Group I see references (14, 15, 28); Group II (4, 22, 28, 40, 140, 147); Group III (28, 45, 58, 73, 139, 158, 161); exchange reactions (34, 73, 90)].

The techniques for measuring exchange rates by nuclear magnetic resonance spectroscopy have been well documented in the leading texts in this field (37, 71, 118).

II

ORGANOLITHIUM COMPOUNDS, (LiR)$_n$

Although the subject of exchange reactions of organolithium compounds has recently been dealt with in detail (14, 15), a brief account is included here because of the many reactions between these compounds and the alkyl derivatives of Groups II and III.

Organolithium compounds occur in solution as dimeric, tetrameric, or hexameric aggregates held together by electron-deficient bridge bonds (14). The actual degree of association depends on the alkyl group involved and the solvent. The nature of the association in these derivatives permits two types of exchange:

(a) group exchange between aggregates (intermolecular);
(b) migration of groups within an aggregate (intramolecular).

Examination of work done primarily by Brown and co-workers shows that in cyclopentane solution, intermolecular exchange of alkyl groups between (Li tert-Bu)$_4$ and [LiCH$_2$Si(CH$_3$)$_3$]$_4$ is relatively slow and proceeds with an activation energy of 24 kcal/mole (15, 52). The rate is enhanced by a factor of 20 when the compounds are placed in toluene solvent. Once the mixed products are formed, however, the intramolecular rearrangements occur rapidly and approximately at the same rate in both solvent systems.

These studies indicate that two quite different mechanisms are needed to explain these processes. The intermolecular exchange has been reported to occur because of the equilibrium which exists between the tetramer (hexamer) and dimer with exchange being effected by recombination of

$$(LiR)_4 \rightleftharpoons 2 (LiR)_2 \tag{1}$$

$$(LiR^*)_4 \rightleftharpoons 2 (LiR^*)_2 \tag{2}$$

$$(LiR)_2 + (LiR^*)_2 \rightleftharpoons Li_4R_2R^*_2 \tag{3}$$

the dimeric units. The difference between cyclopentane and toluene solutions can be accounted for on the basis of the coordinating ability of toluene toward the highly electrophilic $(LiR)_2$ aggregate. No solvent dependence is found for the intramolecular rearrangements.

The exchange between $(LiCH_3)_4$ and $(LiC_2H_5)_4$ has also been examined, but in ether solution (15). In this solvent, there can be no differentiation between the types of exchange, i.e., either both exchange processes are occurring or none. These studies indicate that the predominant factor is the interaction of the ether solvent, which lowers the dissociation energy of the tetramer, and speeds up the exchange process. In fact, the activation energy for the exchange is approximately 11 kcal/mole, which is less than the energy required for intramolecular exchange in cyclopentane.

It is also interesting to note that allyllithium has been shown to exist in a "stable" configuration in THF at low temperature (150). As the temperature is raised, rapid rearrangements with an activation energy of approximately 10.5 kcal/mole occur which give rise to the magnetic equivalence of the two CH_2 groups resulting in an AX_4 PMR spectrum. The exact structure and mechanism of exchange for allylic species of this type are not well understood.

Other studies have shown that rapid exchange occurs at room temperature between an alkyllithium and a lithium halide in ether or tetrahydrofuran solutions (88, 146). This exchange can be stopped at low temperature with the formation of mixed alkyllithium–lithium halide complexes. Further studies have shown that when these systems are enriched with ^{13}C, $^7Li^{-13}C$ coupling can be observed at low temperatures (38). While this clearly shows the interaction which occurs between the metal and carbon atoms, studies using ^{6}Li-enriched samples show no ^{6}Li–^{7}Li coupling (*16*). This has been interpreted to mean that little direct interaction between the metal atoms occurs.

On the basis of these studies, we can conclude that while both types of exchanges outlined above can occur in organolithium systems, the specific details and relative rates of reactions are very dependent on the solvent and the alkyl group involved.

III

GROUP II ORGANOMETALLIC DERIVATIVES

A. BeR₂ and RBeX

Only limited studies have been reported on exchange reactions of dialkylberyllium compounds and alkylberyllium halides due to both their reactivity and toxicity. Some early studies suggested that RBeX did not exist and that alkyl exchange did not occur (*32*). More recently, Ashby and co-workers have carefully reinvestigated the $Be(CH_3)_2$–$BeBr_2$ system and have demonstrated that exchange occurs in several ways (*5*). Of particular importance here is the fact that in solutions of $Be(CH_3)_2$ and $BeBr_2$ mixtures, low-temperature PMR ($-70°$ C) reveals the existence of two methyl-containing species, assigned to $Be(CH_3)_2$ and $Be(CH_3)Br$. Room-temperature spectra show only one methyl resonance and establish that alkyl exchange does take place. It was suggested that the exchange occurs

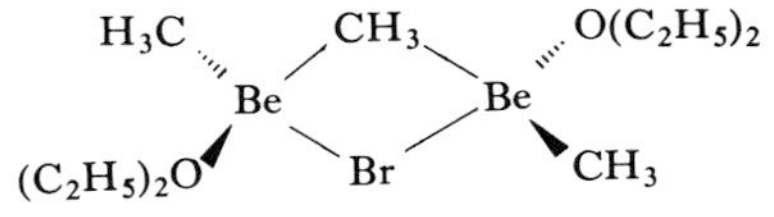

(I)

through a four-centered bridged transition state (I). Quantitative data are not available, however, to test this postulate.

In addition, while no studies have been reported which provide quantitative kinetic information on the exchange of alkyl groups between dialkylberyllium compounds, Coates and Roberts have provided qualitative evidence for the exchange between bridge and terminal positions in

dimeric diorganoberyllium derivatives (27). Their findings, in agreement
with the known formation of stable Be—C—Be bridge bonds, show that
exchange is slow and appears to be stopped on the PMR time scale just
below room temperature. Further studies are needed to provide the rate
data necessary for the postulation of an exchange mechanism.

Coates and co-workers have also investigated a wide variety of compounds of the type $(RBeX)_n$, where $R = CH_3$ or C_2H_5 and $X = NMe_2$,
OMe, SeEt, SeO, $SC_2H_4NMe_2$, etc., and $n = 2$ or 3 (9, 10, 23–25). In a
number of the trimeric derivatives, they have postulated that the ring
structure can be present in two conformations with rapid conversion from
one to the other at room temperature. In some instances, these conformations
may be slowed to give resolvable methyl resonances corresponding to the
nonequivalent methyl positions. No quantitative data have been presented
on these systems.

B. MgR_2 and $RMgX$

While many investigations have dealt with the structure and reactivities
of dialkylmagnesium derivatives and Grignard reagents, it has only been
within the past year that definitive data regarding alkyl exchange reactions
have been available. House et al. have studied a variety of $Mg(CH_3)_2$–
$RMgCH_3$ systems and have obtained kinetic data for the exchange of alkyl
groups (64). Some of the pertinent data are collected in Table I. These
studies also demonstrated that the exchange process was overall second
order, first order with respect to each dialkylmagnesium reagent. From this
and the data showing the solvent dependence given in Table I, a mechanism
for the transfer of the alkyl groups was presented, which is given in Eqs.
(4)–(6).

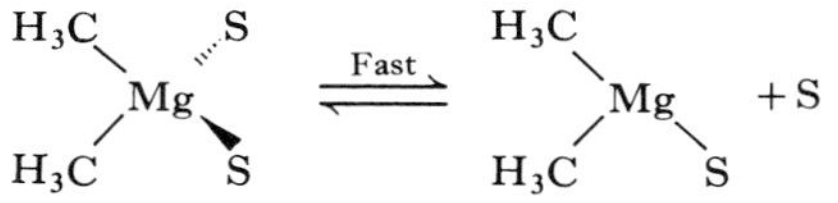

$$\begin{array}{cc} \text{(4)} \end{array}$$

$$\begin{array}{cc} \text{(5)} \end{array}$$

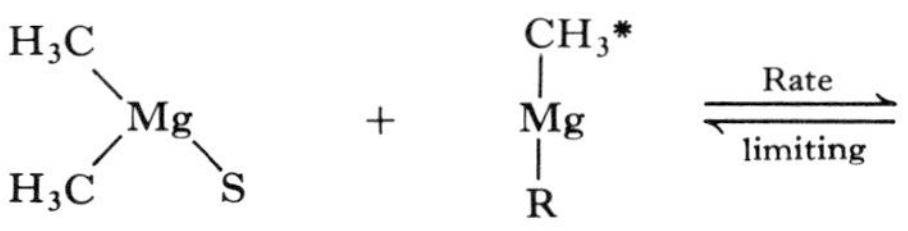

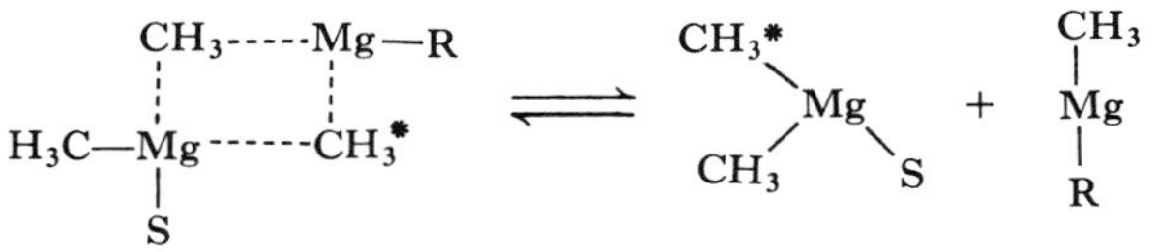

$$CH_3\text{-----}Mg\text{—}R \qquad CH_3^* \qquad\qquad CH_3$$

$$\begin{array}{cccc}
\text{CH}_3\text{-----Mg—R} & & \text{CH}_3^* & \text{CH}_3 \\
\text{H}_3\text{C—Mg-----CH}_3^* & \rightleftharpoons & \text{Mg} & + \quad \text{Mg} \\
& & \text{CH}_3 \quad \text{S} & \text{R}
\end{array} \qquad (6)$$

This mechanism can be used to account for the solvent dependence, which shows that a strongly coordinating solvent slows or prevents step (4) from occurring and thus retards the rate of exchange. This is similar to the effect observed in Group III alkyl derivatives discussed in Section IV,A. However, if complete coordination of the magnesium alkyl does not occur, it becomes necessary to break a

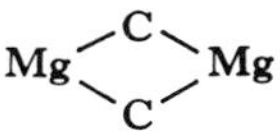

$$Mg \begin{array}{c} \diagup C \diagdown \\ \diagdown C \diagup \end{array} Mg$$

bridge bond before exchange can take place. This indicates a solvent dependence where one might expect a rate enhancement when a "weak base," diethyl ether, is present and a subsequent decrease when a stronger base, such as N,N,N',N'-tetramethylethylenediamine, is present.

TABLE I

RATES OF EXCHANGE OF METHYL GROUPS IN ORGANOMAGNESIUM DERIVATIVES[a]

Reagent	Solvent	E_a (kcal/mole)	Rate constants at $-55°$ C, liter mole^{-1} sec^{-1}
$C_5H_5MgCH_3 + (CH_3)_2Mg$	THF-ether (72:28)	13.8	124
$C_5H_5MgCH_3 + (CH_3)_2Mg$	THF-ether (64:36)	12.2	200
$C_5H_5MgCH_3 + (CH_3)_2Mg$	THF-ether-triethylamine (50:26:24)	11.4	500
$C_5H_5MgCH_3 + (CH_3)_2Mg$	Ether-THF-amine[b] (44:30:26)	12.1	0.63
$C_5H_5MgCH_3 + (CH_3)_2Mg$	DME[c]-ether (72:28)	7.4	25
$PhMgCH_3 + (CH_3)_2Mg$	THF-ether-amine[b] (56:28:16)	8.6	40

[a] From House et al. (64).
[b] N,N,N',N'-Tetramethylethylenediamine.
[c] 1,2-Dimethoxyethane.

Evans and Khan have observed quite similar results to those cited above in their studies on the exchange between fluorinated Grignard reagents and bis(fluoroalkyl)magnesium compounds (38, 39). In their work, relative orders of exchange were presented but specific rates of reaction were not. They did observe that the rate of exchange was greatly decreased on addition of a bidentate amine, which again served to complex the magnesium strongly and prevent its entrance into the formation of a bridged transition state.

It is interesting to note that in comparison to these rapid, low-energy, second-order exchange processes Witanowski and Roberts have shown that inversion of bis(neohexyl)magnesium is a relatively slow first-order process with an activation energy of 20 kcal/mole (154). It therefore appears that exchange and inversion go through different mechanisms.

C. ZnR₂ and RZnX

Several papers have appeared which indicate that unsymmetrical zinc derivatives, RZnR′, can be prepared and separated (1, 129). This implies that for those systems studied, facile exchange does not exist between the dialkylzinc compounds. While these studies were done with larger alkyl groups, a PMR study also indicates the apparent absence of exchange between $Zn(CH_3)_2$ and $Zn(C_2H_5)_2$ (11). This could be misleading, however, because the chemical shifts in the dialkyl compounds and the mixed alkyl compounds, RZnR′, could be very close together and exchange might not be detected. $^{13}C-^{1}H$ coupling studies by Roberts and co-workers support the existence of exchange between molecules of $Zn(CH_3)_2$ (148a). In addition, a preliminary investigation of the $Zn(CH{=}CH_2)_2{-}Zn(CH_3)_2$ system appears to imply rapid exchange in diethyl ether solution (109). The fact that certain zinc compounds can form the necessary four-coordinate transition state is shown through studies of Jeffery and Mole indicating that $Zn(C{\equiv}CPh)_2$ and $Zn[C{\equiv}C(CH_2)_5CH_3]_2$ are associated (69).

Witanowski and Roberts have shown that the inversion process in bis(neohexyl)zinc is relatively slow and proceeds with an activation energy of 26 kcal/mole (154). This process is thought to go through a carbanion intermediate rather than a four-centered transition state. Allylzinc exhibits an AX_4 PMR spectrum which has also been interpreted in terms of an ionic structure (141, 151). With this in mind, it is of interest to note that the rates of reaction of PhHgCl with ZnR_2 appear to follow the polarity of the Zn—C bond rather than the ability of the group to form a bridge bond (2).

PMR studies of $Zn(CH_3)_2$ with $N(C_2H_5)_3$ have shown the formation of a 1:1 complex, but a stable complex was not separated (137). These studies further indicate that exchange occurs between complexed and uncomplexed base.

Boersma and Noltes report that $R_2Zn + ZnX_2$ (X = Cl, Br, I) participate in a Schlenk equilibrium

$$2 \, RZnX \rightleftharpoons R_2Zn + ZnX_2 \tag{7}$$

and that alkylzinc halides catalyze the exchange of alkyl groups (11).

They also report exchange when other groups such as –OR, –NR_2, or –SR replace the halides, and state a rate dependence on the group used (11). Exchanges of this type have been studied by others (3, 12, 20, 113), as has self-exchange within the tri- or tetrameric species $(RZnOR')_n$. The results of most of these studies are inconclusive, but a recent paper of Jeffery and Mole indicates that a fast exchange occurs in the $(CH_3)_2Zn + [Zn(CH_3)(OCH_3)]_4$ system (66). This exchange is stopped on the PMR time scale at $+30°C$ in toluene solvent and at $+5°C$ in pyridine. It appears, then, that the rate of the exchange process is directly dependent on the ability of the solvent to coordinate with either the ZnR_2 or the $(RZnOR')_n$ group.

Further work in this whole area is needed to clarify the numerous conflicting reports and to provide sufficient data on which to base proposed mechanisms.

D. CdR_2 and RCdX

In the study of exchange reactions of closely related organometallic derivatives, such as $ZnR_2 + RZnX$, by PMR spectroscopy, the problem often arises that chemical shift differences are not large enough to provide a convenient range of study. This difficulty is overcome for cadmium derivatives since cadmium possesses three principal isotopes, ^{111}Cd, ^{112}Cd, and ^{113}Cd. The ^{111}Cd (12.9%) and ^{113}Cd (12.3%) both possess spins of $\frac{1}{2}$ which couple strongly with protons in alkyls. The PMR spectrum for dimethylcadmium therefore consists of five lines, one corresponding to the methyl groups attached to the ^{112}Cd isotope and four lines arising from the coupling of the ^{111}Cd and ^{113}Cd with the protons. Because of this, it is particularly easy to determine if self-exchange occurs by observation of the collapse of this spectrum to a single line. Several studies have established that rapid exchange does not take place either in ether or hydrocarbon solutions (*33, 48, 51, 108*) but does occur in more strongly coordinating solvents (*33, 51, 108*). Attempts have been made to examine the rates of these exchange reactions quantitatively, but no satisfactory results have been obtained. Mole and co-workers have shown that the presence of any alkoxide ion catalyzes the exchange, and have estimated the activation energies for the exchange with alkoxide present to be 16 kcal/mole in toluene and 13 kcal/mole in pyridine (*48*).

On the basis of the above studies, one may again postulate that exchange proceeds through a four-centered transition state. The catalytic effect of the basic solvent probably arises from the formation of the complex

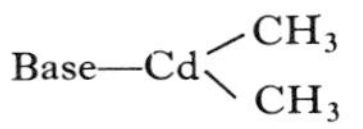

$$\text{Base}\!-\!\text{Cd}\!\!\begin{array}{c}\nearrow\text{CH}_3\\ \searrow\text{CH}_3\end{array}$$

(II)

which can more readily form a bridged transition state

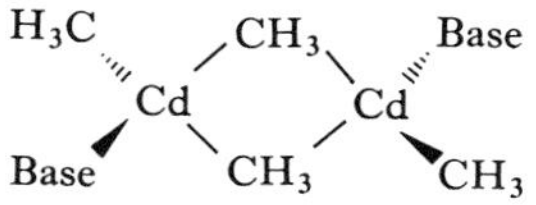

(III)

Thus, participation of solvent may be critical in this type of system, as in the magnesium and zinc systems.

The effect of the alkoxide appears also to be determined by the added ability of the –OR group to form bridged intermediates. Thus, one might visualize the important step as formation of the intermediate (IV). The

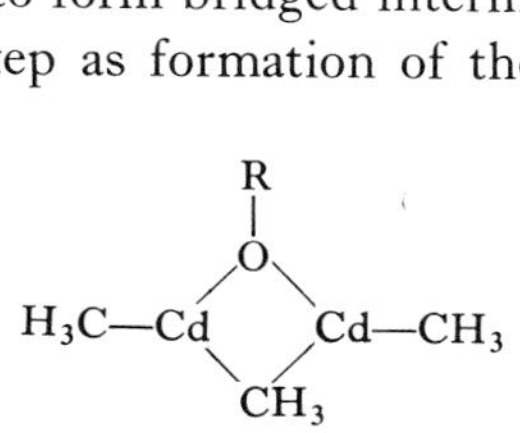

(IV)

intermediate may be further coordinated by solvent but the data are insufficient to determine this.

E. HgR₂ and RHgX

E. HgR$_2$ and RHgX

Dialkylmercury compounds do not for the most part undergo rapid exchange, although the slow transfer of alkyl groups has been extensively studied. This is easily illustrated by the observation of PMR spectrum, which shows the clearly resolved ^{199}Hg satellite (^{199}Hg, $I = \frac{1}{2}$, 16.9%). The only cases which have been reported to undergo exchange are the cyclopentadienide and the allyl derivatives, neither of which is clearly understood (*86, 104a, 117*).

In the case of alkylmercuric halides (RHgX), two possible exchange processes might occur, having different effects on the ^{199}Hg—^{1}H satellite spectra. If an alkyl group was rapidly transferred, the satellites would broaden and collapse as in the case of the cadmium systems. If, instead, the X group was transferred between RHgX and RHgY, the satellites would be averaged between the two values. Experimentally, the satellites have been observed to broaden in a number of cases where X and Y are either halides or pseudohalides. While early studies suggested that this broadening was due to alkyl exchange (*55, 149*), more recent studies have indicated that it is due either to exchange of the X and Y groups or to quadrupolar interaction of the ^{199}Hg nucleus with X, where X is Cl, Br, or I (*41, 49, 104, 130*). This quadrupolar effect would result in broadening because of the rapid spin relaxation of the ^{199}Hg nucleus.

F. MR₂–M′R₂

F. MR$_2$–M′R$_2$

It has been shown that rapid exchange occurs between $Mg(CH_3)_2$ and $Zn(CH_3)_2$ and between $Mg(CH_3)_2$ and $Cd(CH_3)_2$, but that exchange is slow between $Mg(CH_3)_2$ and $Hg(CH_3)_2$ (*33*). In the Zn–Mg system exchange proceeds rapidly even at $-107°$ C (*125*).

In addition, studies indicate that $Zn(CH_3)_2$ undergoes rapid exchange with $Cd(CH_3)_2$, and various postulates concerning the mechanism of this reaction have been made (*57, 87*). Recent studies of this system in methylcyclohexane solvent have clearly shown that the reaction is first order in each of the components and proceeds with an activation energy of 17 kcal/mole (*57*). This study indicates that the exchange process proceeds through a four-centered transition state

$$Zn(CH_3^*)_2 + Cd(CH_3)_2 \rightleftharpoons H_3C-Cd \underset{CH_3}{\overset{CH_3^*}{\diagup\diagdown}} Zn-CH_3^*$$

$$\longrightarrow Cd(CH_3)(CH_3^*) + Zn(CH_3)(CH_3^*) \quad (8)$$

where the high energy of activation represents the difficulty in formation of the M—C—M′ bridge when both metals belong to Group II (57). Further justification for this is provided by the fact that neither $Zn(CH_3)_2$ or $Cd(CH_3)_2$ exchanges readily with $Hg(CH_3)_2$, where bridge formation would be especially difficult (33).

No reports have been presented concerning the reactions of BeR_2 with other Group II organometallic compounds, but it would be anticipated from the foregoing observations that these derivatives would undergo rapid exchange with MgR_2, with ZnR_2, and with CdR_2. The mechanism for these systems would likely be similar to that for $Zn(CH_3)_2$–$Cd(CH_3)_2$ exchange, but the rate would clearly depend on the ease with which the BeR_2 aggregate is broken up. This might either make the dissociation of this aggregate the rate-determining step, as in the Al_2R_6 exchange process (Section IV,A), or lead to a more complex mechanism as outlined for the exchange of lithium alkyls with Group II derivatives (Section III,G).

G. LiR–MR₂

Several studies have shown that rapid exchange and complex formation occur when lithium alkyls are mixed with alkyls of various Group II derivatives. Seitz and Brown (125–127) have shown in a series of papers that rapid exchange occurs in the $LiCH_3$–$Mg(CH_3)_2$ and $LiCH_3$–$Zn(CH_3)_2$ systems, and they have carefully examined both the ^1H and ^7Li NMR. They have also measured the equilibrium constants at low temperature and have examined the exchange in some detail. On the basis of these studies they have proposed the following equilibrium steps (125):

$$\tfrac{1}{2}[Li(CH_3)]_4 + M(CH_3)_2 \;\rightleftharpoons\; Li_2M(CH_3)_4 \;\rightleftharpoons\; Li_3M(CH_3)_5 \qquad (9)$$

When the Li/Mg (Zn) ratio is 2 only a single resonance line is observed at all temperatures. When the ratio is greater than 2, an absorption from $(LiCH_3)_4$ is observed in both the ^7Li and ^1H spectra at low temperatures. When the ratio of Li/Mg (Zn) is less than 2, a proton line associated with $Mg(CH_3)_2$ appears at about $-60°C$ but the exchange remains rapid at $-107°$ for the $Zn(CH_3)_2$ system.

Examination of the data for the system containing excess $(LiCH_3)_4$ suggests the following path:

$$Li_2[M(CH_3)_4] + Li_3M(CH_3)_5 \;\rightleftharpoons\;$$

$$Li_2[M(CH_3)_4]\ldots LiCH_3 \ldots Li_2[M(CH_3)_4] \qquad (10)$$

in which a $LiCH_3$ group is transferred between the two complexes. This exchange proceeds more readily than exchange between $(LiCH_3)_4$ and either of the complexes. The activation energy for the latter process has been estimated and is included in Table II. The possible paths for this are as follows:

$$(^+CH_3Li^*)_4 \rightleftharpoons 2\,(^+CH_3Li^*)_2 \tag{11}$$

$$(^+CH_3Li^*)_2 + Li_2M(CH_3)_4 \rightarrow Li^*LiM(CH_3)_3(^+CH_3) + (LiCH_3)_2 \tag{12}$$

and/or

$$(^+CH_3Li^*)_2 + Li_3M(CH_3)_5 \rightarrow Li^*Li_2M(CH_3)_4(^+CH_3) + (LiCH_3)_2 \tag{13}$$

It has also been found that ^7Li and the methyl group exchange between methyllithium and the complexes at comparable rates. While a distinction between the two above mechanisms cannot be made directly, Seitz and Brown indicate that the second mechanism is the predominant path for exchange. The data for these systems are collected in Table II.

TABLE II

$LiCH_3-Mg(CH_3)_2$ AND $LiCH_3-Zn(CH_3)_2$ SYSTEMS IN ETHER[a]

System	Nucleus	E_a (kcal/mole)
$LiCH_3-Li_2Zn(CH_3)_4$	^7Li	10.9
$LiCH_3-Li_2Zn(CH_3)_4$	^1H	8.5
$LiCH_3-Li_2Mg(CH_3)_4$	^7Li	10.3
	^1H	10
$Mg(CH_3)_2-Li_2Mg(CH_3)_4$	^1H	15

[a] Seitz and Brown (125).

Similar studies have been carried out on $LiPh-MgPh_2$ and $LiPh-ZnPh_2$ systems (126). Although complex formation is present, differences between the rate of ^7Li exchange and Ph exchange are reported. It has been suggested in this case that the rate-determining step for Ph exchange is the dissociation of the complex. In another paper the exchange reactions and complex formation of mixed methyl–Ph systems are also reported (127). Studies on ethyl derivatives of zinc and cadmium with LiR show similar complex formation and exchanges, but less detailed information has been provided (*142*). It has been shown, however, that stable 1:1 complexes (LiMEt$_3$) are possible where the fourth and/or fifth coordination sites are occupied by either THF or DME.

Several studies have been carried out on the interaction of lithium alkyls with dialkylmercury derivatives. In benzene it has been shown that LiC$_2$H$_5$ + Hg(C$_2$H$_5$)$_2$ do not undergo rapid exchange but with added THF exchange occurs (*142*). In diethyl ether, rapid exchange has been reported between Li(CH$_3$) and Hg(CH$_3$)$_2$ (*128*), and exchange has also been observed in the vinyllithium–divinylmercury system in diethyl ether (*109*). These studies are incomplete and do not provide sufficient evidence to confirm a mechanism for exchange, although it was suggested that a complex formation step might be involved.

Evidence supporting this has been obtained in the LiSi(CH$_3$)$_3$–Hg[Si(CH$_3$)$_3$]$_2$ system, in which both rapid exchange of the trimethylsilyl group and formation of LiHg[Si(CH$_3$)$_3$]$_3$ · Base and Li$_2$Hg[Si(CH$_3$)$_3$]$_4$ have been observed in DME solution (*107a*).

IV

GROUP III ORGANOMETALLIC DERIVATIVES

A. AlR$_3$ and AlR$_3$–MR$_3$

1. *Saturated Alkyl Derivatives*

A number of studies have been carried out which establish that tri-methylaluminum exists as a dimer (*77*) having the structure (*78, 145*):

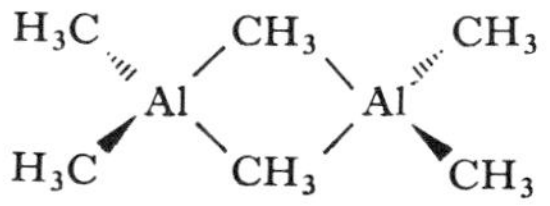

(V)

While this suggests that there should be two resonance signals observed in the PMR spectrum, one corresponding to the bridge methyl groups and the other to the terminal methyl groups, the initial investigation of the PMR spectrum at room temperature showed only a single resonance line (*103*). Upon cooling to $-75°C$ in cyclopentane solution, the expected two lines appeared.

Several studies (*103, 116, 152*) have been carried out to determine the mechanism of this self-exchange. Muller and Pritchard (*103*) found the activation energy for the process to be approximately 15 kcal/mole, which was confirmed by Ramey *et al.* (*116*) as 15.6 kcal/mole. On the basis of this and the known gas-phase dissociation energy for the dimer of $Al_2(CH_3)_6$ (20.4 kcal/mole) (*58, 77*), a simple dissociation process was ruled out (*116*). To account for the exchange, two alternate intramolecular paths were proposed in which rearrangement of the bridge bonds provides a mechanism for the transfer of the methyl groups. In the first of these, given by Eq. (14), the transition state contains four methyl bridge bonds at the corners of a square which may break up in several different ways to give methyl exchange.

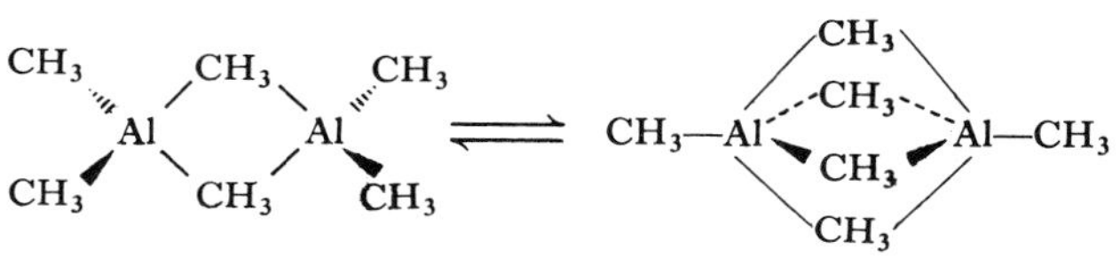

$$\tag{14}$$

The second mechanism, Eq. (15), proceeds through a partial dissociation in which one of the Al—C bonds is broken, followed by rotation about the other, resulting in exchange.

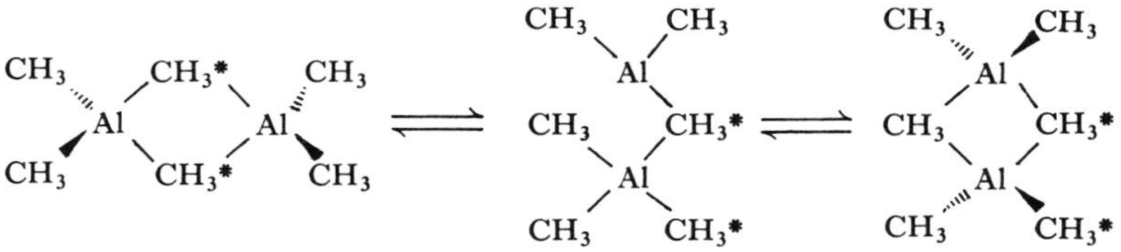

$$\tag{15}$$

Ramey *et al.* (*116*) discounted the mechanism given by Eq. (14) since the formation of a transition state with four bridge bonds would be expected to require a large reorganization energy. They supported the reaction sequence given by Eq. (15) as the most likely process. This also has the advantage that it may be used to account for intermolecular exchange if the intermediate singly bridged species has a sufficiently long lifetime.

As an alternative to this mechanism involving a singly bridged species, a number of groups (*152, 160, 161*) have proposed that alkyl exchange in AlR_3 systems proceeds through the monomeric unit which is formed by the dissociation of the dimeric species in the case of trimethyl- or triethyl-aluminum. For this mechanism, the activation energy of the process must be equal to or greater than the dissociation energy of the dimeric unit. This is a point of real concern, since the gas-phase dissociation energy (20.4 kcal/mole) is substantially greater than the activation energy (15.4 ± 2 kcal/mole) for the exchange process. It has been suggested, however, that the dissociation energy decreases in going to hydrocarbon solutions and on this basis it has been argued that the activation energy for exchange represents the upper limit for the dissociation (*58, 59, 152*). Therefore it, will be useful to detail the work done assuming this mechanism, since it has been well documented and offers a convenient explanation for inter-molecular exchange.

Recently, Brown and Williams (*152*) have carefully examined this approach and have carried out additional experiments on the $Al_2(CH_3)_6$ system to provide further tests of the proposed dissociation mechanism. Their findings are in full agreement with those of Ramey *et al.* (*116*), but they have additional data which have been interpreted to show that the rate and activation energy for bridge–terminal exchange remain the same ($E = 15.4 ± 2$ kcal/mole) in cyclopentane and in toluene solutions. They also found that in the exchange of methyl groups between trimethylgallium or trimethylindium and trimethylaluminum the activation energy was within experimental error of that observed for the bridge–terminal exchange. The rates of exchange were the same in toluene solution but the methyl group exchange was a factor of ten slower than bridge–terminal exchange in cyclopentane solution. The data are collected in Table III. In order to account for these findings, they proposed the following reaction sequence:

$$Al_2(CH_3)_6 \underset{k_{-1}}{\overset{k_1}{\rightleftharpoons}} [2\ Al(CH_3)_3] \underset{k_{-2}}{\overset{k_2}{\rightleftharpoons}} Al(CH_3)_3 \tag{16}$$

Dimer D	Solvent-caged monomer SCM	Solvent-separated monomer SSM

$$\underset{SSM}{Al(CH_3)_3} + E(CH_3)_3 \overset{k_3}{\longrightarrow} Product \tag{17}$$

The bridge–terminal exchange can be accounted for by this process, where $k_{-1} \gg k_1$, i.e., the rate-determining step is the dissociation of D to SCM and the activation energy for the process must be the dissociation energy corresponding to the process D → 2 SCM. In order to also account for the reaction with $Ga(CH_3)_3$ and $In(CH_3)_3$ in cyclopentane and toluene

TABLE III

RATE CONSTANTS FOR BRIDGE–TERMINAL EXCHANGE IN $Al_2(CH_3)_6$
AND FOR METHYL EXCHANGE BETWEEN $Al_2(CH_3)_6$ AND $Ga(CH_3)_3$
IN CYCLOPENTANE AND TOLUENE SOLUTIONS[a]

Temp. (°C)	k (sec^{-1})			
	$Al_2(CH_3)$		$Al_2(CH_3)_6 + Ga(CH_3)_3$	
	Cyclopentane	Toluene	Cyclopentane	Toluene
−55°	7.65	12.3	0.9	18.8
−50°	18.8	26.3	2.1	43.2
−40°	46.2	52.5	4.5	92.0

[a] All values are taken from Williams and Brown (152).

solutions they considered two limiting cases. The first of these is for cyclopentane solution, where $k_3[E] \gg k_2[SSM]$ and $k_{-1} \gg k_2$. With these conditions, the predicted rate of exchange between $Al_2(CH_3)_6$ and $E(CH_3)_3$ will be dependent on the ratio of SCM which dissociates to that which dimerizes. The rate of exchange of $E(CH_3)_3$ will be less than the bridge–terminal exchange by this factor, with the appropriate expression for the lifetime of E given by

$$\frac{1}{\tau_E} = \frac{2}{3}\left(\frac{k_2}{k_{-1}}\right)k_1 \frac{[D]}{[E]} \tag{18}$$

In the second case (toluene solution) the solvent has an appreciable interaction with the SCM, and the limiting conditions become $k_3[E] \gg k_{-2}[SSM]$ and $k_2 \gg k_{-1}$, i.e., all SCM dissociates to SSM, where it exchanges with $E(CH_3)_3$. The correct expression for the lifetime of E then becomes

$$\frac{1}{\tau_E} = \frac{2}{3}k_1 \frac{[D]}{[E]} \tag{19}$$

Under these conditions all of D which dissociates undergoes exchange with $E(CH_3)_3$ and rates comparable to bridge–terminal exchange are observed.

Comparison of the different proposed mechanisms for self-exchange in the trimethylaluminum system suggests that either the singly bridged mechanism of Ramey *et al.* (*116*) or the complete dissociation process described by Williams and Brown (*152*) can account for the kinetics of the bridge–terminal exchange. It is difficult to distinguish between them, but one or the other of the two mechanisms appears to control the self-exchange of trimethylaluminum and play an important rôle in the intermolecular exchange reactions between $Al_2(CH_3)_6$ and other organometallic species.

This intermolecular exchange may be readily accounted for by the process represented in Eqs. (16) and (17). For example, in the studies by Hoffmann (*60, 61, 63*), Poole (*115*), and others (*93, 98, 131, 156, 157*), the exchange of larger organic groups between aluminum atoms may be represented as going through a transition state similar in nature to the trimethylaluminum dimer.

In the methyl–isobutylaluminum system studied by Hoffmann (*60, 61*) the chemical shift of the methyl resonance was proportional to the fraction of methyl groups present over the entire range of concentration studied, showing that rapid exchange of isobutyl groups continues even when the isobutyl:methyl ratio is 3:1. This implies that the isobutyl group must occupy a bridge position. This rapid intermolecular exchange favors a dissociative path and appears to rule out any path involving the singly bridged species because of the extreme complexity of the transition state which would be needed.

The formation of the dimeric intermediate with a large bridging group seems reasonable in view of the dimeric nature of $Al_2(C_2H_5)_6$, which has been shown by Yamamoto (*156*), by a careful investigation of the PMR spectrum at low temperature, to be bridged through the methylene group. This group has also established that the ability of ethyl, *n*-propyl, and isobutyl groups to enter bridge positions relative to methyl is $\frac{1}{6}$, $\frac{1}{7}$, and $\frac{1}{17}$, respectively (*157*).

2. *Exchange of Unsaturated and Cyclopropyl Derivatives*

Recent studies on unsaturated derivatives by Mole and co-workers (*65, 67, 68, 70, 97–99*) and of cyclopropyl derivatives by Sanders and Oliver (*122*) have provided results which shed additional light on the factors influencing the structure, bonding, and exchange reactions in dimeric aluminum systems. Studies on $Al(C_6H_5)_3$ show that the compound is dimeric in solution and that it forms discrete molecules in the crystalline state, with phenyl groups occupying bridging positions almost perpendicular to the Al—Al axis (85). This suggests that one should observe PMR resonances for bridging and terminal groups as in the alkyl series, but no reports of this have been made in the literature. However, Mole has shown that in methylphenylaluminum compounds, the phenyl group is preferentially located in the bridge position (70). He has also given evidence for the existence of mixed methyl–phenyl bridged derivatives (68) such as

$$H_3C \quad Ph \quad CH_3$$
$$Al \quad Al$$
$$H_3C \quad CH_3 \quad CH_3$$

(VI)

Furthermore, he has shown that the phenylethynyl group forms stable bridged derivatives which do not undergo exchange with the same facility as the alkyl compounds (97). The main reason proposed for this is that the π-system of the unsaturated group interacts with the vacant nonbonding molecular orbitals of the aluminum atoms and this interaction stabilizes the bridge system (53). In the aluminum case, where activation (dissociation) of the bridge-bonded species is the rate-determining step for exchange, this stabilizing interaction should markedly decrease the rate of exchange. Quantitative rate data are not available for any of these systems, but this conclusion appears to be borne out qualitatively and will be commented on below along with other factors governing rates of reactions.

The compound $Al_2(cyclopropyl)_6$ exists in the dimeric form and does not undergo rapid bridge–terminal exchange (122), as can be seen from the room-temperature PMR spectra shown in Fig. 1. Studies on the temperature dependence of the PMR spectra of $Al_2(cyclopropyl)_6$–$Al_2(CH_3)_6$ mixtures indicate that the activation energy for exchange in these systems is in the 18–22 kcal/mole range (107). This is far above the value reported for the other alkylaluminum compounds and is comparable to that reported for the phenylethynyl derivatives.

In order to account for this high resistance to exchange, a careful consideration of the nature of the cyclopropyl group has been made. It is clear that a simple π-system such as that which occurs in the benzene or phenylethynyl group does not exist for this moiety, but by using a model such as

Walsh's (*148*), one may show that a *p* orbital on the ring has proper symmetry to overlap the nonbonding orbitals of the metal atoms, thereby stabilizing the bridge bond as in the cases of the phenyl or phenylethynyl systems.

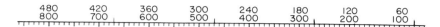

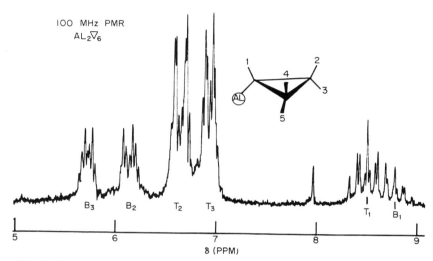

Fɪɢ. 1. The 100-MHz PMR spectrum of tricyclopropylaluminum dimer in benzene (20° C). Protons of the cyclopropyl group are labeled as shown. Regions are labeled B for bridging, T for terminal; subscripts correspond to the individual protons.

3. *Alkyl Exchange in the Presence of Base*

In all of the aforementioned studies on aluminum alkyls, the rate of reaction was dependent on the strength of dimeric molecules or the ability to form them. Jeffery and Mole have undertaken a study of group exchange between $Al(C_6H_5)_3$ and $Al(CH_3)_3$ in the presence of pyridine, where the aluminum alkyls exist in solution as addition compounds with pyridine (*65*). They have found that the mechanism for group exchange is dependent on the amount of pyridine present in excess. When pyridine is present in large excess the rate of exchange is independent of the concentration of pyridine.

This fact is consistent with a bimolecular reaction between the two pyridine adducts, with a transition state as shown in (VII), containing two five-coordinate aluminum atoms.

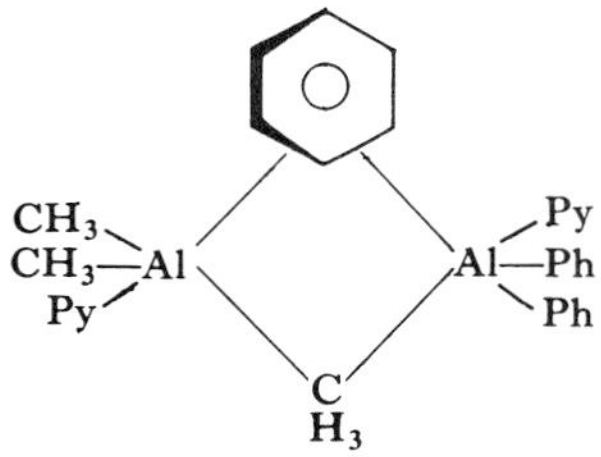

(VII)

When pyridine is present in only slight excess, however, the rate of exchange is dependent on the concentration of excess pyridine and a mechanism is postulated which involves a rate-determining step consisting of attack of a noncomplexed aluminum alkyl, formed in a prior rapid equilibrium, with a complexed aluminum alkyl.

Further evidence of possible formation of five-coordinate aluminum atoms as a pathway for exchange has been reported by Mole and co-workers (50) in a study of alkyl group exchange between trimethylaluminum and dimethylethylaluminum etherates. The exchange rate has an overall second-order dependence on total aluminum concentration and a zero-order dependence on ether concentration, indicating alkyl exchange without prior dissociation of the ether adducts.

B. $AlR_3–MR_2$

Several studies have been reported in which exchange occurs between trimethylaluminum and Group II dimethyl derivatives. McCoy and Allred (87) reported that $Al_2(CH_3)_6$ exchanged rapidly with $Cd(CH_3)_2$ and suggested that this was directly dependent on the concentration of both $Cd(CH_3)_2$ and $Al_2(CH_3)_6$. These findings would support a process in which the rate-determining step is the interaction of the activated species with the Group II alkyl. It has also been found that $Zn(CH_3)_2$ (109) exchanges methyl groups with $Al_2(CH_3)_6$ but $Hg(CH_3)_2$ does not (33).

On the basis of independent studies, Williams and Brown (152) have suggested that the data indicating a bimolecular rate-determining step for the exchange between $Al_2(CH_3)_6$ and $Cd(CH_3)_2$ are incorrect. Supporting this view is the work done on the exchange between $Cd(CH_3)_2$ and the other Group III alkyls, $Ga(CH_3)_3$ and $In(CH_3)_3$ (57). Both of these proceed with relatively low activation energies (7.9 and 8.3 kcal/mole, respectively) and cast doubt on the theory that the bimolecular step is rate-determining in the exchange of $Cd(CH_3)_2$ with $Al(CH_3)_6$. It is clear that a redetermination of the kinetics in this system is in order.

C. GaR₃ and GaR₃–MR₂

Quantitative studies on the exchange reactions of trialkylgallium derivatives have been reported only recently and are quite limited. The bridge–terminal exchange observed for trimethylaluminum is not possible in $Ga(CH_3)_3$ since this compound has been shown to exist as a monomer in solution (102). The possibility of observing self-exchange by observation of proton-gallium coupling also appears unlikely since no ^1H coupling has been reported, though a number of PMR studies have been made (112). The lack of observation of this coupling may either be due to rapid relaxation of the nuclei caused by quadrupolar interactions, or simply to a very small Ga–H coupling constant.

Self-exchange has been reported, however, in trivinylgallium, which is dimeric (134). The proposed structure for this compound is similar to that of trimethylaluminum with bridging vinyl groups:

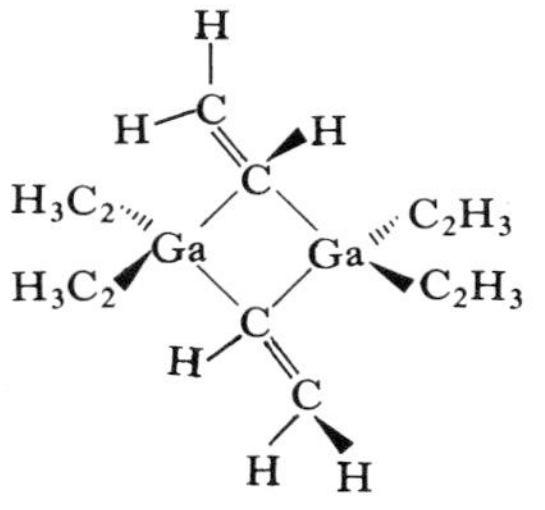

(VIII)

and should give rise to two distinct PMR spectra for the vinyl groups. PMR spectra obtained at $-90°C$, however, show only a single type of vinyl group, which indicates that rapid exchange occurs even at this temperature (100, 144). Rapid exchange has also been suggested for the tripropenylgallium derivatives since they give rise to a single resonance spectrum for the propenyl groups (*100*). In these systems the activation energy for exchange has been estimated to be less than 10 kcal/mole in order to account for the high rates of exchange (*144*).

Studies have also shown that rapid alkyl exchange occurs between $Ga(CH_3)_3$ and $Ga(C_2H_5)_3$ but no additional information is available either on the rate or activation energy of the process (*105*). Extensive studies have recently been reported on the exchange between $Ga(CH_3)_3$ and $Ga(CH=CH_2)_3$ by Visser and Oliver (*144*). Both groups have been shown to undergo rapid exchanges, as indicated by the linear dependence of the chemical shifts of both the methyl and vinyl groups on concentration. These studies provide strong evidence that the vinyl group is preferentially retained in the bridge position and further suggest that they have the orientation proposed for trivinylgallium. These data are best summarized in Figs. 2

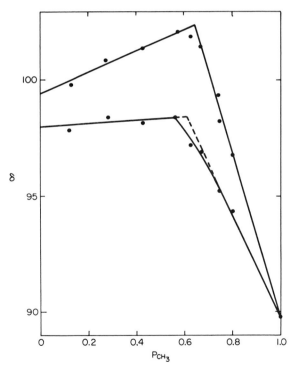

Fig. 2. The chemical shift (in cps, 60 MHz) upfield from cyclopentane of the methyl protons vs. the mole ratio $(CH_3)/[(CH_3)+(C_2H_3)]$. The lower curve was obtained at $+40°$, while the upper one was obtained at $-50°$.

and 3, which show the concentration dependence of the chemical shift of the methyl group and the vinyl group.

From the data available, an upper limit of 10 kcal/mole can be set for the process, suggesting that the dissociation energy for the Ga—CH=CH$_2$—Ga bridge is substantially less than that observed for the aluminum systems. This fact may account for the more rapid exchange reactions observed.

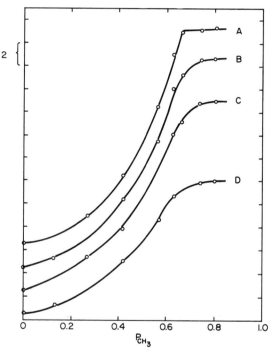

FIG. 3. The chemical shift of the trans proton vs. the mole ratio $(CH_3)/[(CH_3)+(C_2H_3)]$ at various temperatures: A, $-50°$; B, $-30°$; C, $-4°$; and D, $+36°$ C. Each curve is offset by 2 cps (one unit on the graph) on the chemical shift scale (in cps, 60 MHz).

Studies have also been carried out on Ga$_2$(CH$_3$)$_4$(C≡C—Ph)$_2$, which is dimeric in solution and does not undergo rapid exchange, but undergoes slow equilibration with Ga(CH$_3$)$_3$ (69). This is a further indication of the high ability of unsaturated groups to enter into bridge formation, and represents the most stable Ga—C—Ga bridge system so far observed.

Preliminary studies have been carried out to determine if rapid exchange occurs between Ga(CH$_3$)$_3$ and other organometallic compounds. Williams and Brown (*152*) have noted that $In(CH_3)_3$ and $Ga(CH_3)_3$ undergo exchange rapidly even at low temperature, and as previously noted they studied the exchange with $Al(CH_3)_3$. Exchange of $Ga(CH_3)_3$ has also been observed with several Group II alkyl derivatives. For the $Zn(CH_3)_2$–$Ga(CH_3)_3$ system, separate resonances for the zinc and gallium moieties have been observed at $-85°C$ in dichloromethane solvent (*106*). The necessity to study the exchange at such low temperatures indicates that very rapid exchange occurs in this system. Rapid exchange has also been observed for the $Cd(CH_3)_2$–$Ga(CH_3)_3$ system at room temperature, but the reaction is slowed sufficiently at $-50°$ to $-60°C$ so that its rate may be determined by standard PMR line-broadening techniques (*57*). These results indicate that the exchange process has an activation energy of approximately 7.9 kcal/mole with a bimolecular rate-determining step. Mixtures of $Hg(CH_3)_2$ and $Ga(CH_3)_3$ give rise to sharp lines for each component, even at slightly elevated temperatures, indicating that exchange between these two species is slow. Thus, in the exchange of the Group II alkyls with $Ga(CH_3)_3$ we see a marked decrease in the rate of exchange in going from $Zn(CH_3)_2$ to $Cd(CH_3)_2$ to $Hg(CH_3)_2$, which implies that the ease of formation of the activated complex (or metal–carbon–metal bridge bond) decreases in the same order.

It is also interesting to note that alkyl group exchange does not occur rapidly between $Ga(CH_3)_3$ and $Ga(CH_3)_3$—addition compounds at low temperatures. The only reaction observed in these systems is the transfer of the entire $Ga(CH_3)_3$ group (*30, 31*). At temperatures where this takes place rapidly, one cannot tell if alkyl group exchange occurs. This observation is analogous to that for the $Al_2(CH_3)_6$ system, in which alkyl group exchange does not occur with a low activation energy unless the bridge bond can be opened, giving rise to a three-coordinate metal atom.

D. InR_3 and InR_3–MR_2

Only limited efforts have been made to examine the exchange reactions of trialkylindium derivatives. $In(CH_3)_3$ has been shown to exchange with $Al_2(CH_3)_6$ at a measurable rate as previously noted (*152*). It has also been reported that rapid exchange occurs between $In(CH_3)_3$ and either $Ga(CH_3)_3$ (*152*) or $Zn(CH_3)_3$ (*109*) in hydrocarbon solution at $-60°C$. Preliminary studies show a marked solvent dependence for the exchange between $InMe_3$ and $ZnMe_2$. In ether solution, for example, exchange is slowed sufficiently for measurement above $0°$ C (106). Rapid exchange also occurs between In(CH$_3$)$_3$ and Mg(CH$_3$)$_2$ in ether solution at $-60°$ C (106). This should be compared to the In(CH$_3$)$_3$–amine systems (56), in which it has been shown that alkyl group exchange does not occur between free In(CH$_3$)$_3$ and its amine adduct when base exchange is stopped. This is probably a result of the lower dissociation energy for the In(CH$_3$)$_3$–ether complexes.

In(CH$_3$)$_3$ does not exchange with Hg(CH$_3$)$_2$ at a measurable rate but does with Cd(CH$_3$)$_2$ in hydrocarbon solutions. Recent studies on this system show that the reaction proceeds with an activation energy of 8.3 kcal/mole with a rate somewhat less than that observed for the corresponding Ga(CH$_3$)$_3$–Cd(CH$_3$)$_2$ system (57). This may be interpreted as a lessening in the tendency for formation of metal–carbon–metal bridge-bonded transition state.

It has also been shown that In$_2$(CH$_3$)$_4$(C≡C—Ph)$_2$ is dimeric and does not undergo rapid exchange with In(CH$_3$)$_3$. This indicates, therefore, that "stable" In—C—In bridge bonds can be formed with unsaturated derivatives (69).

E. TlR$_3$

Thallium has two isotopes, [203]Tl and [205]Tl, both of which have spins of $\frac{1}{2}$ and have been shown to couple strongly with protons on alkyl groups attached directly to them with coupling constants as high as 1000 cps (81, 83, 84, 143). In view of this, it was startling to note that Tl(CH$_3$)$_3$ showed no Tl–H coupling at room temperature. Upon lowering the temperature, Maher and Evans (84) observed two lines which showed $J_{Tl-H} = 250$ cps. This indicates that very rapid exchange occurs at room temperature, averaging these signals. A complete study of this system shows that this is the case and that the activation energy for the process is 6.3 kcal/mole. In view of the activation energies for Group II–Group III exchange mentioned above, this is a reasonable value for an exchange between two molecules of a monomeric Group III alkyl.

Studies have also been carried out by mixing Tl(CH$_3$)$_3$ with Tl(C$_2$H$_5$)$_3$ in which it is shown that rapid exchange occurs with the formation of the methyl ethyl derivatives (84). No clear attempt has been made to evaluate the mechanism of this reaction but it is certain that all of the mixed thallium derivatives can be prepared. Maher and Evans (82) also examined the

$Tl(CH=CH_2)_{3-n}(CH_3)_n$ system in detail. In this case, rapid exchange occurs, giving the mixed methylvinyl derivatives $Tl(CH=CH_3)(CH_3)_2$ and $Tl(CH=CH_2)_2(CH_3)$. No $Tl(CH=CH_2)_3$ was observed, and it appears that it cannot be prepared either by exchange or by direct preparative procedures. This is the only Group III element for which a trivinyl derivative is unknown.[1]

The kinetics of this system indicate that rapid exchange occurs for both groups and when compared with the studies on pure $Tl(CH_3)_3$, methyl exchange occurs 30 times faster when vinyl groups are present. Another interesting feature is that the vinyl groups exchange faster than the methyl groups by an order of magnitude or more; it is suggested that the vinyl groups can form a bridged transition state more readily than the methyl groups so that one obtains a transition state of type (IX) more readily than

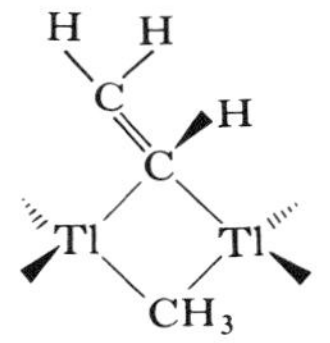

(IX)

one containing two methyl bridges. The slightly increased stability of the metal–carbon–metal bridge bond would give rise to the higher exchange rate observed for the vinyl moiety, and is in full accord with the observed ability of unsaturated derivatives to form bridge bonds more readily than saturated groups.

In examining the $TlPh_3$ system it was noted that the phenyl groups exchanged rapidly in methylene chloride solution while the exchange was considerably slowed in trimethylamine (82). This decrease in rate implies that a stable complex $[Ph_3Tl\cdot N(CH_3)_3]$ is formed which must undergo dissociation prior to phenyl exchange and is similar in this respect to the $Ga(CH_3)_3$ and $In(CH_3)_3$ systems. In contrast to this, methyl exchange in $Tl(CH_3)_3$ is not appreciably slowed by addition of a strong Lewis base, which indicates that no stable complexes are formed. The kinetic data for the $TlPh_3$ and $Tl(CH_3)_3$ systems are collected in Table IV.

---

[1] Trivinylindium has been prepared by C. Beer and J. P. Oliver by exchange of $Hg(CH=CH_2)_2$ with In metal.

TABLE IV

LIFETIMES FOR Tl(CH$_3$)$_3$ AND TlPh$_3$ IN VARIOUS SOLVENTS AT 26° C[a]

	Solvent	$\tau \times 10^{-3}$ sec	Tl(CH$_3$)$_3$ (moles/liter)
Tl(CH$_3$)$_3$	Deuterobenzene	1.30	0.46
		0.78	0.96
		0.54	1.22
	Trimethylamine	3.86	0.74
	Dimethylether	3.25	1.65
TlPh$_3$	Trimethylamine	$> 5 \times 10^{-5}$	—
	Dichloromethane	$< 3 \times 10^{-7}$	—

[a] All data are from Maher and Evans (82).

Studies on phenylethynylthallium derivatives have been very limited, but no evidence for a bridged compound has been found in this system (76).

V

REACTIONS OF MR$_n$X$_{3-n}$ DERIVATIVES

Many Group III derivatives formed by the reaction of MR$_3$ and MX$_3$, where M = Al, Ga, or In and X = Cl, Br, I, CN, NR$_2$, OR, SH, etc., or by cleavage reactions of the various alkyls have been characterized for some time (8, 26, 44, 72, 75, 79, 92, 94, 135, 159). Structural studies and the physical properties of these materials indicate that they all exist as associated aggregates. In systems for which quantitative information is available, the X group is located in the bridging position (46, 62).

Early PMR studies established that for AlR$_m$Cl$_{3-m}$ derivatives, rapid intermolecular exchange occurred at room temperature between all sites in the system (19, 131). Even though not quantitative, these studies indicated a preference of the halide for the bridging position and established that the reactions proceed at approximately the same rate as the self-exchange of trialkylaluminum compounds.

More recent studies have shown that exchange occurs between AlR$_3$ and AlR′$_2$X derivatives where X = Cl, OR″, etc. (54, 91). In the study with alkoxide derivatives, it is found that the rate of alkyl exchange (R for R′) depends on the alkoxide present, with the rate decreasing in the order EtO– > iso-PrO– > tert-BuO– (91). All of these reactions are relatively slow.

On the basis of these and other studies carried out on the exchange reactions of trialkylaluminum derivatives, three mechanisms for exchange have been proposed. These are:

(a) A bimolecular reaction of dimers:

$$R_2Al \underset{X}{\overset{X}{\diamond}} Al\text{-}R_2 + R_2\text{-}Al \underset{Y}{\overset{Y}{\diamond}} AlR_2 \rightleftharpoons \text{Products} \qquad (20)$$

(b) A sequential series of reactions with the first step being the dissociation of one of the species followed by reaction of the monomer with a dimer of the other species:

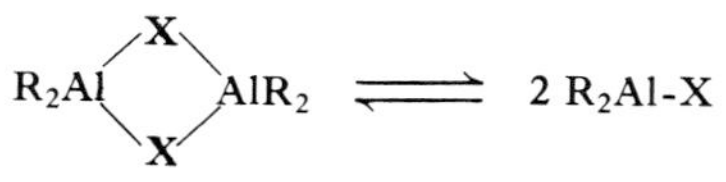

$$R_2Al \underset{X}{\overset{X}{\diamond}} AlR_2 \rightleftharpoons 2\,R_2Al\text{-}X \qquad (21)$$

$$R_2AlX + R_2\text{-}Al \underset{Y}{\overset{Y}{\diamond}} Al\text{-}R_2 \rightleftharpoons \text{Products} \qquad (22)$$

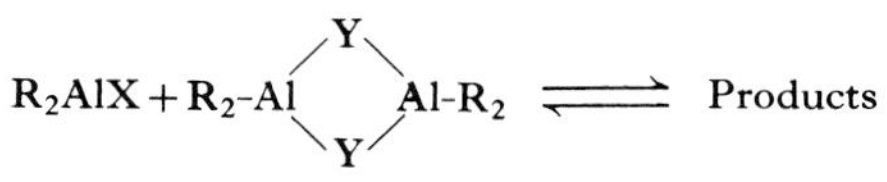

(c) A sequential series of reactions with both reactants first being dissociated followed by reaction of the monomers:

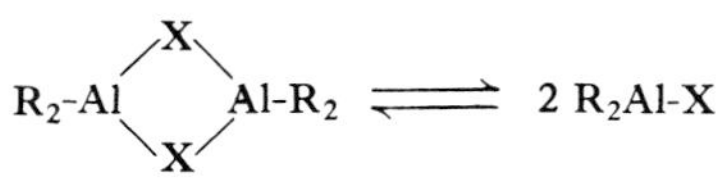

$$R_2\text{-}Al \underset{X}{\overset{X}{\diamond}} Al\text{-}R_2 \rightleftharpoons 2\,R_2Al\text{-}X \qquad (23)$$

$$R_2\text{-}Al \underset{Y}{\overset{Y}{\diamond}} Al\text{-}R_2 \rightleftharpoons 2\,R_2Al\text{-}Y \qquad (24)$$

$$R_2Al\text{-}X + R_2AlY \rightleftharpoons R_2Al \underset{Y}{\overset{X}{\diamond}} Al\text{-}R_2 \rightleftharpoons \text{Products} \qquad (25)$$

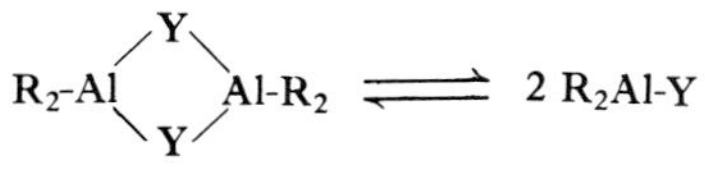

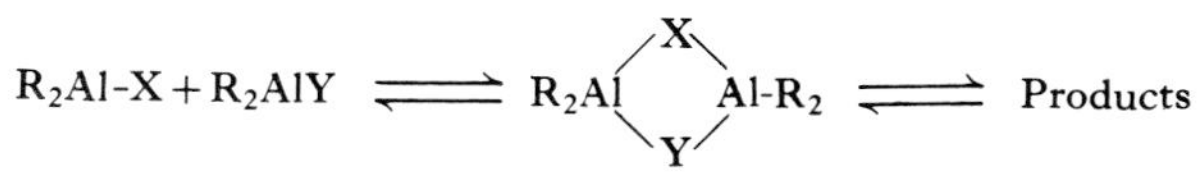

The features that should be noted in these mechanisms are that (a) and (b) suggest the formation of complex transition states and (b) and (c) both postulate a dissociation step prior to exchange. Mechanism (c) has at least in part been supported by the PMR detection of mixed bridged derivatives of the type R_4Al_2XY (67).

Recently, Smith and Wallbridge have examined several of the $(AlEt_2X)_2$ systems and have shown that rapid exchange occurs in hydrocarbon solutions when $X = Cl$, Br, or I (132, 133). They noted further, by substituting a methyl for the ethyl group, that the rate of exchange was dependent on the alkyl group used. Finally, they indicate that the rate of alkyl group exchange is greater than that of the bridged halides. While these studies have not been reported in detail and the mechanism is still uncertain, they suggest that at least two different exchange paths are available for group transfer. One of these involves the dissociation of the bridge bond, as clearly evidenced by the fact that the exchange reactions of compounds with "strong" bridges, such as $-NR_2$, undergo exchange on the PMR time scale at or above room temperature, whereas "weakly" bridged systems, such as $-Cl$ or $-I$, undergo rapid exchange even at low temperatures. It has also been reported that the strength of the bridging groups determines the order of the reaction. When one group is weakly bridging, the reaction is first order in the more strongly bridged species and when both are strong, the reaction has a higher order.

The apparent difference in exchange rates between bridge (X) and terminal (R) groups suggests that some mechanism of exchange is involved which allows transfer of R groups without exchange of X groups. Such a mechanism could be similar to (b) described above. This has not been established, however, and must remain speculative until more data are available.

Finally, work has been done on the exchange of alkyl groups between R_2AlX and R_2AlY molecules in the presence of donor base molecules (51, 132, 133). It has been postulated that, depending on the base involved and the nature of the aluminum derivatives, two dominant pathways for exchange are possible. These are the same as those mentioned in Section IV,A,3 and will not be discussed further here.

Magnuson and Stucky have recently reported both the crystal structure and preliminary PMR data on an alkyl–amino mixed-bridge system (80). Alkyl exchange, which can be stopped at $-50°C$, occurs between bridge and terminal positions. This is of particular interest since the Al—N—Al bridge is very difficult to break, implying that exchange proceeds via a single bridged transition state

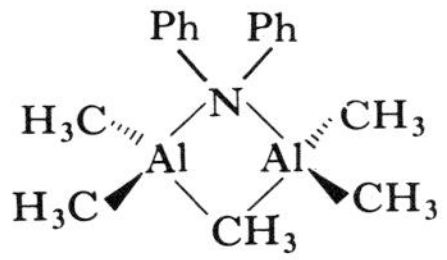

(X)

with exchange occurring by rotation about the Al—N bond. This mechanism has not been established experimentally, but if it is borne out this will represent the first case in which a singly bridged exchange reaction is clearly established.

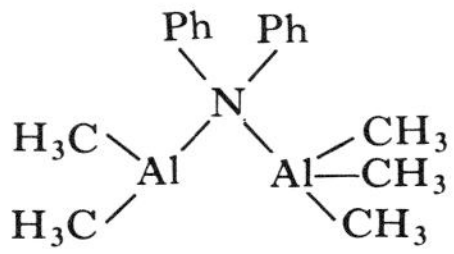

(XI)

VI

REACTIONS OF THE MR$_4^-$ GROUP

Reactions between a Group III alkyl, MR$_3$, and a Group I (or II) metal alkyl, M'R, proceed to completion as follows:

$$M'R + MR_3 \rightarrow M'MR_4 \tag{26}$$

with the formation of the tetraalkylanion of the Group III element (22). The crystal structure of LiAl(C$_2$H$_5$)$_4$ has been determined (42) and may be interpreted either in terms of Li$^+$ and Al(C$_2$H$_5$)$_4^-$ ion or a covalently bonded structure possessing two Li—C—Al bridge bonds. Infrared studies on LiAl(CH$_3$)$_4$ in the solid state and in weakly interacting solvents, such as cyclopentane or diethyl ether, indicate the existence of Li—C bonds at least on the time scale of the infrared experiment. This was shown by a marked ^{6}Li isotope effect (112). PMR studies are sketchy in these solvents but have been extensively carried out in tetrahydrofuran solution (47, 112). These investigations have shown an equilibrium between contact and solvent-separated ion pairs. Through ^{27}Al nuclear resonance, Gutowsky and Gore have obtained equilibrium constants for the "exchange" between solvent-separated and solvent-caged ion pairs in the LiAl(CH$_3$)$_4$ and NaAl(CH$_3$)$_4$ systems (47). These values are 100 and 10, respectively. This equilibrium has also been observed through proton resonance studies where the degree of ion interaction can be approximated by the amount of quadrupolar interaction observed. For complete dissociation, as in dimethoxyethane solvent, a six-line proton spectrum is observed due to the ^{27}Al–^{1}H coupling (111). When the ions are closely associated, this spectrum collapses due to the quadrupolar interaction of the ^{27}Al nucleus with the unsymmetrical field gradient. The extent of this collapse has been calculated and interpreted in terms of an associated-ion-pair–solvent-separated-ion-pair equilibrium (47).

Studies on other exchange reactions of the tetraalkyl anions are fairly limited and are summarized in Table V. Williams and Brown have provided the only quantitative data and have proposed mechanisms for both lithium and alkyl exchange (153). They have suggested that the lithium exchange between (LiR)$_4$ and LiMR$_4$ proceeds with the rate-determining step

TABLE V

EXCHANGE REACTIONS INVOLVING ORGANOMETALLIC ANIONS

System	Lithium exchange relative rate E_a^* (kcal/mole)	Alkyl exchange relative rate E_a^* (kcal/mole)
LiCH$_3$–LiB(CH$_3$)$_4$[a]	Intermediate, 11.9	Slow
LiCH$_3$–LiAl(CH$_3$)$_4$[a]	Intermediate, 10.5	Slow
LiAl(CH$_3$)$_4$–LiB(CH$_3$)$_4$[a]	Fast	Slow
Al(CH$_3$)$_3$–LiAl(CH$_3$)$_4$[a,b]	—	Fast, 9.1
Al(CH$_3$)$_3$–LiB(CH$_3$)$_4$[a]	—	Slow
LiC$_2$H$_5$–LiAl(C$_2$H$_5$)$_4$[a]	Fast	Slow
LiC$_2$H$_5$–LiGa(C$_2$H$_5$)$_4$[a]	Fast	Slow
LiAl(C$_2$H$_5$)$_4$–LiGa(C$_2$H$_5$)$_4$[a]	Fast	Slow
Al(CH$_3$)$_3$–LiGa(CH$_3$)$_4$[b]	—	Fast
Al(CH$_3$)$_3$–LiIn(CH$_3$)$_4$[b]	—	Fast
Al(CH$_3$)$_3$–LiTl(CH$_3$)$_4$[b]	—	Slow

[a] Williams and Brown (153).
[b] Oliver and Wilkie (112).

being the dissociation of the lithium tetramer to the dimer as described in Section II. This is followed by

$$(LiCH_3)_2 + Li \parallel M(CH_3)_4 \rightleftharpoons \left[Li \begin{array}{c} CH_3 \\ Li \\ \\ Li \\ CH_3 \end{array} \right]^+ M(CH_3)_4^- \qquad (27)$$

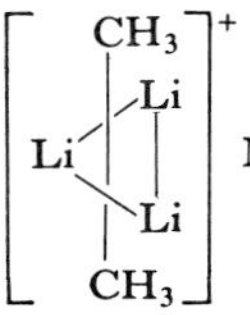

For lithium exchange between Li $M(CH_3)_4$ and Li $M'(CH_3)_4$ it was suggested that a direct bimolecular reaction occurred, but insufficient data were available to test this.

The exchange of alkyl groups between anions MR_4^- and $M'R_4^-$ does not proceed readily, but reaction between $M(CH_3)_3$ and $M(CH_3)_4^-$ has been observed in several instances (110, 112, 153). On the basis of their studies of the $Al(CH_3)_3$–$LiAl(CH_3)_4$ systems, Williams and Brown suggested that the rate-determining step is the formation of solvent-separated ion pairs,

$$LiAl(CH_3)_4 \rightleftharpoons Li^+ \parallel Al(CH_3)_4^- \qquad (28)$$

followed by a rapid reaction

$$Li^+ \parallel Al(CH_3)_4^- + Al(\overset{*}{C}H_3)_3 \rightleftharpoons Li^+ \parallel Al(CH_3)_3\overset{*}{C}H_3^- + Al(\overset{*}{C}H_3)_2CH_3 \qquad (29)$$

It was suggested that this must proceed through the single bridged transition

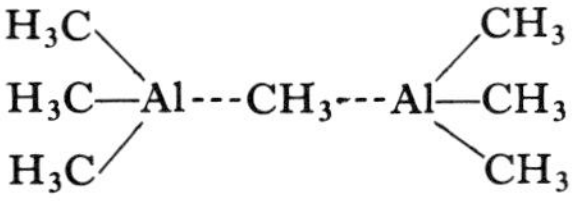

$$\begin{array}{c} H_3C \\ H_3C-Al\cdots CH_3\cdots Al-CH_3 \\ H_3C \end{array} \begin{array}{c} CH_3 \\ \\ CH_3 \end{array}$$

(XII)

state shown in (XII). It has been shown that $LiB(CH_3)_4$ does not exchange rapidly with $Al(CH_3)_3$ but that the reverse reaction takes place quantitatively (110). Therefore, the lack of exchange observed in both this case and that of

$$LiAl(CH_3)_4 + B(CH_3) \rightarrow LiB(CH_3)_4 + Al(CH_3)_3 \qquad (30)$$

$Tl(CH_3)_4^-$ may be due to the high stability of the complex formed. Further studies in this area are clearly necessary to provide details concerning alkyl group exchange and relative stabilities of compounds.

VII

ADDITION COMPOUNDS

A. BF_3–Base and BR_3–Base Systems

A large amount of work has been done on exchange involving addition compounds of Group III alkyl derivatives, but before treating these it would be useful to discuss studies carried out on BF_3 adducts since these studies lay the groundwork for most of the later discussion.

In systems containing BF_3 and two ethers, two routes for exchange have been proposed (*119–121*). The first of these, predominating in solutions containing a large excess of uncomplexed ether, is a nucleophilic displacement described by Eq. (31).

$$BF_3 \cdot Base(1) + Base(2) \rightleftarrows BF_3 \cdot Base(2) + Base(1) \qquad (31)$$

In solutions containing little or no free ether, a different bimolecular exchange appears to be operative:

$$BF_3 \cdot Base(1) + *BF_3 \cdot Base(2) \rightleftarrows BF_3 \cdot Base(2) + *BF_3 \cdot Base(1) \qquad (32)$$

Both of these reactions proceed with an activation energy less than the dissociation energy of the BF_3 adducts. In solutions containing an intermediate amount of free ether, a complex rate law appears to describe the exchange, because one must also include terms for the relative stabilities of the respective BF_3-Base complexes to describe adequately the rate of exchange observed in the system. In these studies, no mention was made of the steric requirements of the transition state for exchange by either mechanism, and no mention was made of the possibility of the exchange reaction occurring through a dissociation process.

Diehl (*35, 36*) has observed the exchange reactions in BF_3–alcohol systems by ^{19}F resonance. The exchange in these systems, where the ratio of BF_3 to alcohol is less than one, was attributed to the exchange of alcohol in a hydrogen bonded dicomplex with structure (XIII).

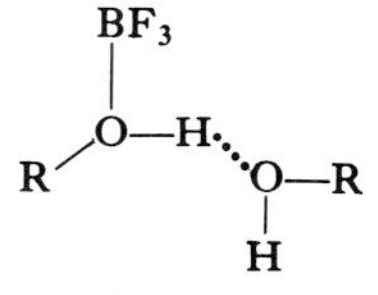

$$
\begin{array}{c}
BF_3 \\
| \\
\underset{R}{\diagdown} O\text{---}H\cdots \\
 \cdots O\text{---}R \\
 | \\
 H
\end{array}
$$

(XIII)

The activation energy of 7.3 kcal/mole remains constant for several concentration ratios, but the concentration dependence of the lifetime of the complex species shows a sharp change in value at $BF_3/ROH \simeq 0.5$. This can probably be explained by the fact that the dicomplex above is a completely solvated form of the BF_3-ROH complex. When $BF_3/ROH > 0.5$, Eq. (33) will describe the exchange (43).

$$BF_3-ROH(1) + ROH(2) \rightleftharpoons BF_3-ROH(2) + ROH(1) \tag{33}$$

Therefore, the lifetime of the BF_3-ROH complex (τ_{AB}) will be related directly to the free alcohol present.

$$\frac{1}{\tau_{AB}} = k[ROH] \tag{34}$$

When $BF_3/ROH < 0.5$, the species in solution will include the dialcohol complex and exchange will be complicated by this intramolecular path.

$$[BF_3-ROH(1)][ROH(2)] \rightleftharpoons [BF_3-ROH(2)][ROH(1)] \tag{35}$$

Thus, for this reaction, τ_{AB} will be independent of the concentration of alcohol. Therefore, while there should be a drastic change in concentration dependence, the activation energy should be changed only slightly because the basic displacement reaction of ROH(1) by ROH(2) is still the same.

Studies have also been carried out in systems containing excess BF_3 (17, 18). The results (18) show that when the base is dimethyl ether, anisole, tetrahydrofuran, or pyridine, the exchange of BF_3 is rapid and probably proceeds through an electrophilic displacement reaction in which the excess BF_3 attacks the complex. These reactions all have activation energies of less than 10 kcal/mole, eliminating the possibility of a dissociation process. The data available, however, do not allow a complete evaluation of the reaction mechanism. Studies carried out on BF_3–methanol complexes by ^{19}F NMR (17) indicate displacement reactions having an activation energy of 5.3 kcal/mole.

The only evidence for a dissociation process in BF_3 adduct systems appears for $BF_3 \cdot N(C_2H_5)_3$, in which slow exchange was observed by Brownstein (18). The difference in mechanism for this system is thought to lie in the steric nature of the triethylamine ligand (13). The steric crowding around the amine in the complex could prevent formation of the five-coordinate transition state necessary for bimolecular displacement. Further support for this is found through the observation of ^{11}B–^{1}H coupling in trimethyl- and triethylamine adducts of BCl_3 and BBr_3 (89, 124). This observation shows that the B—N bond must have a reasonably long lifetime so that a low-energy pathway for exchange is not present.

More recent studies by Cowley and Mills (29) postulate that in a solution of $B(CH_3)_3 \cdot N(CH_3)_3$ with either excess $B(CH_3)_3$ or $N(CH_3)_3$, group exchange proceeds through a dissociative mechanism. As will be seen below, the dissociative mechanism has not been found for any of the gallium or indium systems in the presence of excess base. Its appearance in the $B(CH_3)_3$ system could be attributed to the amount of steric crowding around the boron atom, making the transition state for a bimolecular mechanism too crowded to be stable.

B. AlR₃–Base Systems

Although aluminum alkyls have been extensively studied, as we have already seen, little quantitative information has appeared in the literature regarding their exchange with addition compounds. Mole (96) has suggested that the single resonance due to $(CH_3)_3N$ protons in the $(CH_3)_3Al$–NMe_3 system with excess $N(CH_3)_3$ present can be explained by rapid exchange of $N(CH_3)_3$. It is suggested that the probable mechanism for exchange would be a nucleophilic displacement of the $(CH_3)_3N$ in the adduct. This type of exchange must also occur in ether solutions of aluminum alkyls, but only qualitative evidence is available to support this mechanism (95).

Several other qualitative studies have been carried out. Takashi has shown that a number of bases undergo rapid exchange with $Al(C_2H_5)_3$ when excess base is present (136). These are listed in Table VI. Other reports have also mentioned fast exchange of excess base with AlR_3 addition compounds (93, 95, 96).

It is anticipated that all of the complex reactions observed in the boron system and in the studies on indium and gallium derivatives to be discussed below will be found in alkylaluminum systems. There also will be some additional complications which arise from the fact that the AlR_3 moiety often occurs in a dimeric form which will alter the energetics of the reaction and may in some instances govern the mechanism of exchange as observed for alkyl group transfer.

TABLE VI

EXCHANGE OF BASE WITH TRIALKYLALUMINUM ADDITION COMPOUNDS

Adduct	Rate of exchange with excess		References
	AlR_3	Base	
$Al(C_2H_5) \cdot N(C_3H_5)_3$	Slow	Fast	136
$Al(C_2H_5)_3 \cdot NC_5H_5$	Slow	Fast	136
$Al(C_2H_5)_3 \cdot O(C_2H_5)_2$	Fast	Fast	136
$Al(C_2H_5)_3 \cdot OC_4H_8$	Fast	Fast	95, 136
$Al(C_2H_5)_3 \cdot Sb(C_3H_5)_3$	Fast	Fast	136
$Ph_3Al \cdot NCPh$	Slow	Slow	95
$Ph_3Al \cdot O(C_2H_5)_2$	—	Fast	95
$Ph_3Al \cdot N(C_2H_5)_3$	Slow	Slow	95

C. GaR₃–Base and InR₃–Base Systems

The results of exchange studies on addition compounds of alkylgallium and -indium derivatives are summarized in Table VII, which shows the

TABLE VII

PARAMETERS FOR EXCHANGE OF GaR_3 AND InR_3 WITH THEIR ADDITION COMPOUNDS

System	E_a^* (kcal/mole)	S (e.u.)	Mechanism	References
$Ga(CH_3)_3 \cdot N(CH_3)_3 + Ga(CH_3)_3$	23.8	$+30$ (293° K)	Dissociation	30
$Ga(CH_3)_3 \cdot N(CH_3)_2H + Ga(CH_3)_3$	19.1	$+12$ (283° K)	Dissociation	31
$Ga(CH_3)_3 \cdot N(CH_3)H_2 + Ga(CH_3)_3$	9.9	-13 (264° K)	Bimolecular	31
$Ga(CH_3)_3 \cdot NH_3 + Ga(CH_3)_3$	8.5	-15 (233° K)	Bimolecular	31
$Ga(CH_3)_3 \cdot P(CH_3)_3 + Ga(CH_3)_3$	16.8	$+12$ (251° K)	Dissociation	106
$Ga(CH_3)_3 \cdot NC_5H_5 + Ga(CH_3)_3$	8	-8 (196° K)	Bimolecular	15
$In(CH_3)_3 \cdot N(CH_3)_3 + In(CH_3)_3$	19.7	$+21$ (256° K)	Dissociation	56
$In(CH_3)_3 \cdot N(CH_3)_2H + In(CH_3)_3$	14.9	$+5$ (256° K)	Dissociation	56
$In(CH_3) \cdot N(CH_3)H_2 + In(CH_3)_3$	11.6	-1 (236° K)	Bimolecular	56

activation parameters and mechanisms for the various reactions. All of these systems contain an excess of the free organometallic moiety. From these results it is quite clear that two processes are operative for the exchange reactions. These are

(1) $M(CH_3)_3 \cdot Base \rightleftharpoons M(CH_3)_3 + Base$ (36)

and

(2) $M(CH_3)_3 \cdot Base + M^*(CH_3)_3 \rightleftharpoons M^*(CH_3)_3 \cdot Base + M(CH_3)_3$ (37)

In case (1), the rate is determined by dissociation of the addition compound. In case (2) the limiting factor is the formation of the transition state caused by attack on the adduct by a molecule of free $M(CH_3)_3$.

It has been suggested, and these data confirm, that the factor governing the mechanism of exchange for these systems is the steric interaction involved in the formation of the bimolecular transition state. If this interaction is too great, the exchange will proceed through the dissociative mechanism. For the systems studied for $Ga(CH_3)_3$ and $In(CH_3)_3$, in no instance did exchange with excess base proceed through dissociation. This is in contrast to the system involving $B(CH_3)_3$ and $N(CH_3)_3$ mentioned earlier, and is also probably due to steric effects.

The geometry required for the bimolecular transition state is probably best illustrated by the recent work of Stucky (6), which shows the AlR_3 group remaining almost planar when an addition compound is formed.

The studies of Schmidbaur et al. are also of interest in considering the possible mechanism of reaction of addition compounds (123). In this work

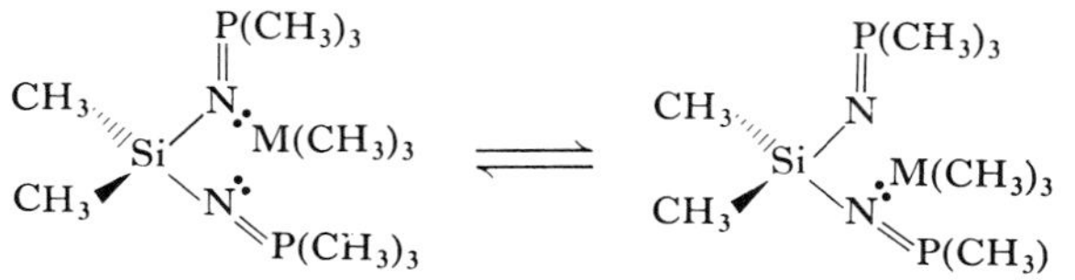

it was shown that reaction (38) occurs and can be stopped at $-60°\,C$ when $M = Ga$. Analysis of the data indicates that this process proceeds with an activation energy of 12.8 kcal/mole. When $M = In$ the reaction either proceeds much more rapidly or the indium is pentacoordinate, as only one set of PMR resonances are observed at all temperatures.

VIII

SUMMARY AND CONCLUSIONS

From the foregoing material it is quite clear that transfer of an organic group from one organometallic compound to another proceeds through a bridged transition state. The rate of this reaction is determined by one of two factors:

(a) dissociation of a dimer or higher polymer
(b) formation of the bridged transition state

The first of these is found to control reaction rates for the strongly bridged derivatives of lithium and aluminum. The second predominates in the exchange of Group II alkyls with one another or in the exchange of the monomeric Group III alkyls with the Group II derivatives. In these studies the relative rates of reaction indicate that bridge formation occurs in the order $Zn > Cd \gg Hg$. Studies on different organic substituents show that unsaturated and cyclopropyl derivatives normally form more stable bridges than do alkyl groups. This in fact has been shown to cause the rate-determining step to be shifted from (b) to (a), where the bridging becomes so strong that stable dimeric species are present. In other instances, however, this increased tendency to form bridge bonds accelerates the reaction, as in the case of the vinyl moiety.

Among the Group III alkyl addition compounds three mechanisms have been observed:

(a) dissociation
(b) electrophilic attack of the free Group III alkyl
(c) nucleophilic attack of free base

The factors governing which mechanism is operative are the dissociation energy of the adduct involved and the steric effects of the organic substituents.

The studies on X-bridged systems ($X = Cl$, Br, I, NR, etc.) indicate that they generally are slower and go through the same paths described for other bridged systems. There is, however, growing evidence that some of these may also go through a reaction involving the dimeric or polymeric species without prior dissociation.

Finally, it should be pointed out that there are many areas which have not been adequately studied and have not been included here. These include studies on the interaction of organometallic compounds with transition metal derivatives, such as those described by Brunner *et al.* in which it

was shown that the adducts undergo rapid exchange with $Al(CH_3)_3$ *(21)*. Other studies also indicate that rapid exchange may occur between Group III and Group IV derivatives of the type MR_nX_{3-n} and $M'R_nX_{4-n}$ *(7)*. This area should provide further insight into the mechanisms of organometallic exchange reactions, but has not been developed sufficiently to warrant discussion at this time.

Acknowledgments

I would like to express my thanks to my former and current graduate students who have contributed both to the work reported in this manuscript and to discussions concerning this topic. I would particularly like to express my thanks to Dr. J. B. De Roos who initiated work in this area in my research group and to Mr. K. L. Henold for aiding in the preparation of the manuscript. Finally, I would like to express my thanks to the National Science Foundation for support of my research in this area.

References

1. Abraham, M. H., and Hill, J. A., *J. Organometal. Chem. (Amsterdam)* **7**, 23 (1967).
2. Abraham, M. H., and Rolfe, P. H., *J. Organometal. Chem. (Amsterdam)* **8**, 395 (1967).
3. Allen, G., Bruce, J. M., Farren, D. W., and Hutchinson, F. G., *J. Chem. Soc., B (London)* p. 799 (1966).
4. Ashby, E. C., *Quart. Rev. (London)* **21**, 259 (1967).
5. Ashby, E. C., Sanders, R., and Carter, J., *Chem. Commun.* p. 997 (1967).
6. Atwood, J. L., and Stucky, G. D., *J. Am. Chem. Soc.* **89**, 5362 (1967).
7. Badin, E. J., *J. Phys. Chem.* **63**, 1791 (1959).
8. Beachley, O. T., and Coates, G. E., *J. Chem. Soc.* p. 3241 (1965).
9. Bell, N. A., Coates, G. E., and Emsley, J. W., *J. Chem. Soc., A* p. 49 (1966).
10. Bell, N. A., Coates, G. E., and Emsley, J. W., *J. Chem. Soc., A* p. 1360 (1966).
11. Boersma, J., and Noltes, J. G., *J. Organometal. Chem. (Amsterdam)* **8**, 551 (1967).
12. Boersma, J., and Noltes, J. G., *J. Organometal. Chem. (Amsterdam)* **13**, 291 (1968).
13. Brown, H. C., Stehle, P. F., and Tierney, P. A., *J. Am. Chem. Soc.* **79**, 2020 (1957).
14. Brown, T. L., *Advan. Organometal. Chem.* **3**, 365–393 (1965).

15. Brown, T. L., *Accounts Chem. Res.* **1**, 23 (1968).
16. Brown, T. L., Seitz, L. M., and Kimura, B. Y., *J. Am. Chem. Soc.* **90**, 3245 (1968).
17. Brownstein, S., and Paasivirta, J., *J. Am. Chem. Soc.* **87**, 3593 (1965).
18. Brownstein, S., Eastham, A. M., and Latremouille, G. A., *J. Phys. Chem.* **67**, 1028 (1963).
19. Brownstein, S., Smith, B. C., Erlich, G., and Laubengayer, A. W., *J. Am. Chem. Soc.* **82**, 1000 (1960).
20. Bruce, J. M., Cutsforth, B. C., Farren, D. W., Hutchinson, F. G., Rabagliati, F. M., and Reed, D. R., *J. Chem. Soc.*, *B* p. 1020 (1966).
21. Brunner, H., Wailes, P. C., and Kaesz, H. D., *Inorg. Nucl. Chem. Letters* **1**, 125 (1965).
22. Coates, G. E., *Record Chem. Progr.* (*Kresse-Hooker Sci. Lib.*) **28**, 3 (1967).
23. Coates, G. E., and Fishwick, A. H., *J. Chem. Soc.*, *A* p. 477 (1968).
24. Coates, G. E., and Fishwick, A. H., *J. Chem. Soc.*, *A* p. 635 (1968).
25. Coates, G. E., and Fishwick, A. H., *J. Chem. Soc.*, *A* p. 640 (1968).
26. Coates, G. E., and Mukherjee, R. N., *J. Chem. Soc.* p. 229 (1963).
27. Coates, G. E., and Roberts, P. D., *J. Chem. Soc.*, *A* p. 2651 (1968).
28. Coates, G. E., and Wade, K., *in* "Organometallic Compounds," 3rd ed., Vol. I, Chapters 1, 2, and 3. Methuen, London, 1967.
29. Cowley, A. H., and Mills, J. L., private communication (1968).
30. De Roos, J. B., and Oliver, J. P., *Inorg. Chem.* **4**, 1741 (1965).
31. De Roos, J. B., and Oliver, J. P., *J. Am. Chem. Soc.* **89**, 3970 (1967).
32. Dessy, R. E., *J. Am. Chem. Soc.* **82**, 1580 (1960).
33. Dessy, R. E., Kaplan, F., Coe, G. R., and Salinger, R. M., *J. Am. Chem. Soc.* **85**, 1191 (1963).
34. Dessy, R. E., and Kitching, W., *Advan. Organometal. Chem.* **4**, 267–351 (1966).
35. Diehl, P., *Helv. Phys. Acta* **31**, 686 (1958).
36. Diehl, P., and Ogg, R. A., Jr., *Chem. Coord. Compounds Symp.*, *Rome*, 1957 pp. 468–475. Pergamon Press, Oxford, 1958.
37. Emsley, J. W., Feeney, J., and Sutcliff, L. H., "High Resolution Nuclear Magnetic Resonance Spectroscopy," Chapter 9. Pergamon Press, Oxford, 1965.
38. Evans, D. F., and Khan, M. S., *J. Chem. Soc.*, *A* p. 1643 (1967).
39. Evans, D. F., and Khan, M. S., *J. Chem. Soc.*, *A* p. 1648 (1967).
40. Fetter, N. R., *Organometal. Chem. Rev.* **3**, 1 (1968).
41. Ford, D. N., Wells, P. R., and Lauterbur, P. C., *Chem. Commun.* p. 616 (1967).
42. Gerteis, R. L., Dickerson, R. E., and Brown, T. L., *Inorg. Chem.* **3**, 872 (1964).
43. Gillespie, R. J., and Hartman, J. S., *Can. J. Chem.* **45**, 2243 (1967).
44. Glick, R. E., and Zwickel, A., *J. Inorg. & Nucl. Chem.* **16**, 149 (1960).
45. Greenwood, N. N., *Advan. Inorg. Chem. Radiochem.* **5**, 91 (1963).
46. Groenewege, M. P., Smidt, J., and DeVries, H., *J. Am. Chem. Soc.* **82**, 4425 (1960).
47. Gutowsky, H. S., and Gore, E. S., private communication (1968).
48. Ham, N. S., Jeffery, E. A., Mole, T., Saunders, J. K., and Stuart, S. N., *J. Organometal. Chem.* (*Amsterdam*) **8**, P7 (1967).
49. Ham, N. S., Jeffery, E. A., Mole, T., and Stuart, S. N., *Chem. Commun.* p. 254 (1967).
50. Ham, N. S., Jeffery, E. A., Mole, T., and Saunders, J. K., *Australian J. Chem.* **20**, 2641 (1967).
51. Ham, N. S., Jeffery, E. A., Mole, T., and Saunders, J. K., *Australian J. Chem.* **21**, 659 (1968).
52. Hartwell, G. E., and Brown, T. L., *J. Am. Chem. Soc.* **88**, 4625 (1966).
53. Hata, G., *Chem. Commun.* p. 7 (1968).

54. Hatada, K., and Yuki, H., *Tetrahedron Letters* p. 5227 (1967).
55. Hatton, J. V., Schneider, G. W., and Siebrand, W., *J. Chem. Phys.* **39**, 1330 (1963).
56. Henold, K. L., and Oliver, J. P., *Inorg. Chem.* **7**, 950 (1968).
57. Henold, K. L., Soulati, J., and Oliver, J. P., *J. Am. Chem. Soc.* **91**, 3171 (1969).
58. Henrickson, C. H., and Eyman, D. P., *Inorg. Chem.* **6**, 1461 (1967).
59. Henrickson, C. H., Duffy, D., and Eyman, D. P., *Inorg. Chem.* **7**, 1047 (1968).
60. Hoffmann, E. G., *Bull. Soc. Chim. France* p. 1467 (1963).
61. Hoffmann, E. G., *Trans. Faraday Soc.* **58**, 642 (1962).
62. Hoffmann, E. G., *Z. Elektrochem.* **64**, 616 (1960).
63. Hoffmann, E. G., *Z. Elektrochem.* **64**, 144 (1960).
64. House, H. O., Lathan, R. A., and Whitesides, G. M., *J. Org. Chem.* **32**, 2481 (1967).
65. Jeffery, E. A., and Mole, T., *Australian J. Chem.* **21**, 1497 (1968).
66. Jeffery, E. A., and Mole, T., *Australian J. Chem.* **21**, 1187 (1968).
67. Jeffery, E. A., Mole, T., and Saunders, J. K., *Australian J. Chem.* **21**, 649 (1968).
68. Jeffery, E. A., Mole, T., and Saunders, J. K., *Australian J. Chem.* **21**, 137 (1968).
69. Jeffery, E. A., and Mole, T., *J. Organometal. Chem. (Amsterdam)* **11**, 393 (1968).
70. Jeffery, E. A., Mole, T., and Saunders, J. K., *Chem. Commun.* p. 696 (1967).
71. Johnson, C. S., Jr., *Advan. Magnetic Resonance* **1**, 33 (1965).
72. Kenny, M. E., and Laubengayer, A. W., *J. Am. Chem. Soc.* **76**, 4839 (1954).
73. Köster, R., and Binger, P., *Advan. Inorg. Chem. Radiochem.* **7**, 263 (1965).
74. Köster, R., and Bruno, G., *Ann. Chem.* **629**, 89 (1960).
75. Kurosawa, H., and Okawara, R., *Inorg. Nucl. Chem. Letters* **3**, 21 (1967).
76. Kurosawa, H., Tanaka, M., and Okawara, R., *J. Organometal. Chem. (Amsterdam)* **12**, 241 (1968).
77. Laubengayer, A. W., and Gilliam, W. F., *J. Am. Chem. Soc.* **63**, 477 (1941).
78. Lewis, P. H., and Rundle, R. E., *J. Chem. Phys.* **21**, 986 (1953).
79. Lloyd, J. E., and Wade, K., *J. Chem. Soc.* p. 2662 (1965).
80. Magnuson, V. R., and Stucky, G., *J. Am. Chem. Soc.* **90** 3269 (1968).
81. Maher, J. P. and Evans, D. F., *J. Chem. Soc.* p. 637 (1965).
82. Maher, J. P., and Evans, D. F., *J. Chem. Soc.* p. 5534 (1963).
83. Maher, J. P., and Evans, D. F., *Proc. Chem. Soc.* p. 176 (1963).
84. Maher, J. P., and Evans, D. F., *Proc. Chem. Soc.* p. 208 (1961).
85. Malone, J. F., and McDonald, W. S., *Chem. Commun.* p. 444 (1967).
86. Maslowsky, E., and Nakamoto, K., *Chem. Commun.* p. 257 (1968).
87. McCoy, R., and Allred, A. L., *J. Am. Chem. Soc.* **84**, 912 (1962).
88. McKeever, L. D., Waack, R., Doran, M. A., and Baker, E. B., *J. Am. Chem. Soc.* **90**, 3244 (1968).
89. Miller, J. M., and Onyszchuk, M., *Can. J. Chem.* **42**, 1518 (1964).
90. Moedritzer, K., *Advan. Organometal. Chem.* **6**, 171–262 (1968).
91. Mole, T., *Australian J. Chem.* **19**, 381 (1966).
92. Mole, T., *Australian J. Chem.* **19**, 373 (1966).
93. Mole, T., *Australian J. Chem.* **18**, 1183 (1965).
94. Mole, T., *Australian J. Chem.* **17**, 1050 (1964).
95. Mole, T., *Chem. & Ind. (London)* p. 281 (1964).
96. Mole, T., *Australian J. Chem.* **16**, 801 (1963).
97. Mole, T., and Surtees, J. R., *Australian J. Chem.* **17**, 1229 (1964).
98. Mole, T., and Surtees, J. R., *Australian J. Chem.* **17**, 310 (1964).
99. Mole, T., and Surtees, J. R., *Chem. & Ind. (London)* p. 1727 (1963).
100. Moy, D., Emerson, M. T., and Oliver, J. P., *J. Am. Chem. Soc.* **86**, 371 (1964).

101. Muller, N., and Otermat, A. L., *Inorg. Chem.* **2**, 1075 (1963).
102. Muller, N., and Otermat, A. L., *Inorg. Chem.* **4**, 296 (1965).
103. Muller, N., and Pritchard, D. E., *J. Am. Chem. Soc.* **82**, 248 (1960).
104. Murell, L. L., and Brown, T. L., *J. Organometal. Chem. (Amsterdam)* **13**, 301 (1968).
104a. Nesmeyanov, A.N., Fredorov, L.A., Matevikova, R.B., Fedin, E.I., and Kochetova, N.S., *Chem. Commun.* p. 105 (1969).
105. Oliver, J. P., and De Roos, J. B., unpublished observations (1966).
106. Oliver, J. P., and Henold, K. L., unpublished observations (1968).
107. Oliver, J. P., and Sanders, D. A., unpublished observations (1968).
107a. Oliver J. P., and Schaaf, T., *J. Am. Chem. Soc.* **91**, 4327 (1969).
108. Oliver, J. P., and Soulati, J., unpublished observations (1968).
109. Oliver, J. P., and Visser, H. D., unpublished observations (1968).
110. Oliver, J. P., and Weibel, T., unpublished observations (1968).
111. Oliver, J. P., and Wilkie, C. A., *J. Am. Chem. Soc.* **89**, 163 (1967).
112. Oliver, J. P., and Wilkie, C. A., unpublished observations (1967).
113. Noltes, J. G., and Boersma, J., *J. Organometal. Chem. (Amsterdam)* **12**, 425 (1968).
114. Pauling, L., and Laubengayer, A. W., *J. Am. Chem. Soc.* **63**, 480 (1941).
115. Poole, C. P., Jr., Swift, H. E., and Itzel, J. F., Jr., *J. Chem. Phys.* **42**, 2576 (1965).
116. Ramey, K. C., O'Brien, J. F., Hasegawa, I., and Borchert, A. E., *J. Phys. Chem.* **69**, 3418 (1965).
117. Rappoport, Z., Sleezer, P. D., Winstein, S., and Young, W. G., *Tetrahedron Letters* p. 3719 (1965).
118. Redfield, A. G., *Advan. Magnetic Resonance* **1**, 1 (1965).
119. Rutenburg, A. C., Palko, A. A., and Drury, J. S., *J. Am. Chem. Soc.* **85**, 2702 (1963).
120. Rutenburg, A. C., Palko, A. A., and Drury, J. S., *J. Phys. Chem.* **68**, 976 (1964).
121. Rutenburg, A. C., and Palko, A. A., *J. Phys. Chem.* **69**, 527 (1965).
122. Sanders, D. A., and Oliver, J. P., *J. Am. Chem. Soc.* **90**, 5910 (1968).
123. Schmidbaur, H., Klein, H. F., and Eiglmeier, K., *Angew. Chem.* **79**, 821 (1967).
124. Schuster, R. E., Fratiello, A., and Onak, T. P., *Chem. Commun.* p. 1038 (1967).
125. Seitz, L. M., and Brown, T. L., *J. Am. Chem. Soc.* **88**, 4140 (1966).
126. Seitz, L. M., and Brown, T. L., *J. Am. Chem. Soc.* **89**, 1602 (1967).
127. Seitz, L. M., and Brown, T. L., *J. Am. Chem. Soc.* **89**, 1607 (1967).
128. Seitz, L. M., and Hall, S. D., *J. Organometal. Chem. (Amsterdam)* **15**, 7 (1968).
129. Sheverdina, N. I., Paleeva, I. E., and Kocheshkov, K. A., *Izv. Akad. Nauk SSSR, Ser. Khim.* p. 587 (1967).
130. Simpson, R. B., *J. Chem. Phys.* **46**, 4775 (1967).
131. Smidt, J., Groenewege, M. P., and DeVries, H., *Rec. Trav. Chim.* **81**, 729 (1962).
132. Smith, C. A., and Wallbridge, M. G. H., *J. Chem. Soc., A* p. 7 (1967).
133. Smith, C. A., and Wallbridge, M. G. H., *Abstr. 3rd Intern. Symp. Organometal. Chem., Munich,* 1967 p. 346.
134. Stevens, L. G., and Oliver, J. P., *J. Inorg. & Nucl. Chem.* **24**, 953 (1962).
135. Tada, H., Yasuda, K., and Okawara, R., *Inorg. Nucl. Chem. Letters* **3**, 315 (1967).
136. Takashi, Y., *Bull. Chem. Soc. Japan* **40**, 612 (1967).
137. Takashi, Y., *Bull. Chem. Soc. Japan* **40**, 1001 (1967).
138. Taylor, M. J., *J. Chem. Soc., A.* p. 1462 (1966).
139. Thayer, J. S., and West, R., *Advan. Organometal. Chem.* **5**, 169–219 (1967).
140. Thiele, K. H., *Organometal. Chem. Rev.* **1**, 331 (1966).
141. Thiele, K. H., and Zdunneck, P., *J. Organometal. Chem. (Amsterdam)* **4**, 10 (1965).

142. Toppet, S., Slinckx, G., and Smets, G., *J. Organometal. Chem. (Amsterdam)* **9**, 205 (1967).
143. Vatton, J. V., *J. Chem. Phys.* **40**, 933 (1964).
144. Visser, H. D., and Oliver, J. P., *J. Am. Chem. Soc.* **90**, 3579 (1968).
145. Vranka, R. G., and Amma, E. L., *J. Am. Chem. Soc.* **89**, 3121 (1967).
146. Waack, R., Doran, M. A., and Baker, E. B., *Chem. Commun.* p. 1291 (1967).
147. Wakefield, B. J., *Organometal. Chem. Rev.* **1**, 131 (1966).
148. Walsh, A. D., *Trans. Faraday Soc.* **45**, 179 (1949).
148a. Weigert, F. J., Winokur, M., and Roberts, J. D., *J. Am. Chem. Soc.* **90**, 1566 (1968).
149. Wells, P. R., Kitching, W., and Henzell, R. F., *Tetrahedron Letters* p. 1029 (1964).
150. West, P., Purmort, J. I., and McKinley, S. V., *J. Am. Chem. Soc.* **90**, 798 (1968).
151. Wilke, G., Talk presented at the *Robert A. Welch Found. Conf. Chem. Res. IX. Organometal. Compounds, Houston, Texas,* 1965.
152. Williams, K. C., and Brown, T. L., *J. Am. Chem. Soc.* **88**, 5460 (1966).
153. Williams, K. C., and Brown, T. L., *J. Am. Chem. Soc.* **88**, 4134 (1966).
154. Witanowski, M., and Roberts, J. D., *J. Am. Chem. Soc.* **88**, 737 (1966).
155. Worrall, I. J., and Wallwork, S. C., *J. Chem. Soc.* p. 1816 (1965).
156. Yamamoto, O., *Bull. Chem. Soc. Japan* **37**, 1125 (1964).
157. Yamamoto, O., and Hayamizu, K., *J. Phys. Chem.* **72**, 822 (1968).
158. Yasuda, K., and Okawara, R., *Organometal. Chem. Rev.* **2**, 255 (1967).
159. Yasuda, K., and Okawara, R., *Inorg. Nucl. Chem. Letters* **3**, 135 (1967).
160. Ziegler, K., *Bull. Soc. Chim. France* p. 1456 (1963).
161. Ziegler, K., *in* "Organometallic Chemistry" (H. Zeiss, ed.), Chapter 5, p. 194. Reinhold, New York, 1960.

Mass Spectra of Metallocenes and Related Compounds

M. CAIS and M. S. LUPIN

Department of Chemistry, Technion-Israel Institute of Technology,
Haifa, Israel

I

INTRODUCTION

Mass spectrometry of organic molecules has become well established in recent years and both the recording of spectra and the fragmentation patterns of functional groups have been comprehensively discussed (*14, 19, 33, 126, 142*). In the last 2–3 years the study of the behavior of organometallic compounds in the mass spectrometer, previously a neglected field, has become much more widespread, and systematic investigations are now being carried out. Bruce (27) has reviewed the literature up to the beginning of 1967 and has dealt mainly with the mass spectra of metal carbonyl and related compounds of the transition metals.

This review is a survey of the mass spectral behavior of metallocenes and related π-bonded organometallic systems for which data are becoming available and detailed discussions of fragmentation patterns are being reported at a rapidly increasing rate. The effect of the metal on the fragmentation of the organic moiety is receiving considerable attention and the identification of metal-containing fragments is often facilitated by the isotope patterns of the metal. However, for metals having several naturally occurring isotopes of approximately the same relative abundance (e.g., ruthenium, molybdenum, tungsten, and palladium), the complicated patterns observed may make identification of fragment ions more difficult, especially if the loss of hydrogen atoms is suspected.

One of the inherent difficulties in the study of mass spectra of organometallic compounds is that decomposition, either thermally or by electron bombardment, causes contamination of the instrument, and repeated cleaning of the ion source is necessary (199). Also, as many organometallic compounds are not very volatile, the temperature of the inlet system is often high (200°–300° C) in order to create a sufficient pressure of gaseous molecules, and the possibility of thermal decomposition to give species not originally present in the sample must not be overlooked (114, 166). This is illustrated by the mass spectrum of the trimer $[C_5H_5CoCO]_3$, which corresponds to the tetranuclear complex, $(C_5H_5)_4Co_4(CO)_2$, probably formed by pyrolysis in the spectrometer (104).

Appearance potentials of molecular ions and fragment ions have been reported for some compounds (Section VII) and an estimation of heats of formation and of bond strengths has been attempted in several cases, notably by Winters and Kiser (199). These results must be treated cautiously, however, because the appearance potentials may include excess energy due to contributions from excited states.

Metallocenes and related compounds for which mass spectral data have been reported are given in Tables I–XII, but in many cases only the molecular ion is reported. Metal carbonyls, nitrosyls and their derivatives, and fluorocarbon complexes are not discussed in this review, but Table XIX summarizes the compounds for which mass spectral data have been reported since Bruce's review (27).

The spectra of many organometallic compounds of the main group elements have been measured but a discussion of these results does not fall within the scope of this review.

II

TITANIUM, ZIRCONIUM, AND HAFNIUM

Nesmeyanov *et al.* (*155*) have studied several cyclopentadienyl compounds of the type $C_5H_5Ti(OEt)_{3-n}Cl_n$ ($n=0$, 1, 2, 3), and the main features of the spectra are the appearance of the molecular ion, the fragment ions $[M-A]^+$, where $A=Cl$ or OEt, and a strong peak for $C_5H_5{}^+$. No peak is observed for $[C_5H_5Ti]^+$ when $n=0$, 1, or 2, but instead a peak corresponding to the ion $[C_5H_5TiO]^+$ is observed. Also, when $n=1$ or 2, the elimination of HCl is observed from the molecular ion, but in none of the spectra is any fragmentation of the cyclopentadienyl ring observed. Assuming that the formation of an $(M-C_5H_5)^+$ ion would increase with increase in ionic character of the titanium–cyclopentadienyl bond, the data indicate that the ionic character of the bond increases with a decrease in n, so that the most intense peak for the $[M-C_5H_5]^+$ fragment ion is observed in the spectrum of $C_5H_5Ti(OEt)_3$ ($n=0$).

The trimeric nature of $[(C_5H_5)_2TiCN]_3$ has been confirmed by mass spectrometry. The spectrum exhibits the molecular ion at m/e 612 and fragment ions corresponding to the loss of C_5H_5, CN, and $[(C_5H_5)_2TiCN]$ (*52*). The thiocyanate complex $[(C_5H_5)_2TiSCN]_3$ is more volatile than the cyanide analog and shows no peaks above m/e 236, $[(C_5H_5)_2TiSCN]^+$. Loss of both C_5H_5 and SCN is observed from this ion (*52*). The monomeric isocyanate complex $(C_5H_5)_2TiNCO$ exhibits a strong molecular ion and the expected fragment ions $[M-C_5H_5]^+$ and $(M-NCO)^+$, as well as loss of oxygen from the isocyanate group, giving the ion $[TiNC]^+$ m/e 74. The mass spectrum of the isocyanate complex also has weak peaks at masses higher than the molecular ion corresponding to $[(C_5H_5)_2TiNCOC_2H]^+$ and $[(C_5H_5)_2TiNCOC_3H_3]^+$ formed possibly by ion–molecule reactions (*52*) but more probably by thermal decomposition reactions on the probe.

The mass spectrum of dicyclopentadienyltitanium-α,α'-dipyridyl shows the molecular ion $[(C_5H_5)_2TiC_{10}H_8N_2]^+$ as well as the fragment ions $(M-C_5H_5)^+$, $TiC_5H_5{}^+$, $C_{10}H_8N_2{}^+$, and Ti^+. However, the peak due to the dipyridyl ion is very strong, possibly because of thermal decomposition

—a suggestion which is supported by the very low abundance of Ti^+ (*67*).
The mass spectra of $(C_5H_5)_2ZrCl_2$ and $[(C_5H_5)_2ZrCl]_2O$ have been measured (*171*). Fragments containing zirconium and/or chlorine are easily recognized because of characteristic isotopic distributions [$^{90}Zr(51\%)$, $^{91}Zr(11\%)$, $^{92}Zr(17\%)$, $^{94}Zr(17\%)$, $^{35}Cl(75\%)$, $^{37}Cl(25\%)$]. Fragmentation of $(C_5H_5)_2ZrCl_2$ occurs by either loss of chlorine or cyclopentadienyl radicals from the molecular ion, but no peak for $[C_5H_5Zr]^+$ is observed. A predominant feature of the spectrum is fragmentation of the cyclopentadienyl ring from ions containing both zirconium and chlorine, with the formation of ions containing the cyclopropenyl group, e.g., $[C_3H_3ZrCl_2]^+$, $[C_3H_3ZrCl]^+$, or containing acetylene or acetylide groups $[C_2H_2ZrCl_2]^+$, $[C_2HZrCl_2]^+$, and $[C_2HZrCl]^+$. The cyclopropenylium ion $(C_3H_3)^+$ exhibits a strong peak in this spectrum [30% (Relative Abundance)]. [$(C_5H_5)_2ZrCl]_2O$ does not show a peak for the molecular ion and only a weak peak for $[M - Cl]^+$, but a strong peak for $[M - C_5H_5]^+$. The spectrum exhibits several ions in which the Zr—O—Zr unit is present and the presence of two atoms of zirconium makes possible a number of novel structures for which no analogies are found in the spectrum of $C_5H_5ZrCl_2$. In a scheme of possible mechanisms to account for the fragment ions observed, structures are postulated in which a C_5H_4 unit is σ-bonded to one zirconium atom and π-bonded to the other, e.g., (I).

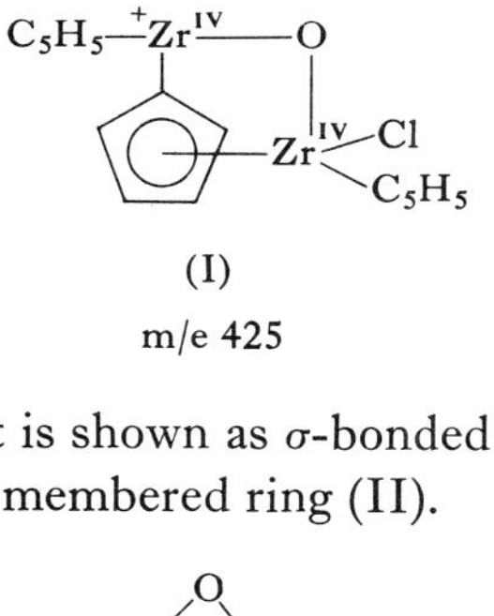

(I)

m/e 425

In other ions a C_5H_3 unit is shown as σ-bonded to both zirconium ions with the formation of a five-membered ring (II).

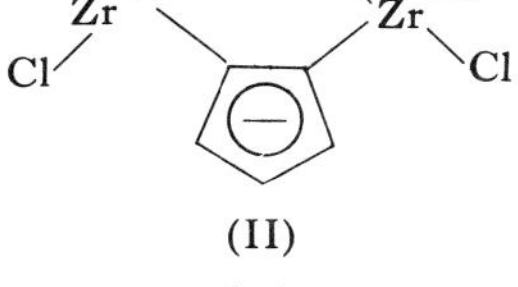

(II)

m/e 359

For several ions plausible structures require trivalent or divalent zirconium, e.g., $[C_5H_5ZrCl]^+$, $[(C_5H_5)_2Zr]^+$, $[ZrCl_2]^+$, and $[(C_5H_5)(Cl)ZrOZrCl]^+$. In the last ion the resonance structure includes contributions in which both zirconium atoms are formally trivalent and others in which one is tetravalent and the other divalent.

$$
\begin{array}{ccc}
\overset{\displaystyle Cl}{\underset{\displaystyle |}{(C_5H_5)Zr^{III}}}\!-\!O\!-\!Zr^{+III}\!-\!Cl & \leftrightarrow & \overset{\displaystyle Cl}{\underset{\displaystyle |}{(C_5H_5)Zr^{III}}}\!-\!O\!-\!Zr^{III}\!\!=\!\!Cl^+ & \leftrightarrow & \overset{\displaystyle Cl}{\underset{\displaystyle |}{(C_5H_5)Zr^{III}}}\!-\!\overset{+}{O}\!\!=\!\!Zr^{III}\!-\!Cl
\end{array}
$$

$$
\begin{array}{ccc}
\overset{\displaystyle Cl}{\underset{\displaystyle |}{(C_5H_5)Zr^{IV}}}\!\!=\!\!\overset{+}{O}\!-\!Zr^{II}\!-\!Cl & \leftrightarrow & \overset{\displaystyle Cl}{\underset{\displaystyle |}{(C_5H_5)Zr^{+IV}}}\!-\!O\!-\!Zr^{II}\!-\!Cl & \leftrightarrow & \overset{\displaystyle Cl^+}{\underset{\displaystyle \|}{(C_5H_5)Zr^{IV}}}\!-\!O\!-\!Zr^{II}\!-\!Cl
\end{array}
$$

The use of mass spectrometry to determine hydride species is becoming well established and has been used to identify the product of the reaction of $(C_5H_5)_2{}^{90}ZrBH_4$ with amine as $(C_5H_5)_2{}^{90}Zr(H)BH_4$, m/e 236 (93).

The observation of the molecular ion for $(C_5H_5)_2ClZrSiPh_3$ has been reported (39) and the mass spectrum of $(C_5H_5)_2ClHfSiPh_3$ shows the molecular ion peak and fragments consistent with the proposed formula (125b).

The mass spectra of the tetraallyl complexes $M(C_3H_5)_4$, where M = Zr, Hf, have been recorded (12). The molecular ion is weak and the dominant metal-containing fragment is $[M - C_3H_5]^+$. This is illustrated by a comparison of the field ionization and electron impact mass spectra of $Zr(C_3H_5)_4$ (12). Interestingly, the electron impact spectrum of $Zr(C_3H_5)_4$ does not show any peaks for the fragments $[Zr(C_3H_5)_2]^+$, $[ZrC_3H_5]^+$, or Zr^+, but rather, the fragment $[Zr(C_3H_5)_3]^+$ eliminates ethylene in a one-step process (metastable peak observed). A possible intermediate for this fragmentation might be (III), from which ethylene can be eliminated, leaving two vinyl groups attached to the metal.

$$
\left[
\begin{array}{c}
CH_2 \qquad\qquad CH_2 \\
\diagdown\quad\cdots M\cdots\quad\diagup \\
CH \qquad\qquad CH \\
\diagdown CH_2\cdots H_2C\diagup
\end{array}
\right]^+
$$

(III)

III

VANADIUM AND NIOBIUM

In one of the first important papers on the mass spectra of organometallic compounds, Friedman *et al.* (*81*) examined the mass spectra of several bis(cyclopentadienyl) compounds, including bis(cyclopentadienyl)vanadium, $(C_5H_5)_2V$. The spectrum is that of a typical covalent sandwich compound, the main ions observed being $(C_5H_5)_2V^+$, $C_5H_5V^+$, and V^+. The molecular ion is the base peak of the spectrum and very little fragmentation of the cyclopentadienyl ring is observed at an ionizing potential of 50 eV. Although the most intense peak corresponds to the molecular ion, the total ion current of this ion is much less than the corresponding ions in the spectra of $(C_5H_5)_2Fe$ and $(C_5H_5)_2Co$. This difference cannot be accounted for simply by weakening of the metal–ring bond, for the total ion current of V^+ is slightly less than for the corresponding metal ions in the iron and cobalt compounds. Therefore, a greater stability of the $C_5H_5V^+$ ion relative to the ions $C_5H_5Fe^+$ and $C_5H_5Co^+$ is postulated. The mass spectrum of $(C_5H_5)_2V$ has recently been reinvestigated by Müller and D'Or (*148*), who observed, in addition to the main peaks observed by Friedman *et al.*, a strong peak for the ion $C_3H_3V^+$ which probably has the cyclopropenyl structure.

The mass spectrum of $C_5H_5VC_7H_7$ (*117, 150*) shows that loss of the seven-membered ring occurs preferentially from the molecular ion. However, besides the simple cleavage processes giving the ions $C_7H_7V^+$, $C_5H_5V^+$, and V^+ (formed in a one-step process from M^+), the molecular ion also eliminates acetylene to give the bis(cyclopentadienyl) ion $(C_5H_5)_2V^+$. A most interesting fragmentation of the molecular ion is by loss of a C_6H_6 fragment, giving an abundant ion $C_6H_6V^+$, which then eliminates a second C_6H_6 unit to give the metal ion. All these fragmentation paths are confirmed by the observations of the appropriate metastable peaks.

The main ions in the spectrum of dibenzenevanadium are $(C_6H_6)_2V^+$, $C_6H_6V^+$, and V^+ (*150*). Some fragmentation of the benzene ring, while it is still bonded to vanadium, is observed, and the main fragments formed are $C_4H_xV^+$ ($x = 1–4$) and C_2HV^+. This latter ion, which is the strongest of the ions formed by breakdown of the benzene ring, is probably the metal-acetylide ion. A metastable peak is observed which corresponds to the fragmentation of the doubly charged molecular ion into singly charged species.

$$(C_6H_6)_2V^{2+} \rightarrow C_6H_6V^+ + C_6H_6^+$$

The mass spectrum of $C_5H_5V(CO)_4$ (*114, 199*) shows that the molecular ion fragments by loss of carbon monoxide rather than by elimination of a cyclopentadienyl radical, and successive removal of carbon monoxide molecules gives the base peak of the spectrum $C_5H_5V^+$. No fragmentation of the carbonyl groups occurs and the only metal carbonyl fragment observed is VCO^+ (7.1% R.A.). However, fragmentation of the cyclopentadienyl ring does occur with elimination of acetylene giving a strong peak for $C_3H_3V^+$. Negative ions were observed for the species $(C_5H_y)V(CO)_x$ ($y=3\text{--}5$, $x=2$, 3) and are formed by both a dissociative electron capture (1) and an ion pair process (2) (*199*). For the measurement of the

$$AB + e \rightarrow A^- + B \qquad (1)$$

$$AB + e \rightarrow A^- + B^+ + e \qquad (2)$$

appearance potentials of the various fragment ions the heat of formation for the process (3) was 139 kcal/mole (*199*) (see Section VII). The reaction

$$e + C_5H_5V(CO)_4 \rightarrow V^+ + C_3H_3 + C_2H_2 + 4\,CO + 2\,e \qquad (3)$$

of $C_5H_5V(CO)_4$ with acetic acid gives cyclopentadienylvanadium diacetate, which is shown to have the monomeric formulation (IV) in the gaseous state by its mass spectrum (*105*).

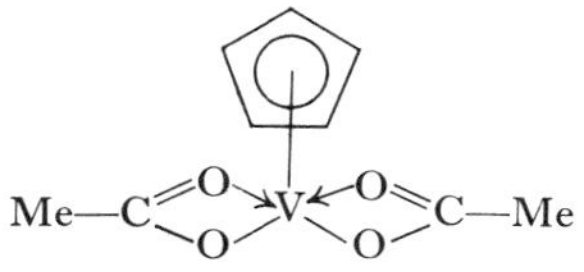

(IV)

The peak at highest mass corresponds to $[C_5H_5V(CO_2CH_3)_2]^+$, and loss of either one or two molecules of ketene gives strong peaks for $[C_5H_5V(OH)CO_2CH_3]^+$ and $[C_5H_5V(OH)_2]^+$. The monomeric formulation corresponds to a vanadium(III) derivative which would have two unpaired electrons and a magnetic moment of about 2.83 μ_B (Bohr magnetons). Magnetic susceptibility measurements, in the solid state, indicate a magnetic moment of 1.49 ± 0.02 μ_B, suggesting spin pairing in the solid state by means of vanadium–vanadium bonds.

Treatment of $C_5H_5V(CO)_4$ with bis(trifluoromethyl)dithietene gives the dimer $[C_5H_5VC_4F_6S_2]_2$ (*110*), and its partial mass spectrum confirms the dimeric nature of the compound, exhibiting a strong molecular ion peak at m/e 684 and several other fragments containing two vanadium atoms (*120*).

An interesting feature in the spectrum of $C_5H_5V(CO)_3P(NMe_2)_3$ is the loss of two carbonyl groups in a one-step process from the $[M-CO]^+$ fragment ion, and a metastable peak is observed for this transition (*113*). Similar eliminations of two carbonyl groups have been observed for other transition metal carbonyl complexes (*131*). $(C_5H_5)_2V_2(CO)_5$ exhibits the molecular ion and fragment ions corresponding to the stepwise loss of the five CO groups; an intense peak occurs at m/e 232 $(C_5H_5)_2V_2^+$ (*77a*).

The only niobium complex for which mass spectra data have been reported is $(C_5H_5)_2NbOCl$, which exhibits the molecular ion peak (*195*).

IV

CHROMIUM, MOLYBDENUM, AND TUNGSTEN

The mass spectra of a considerable number of π-bonded complexes of the group VIA metals have been reported, but in many cases mass spectrometry has only been used to determine the molecular weight, so that a detailed examination of the fragmentation processes involved has not been attempted, and only the molecular ion and perhaps a few other major peaks are reported. Within this section it is more convenient to discuss the compounds in terms of the attached ligands rather than in terms of the central metal atom. The classifications are (A) cyclopentadienyl compounds; (B) arene compounds; and (C) olefin, acetylene, and allyl compounds.

A. Cyclopentadienyl Compounds

1. Bis(cyclopentadienyls)

The mass spectrum of bis(cyclopentadienyl)chromium $(C_5H_5)_2Cr$ shows three main peaks for $(C_5H_5)_2Cr^+$, $C_5H_5Cr^+$, and Cr^+, but very little cleavage of the cyclopentadienyl ring is observed (*81, 148*). Also significant is the high intensity of the Cr^+ ion, which is more abundant than the metal ion fragments in ferrocene, cobaltocene, or nickelocene, suggesting that the metal–ring bond in $(C_5H_5)_2Cr$ is weaker than in ferrocene, etc. However, estimation of the metal–ring bond energies from appearance potential data for the fragment ions does not show the expected difference between $(C_5H_5)_2Cr$ and the other bis(cyclopentadienyl) compounds, and it is therefore possible that the increased yield of Cr^+ ions occurs because of thermal decomposition of the complex rather than by an electron bombardment phenomenon. However, such thermal decomposition would result in a high intensity for $C_5H_5^+$, and the relative abundance of this ion has not been recorded.

The mass spectrum of $(C_5H_5)_2WH_2$ shows a very strong molecular ion peak (148). The fragmentation pattern is very similar to the isoelectronic compounds $(C_5H_5)_2Os$ and $(C_5H_5)_2ReH$, the major fragments being $C_8H_xW^+$ and $C_6H_xW^+$ (exact calculation of the number of hydrogens is complicated by the several naturally occurring isotopes of tungsten).

2. Cyclopentadienyl Carbonyl Compounds

The mass spectrum of $[C_5H_5Cr(CO)_3]_2$ shows no ions containing two chromium atoms, and the peaks at highest mass correspond to $[C_5H_5Cr(CO)_3H]^+$, m/e 202, and $[C_5H_5Cr(CO)_3]^+$, m/e 201, indicating that the chromium–chromium bond is so weak that on vaporization it breaks to give the monomeric $C_5H_5Cr(CO)_3$ vapor (111). By contrast, the mass spectrum of $[C_5H_5Mo(CO)_3]_2$ (111, 131) exhibits a peak for the molecular ion and peaks for the ions $(C_5H_5)_2Mo_2(CO)_n$ ($n=0$–5), showing that the metal–metal bond in the molybdenum complex is considerably stronger than in the chromium complex. The main fragmentation pattern of the molybdenum complex is by successive cleavage of the carbonyl groups, giving as the base peak of the spectrum $[(C_5H_5)_2Mo_2]^+$, but some cleavage of the metal–metal bond does occur, giving the species $[C_5H_5Mo(CO)_n]^+$ ($n=1$–3), in low abundance (131, 182), and very little cleavage of the cyclopentadienyl group is observed (182). Ion–molecule reaction products such as $[(C_5H_5)_3Mo_2(CO)_n]^+$ ($n=4$–6) and $[(C_5H_5)_2MoCH_2(CO)_n]^+$ ($n=0$–4) have been observed by Schumacher and Taubenest (182). The appearance of the molecular ion for $[(C_5H_5Mo(CO)_3]_2Hg$ has been reported (139). The reaction of pentamethylcyclopentadiene with $Mo(CO)_6$ gives the novel compound $[C_5Me_5Mo(CO)_2]_2$, and its mass spectrum shows peaks for the ions $[(C_5Me_5)_2Mo_2(CO)_n]^+$ ($n=0$–4) and also $[(C_5Me_5)_2Mo_2]^{2+}$, indicating a very strong metal–metal interaction (121). The structure proposed by King for this compound involves a metal–metal triple bond.

A similar structure has been proposed for the product of the reaction of 9,10-dihydroindene with $Mo(CO)_6$, which was originally believed to be cyclononatetraenemolybdenum tricarbonyl (*124*), but mass spectrometry now suggests the formula $[C_9H_9Mo(CO)_2]_2$, as the peak at highest mass corresponds to $[(C_9H_9)_2Mo_2(CO)_4]^+$ and successive loss of the carbonyl groups is observed. However, once all the carbonyl groups are removed

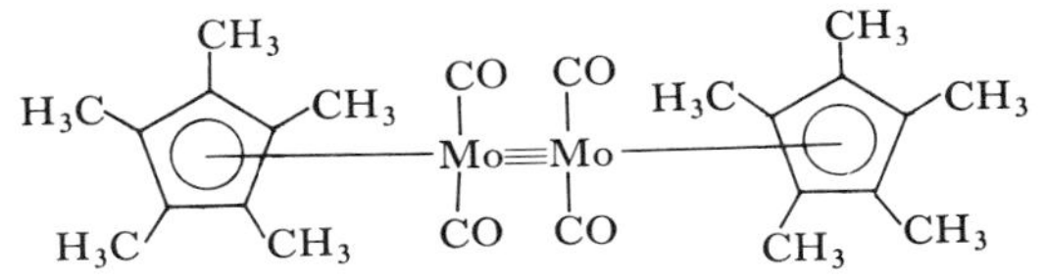

dehydrogenation becomes a much favored step and no $(C_9H_9)_2Mo_2^+$ ions are observed. Instead, the ion $(C_9H_7)_2Mo_2^+$ is observed in greater concentration than any other molybdenum-containing ion. A structure involving a metal–metal bond is postulated for this compound (*108*).

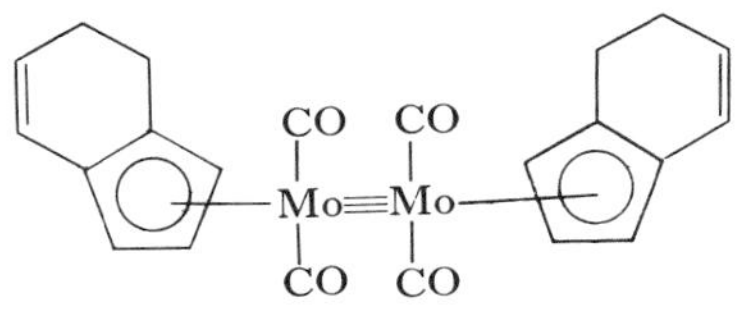

The observation of the molecular ion peaks for the mercury derivatives $[C_5H_5M(CO)_3]_2Hg$ (M = Mo, W) has been reported (*139*). The composition of the bis(trifluoromethyl)ethylenedithiolate complex $[C_5H_5W(CO)S_2C_2(CF_3)_2]_2$ was confirmed by observation of the molecular ion and the $[M - CO]^+$ fragment ion (*120*).

Whereas the dimeric cyclopentadienyl compounds discussed above show very little fragmentation of the cyclopentadienyl ring, the monomeric compounds $C_5H_5Mo(CO)_3Br$ (*181*) and $C_5H_5Mo(CO)_3(CH_2)Br$ (*114*) show strong peaks for $[C_3H_3MoBr]^+$ and $[C_3H_3Mo]^+$ formed by expulsion of acetylene from the cyclopentadienyl ring. For both compounds stepwise loss of the carbonyl groups occurs preferentially to cleavage of bromine or of the alkyl bromide chain, and ions containing molybdenum–bromine bonds are very abundant. The propyl iodide complex $C_5H_5Mo(CO)_3(CH_2)_3I$ (*88*) does not show any fragmentation of the cyclopentadienyl ring, and the base peak of the spectrum is $[C_5H_5MoI]^+$, formed probably via a five-membered cyclic transition state. An alternative fragmentation, involving the elimination of ethylene, gives the fragment ion $[C_5H_5MoCH_2I]^+$.

The highest peak observed in the spectrum of $C_5H_5Mo(CO)_3CH_2$ $OCOC_2H_5$ (*114*) corresponds to the ion $[M-CO]^+$, indicating that decarbonylation occurs readily in the spectrometer. Stepwise loss of the two carbonyl groups is followed by loss of ethylene, giving the fragmentation ion $[C_5H_5MoCH_2OCOH]^+$, and this ion undergoes hydroxyl migration, giving the ions $[C_5H_5MoOH]^+$ and $[C_5H_5MoO]^+$. Similar fragmentations have been observed in the mass spectra of substituted cymantrene derivatives (*37*) and are indicative of strong metal–oxygen bonds. An alternative fragmentation observed in the spectrum of this ester derivative is the loss of ketene from $[M-CO]^+$ ion, with migration of the ethoxy group to the metal; the resulting ion then loses CO, followed by loss of $CO+H_2$, giving a fragment ion m/e 204. This ion then loses acetylene to give the ion $[C_5H_5MoOH]^+$, indicating that the fragment ion m/e 204 probably has the structure $[C_5H_5MoOCH=CH_2]^+$ rather than the isomeric acetyl structure with a molybdenum–carbon bond. A third fragmentation path is by loss of OH from the $[M-CO]^+$ to give an ethoxyacetylene ion.

The mass spectra of the two complexes $C_5H_5Mo(CO)_nCH_2SMe$ (*n* = 2, 3) are essentially the same and correspond to the spectrum expected for the dicarbonyl complex (*115*). This is consistent with the observed facile decarbonylation of tricarbonyl complex to the dicarbonyl complex. Loss of the carbonyl groups is followed by loss of methane, giving the ion $C_6H_6SMo^+$ which is the base peak of the spectrum, or, alternatively, the $[M-2\ CO]$ ion eliminates ethylene to give the ion $C_5H_5MoSH^+$. Cleavage of the methyl group from the molecular ion is observed, but is a much less favored process than loss of CO.

The spectra of $C_5H_5Mo(CO)_3COCF_3$ and $C_5H_5Mo(CO)_3CF_3$ are very similar and correspond to $C_5H_5Mo(CO)_3CF_3$ (*116*). Stepwise loss of the three carbonyl groups is observed, but the M^+ and $[M-2\ CO]^+$ ions also show loss of fluorine. The $[M-3\ CO]^+$ ion then loses either CF_2 with concurrent migration of fluorine to the metal atom, or loses HF. Further fragmentation gives the expected ions $C_5H_5Mo^+$, $C_3H_3MoF^+$, $C_3H_3Mo^+$. The one difference between the two spectra is that $C_5H_5Mo(CO)_3COCF_3$ exhibits the tetracarbonyl ion $C_5H_5Mo(CO)_4^+$ and its decarbonylation products, possibly because some of the tricarbonyl complex is not decarbonylated thermally, but fragments under electron bombardment according to the following scheme:

$$C_5H_5Mo(CO)_3COCF_3 \rightarrow C_5H_5Mo(CO)_4^+ + CF_3^-$$

The tungsten complex $C_5H_5W(CO)_3COCF_3$ does exhibit a molecular ion, and the much lower tendency for the tungsten compound to undergo decarbonylation can be related to the greater stability of tungsten–carbon bonds as compared to analogous molybdenum–carbon bonds (116). The expected cleavage of CF_3 from the molecular ion is observed to give the ion $C_5H_5W(CO)_4^+$, and also observed is the ion $C_5H_5W(CO)_2CF^+$, which can arise by elimination of COF_2 from the $[M-CO]^+$ ion. The compounds $C_5H_5Mo(CO)_3COC_3F_7$ and $C_5H_5Mo(CO)_3C_3F_7$ show essentially the same spectra, namely that of the decarbonylated complex $C_5H_5Mo(CO)_3$ C_3F_7, a situation identical to that of the trifluoromethyl and trifluoroacetyl complexes discussed above. An interesting feature of these spectra is the migration of fluorine atoms to the metal to give the ion $C_5H_5MoF_2^+$, which is the most abundant ion containing molybdenum. The tungsten complex $C_5H_5W(CO)_3COC_3F_7$, as does the trifluoroacetyl complex, exhibits the molecular ion and the ion $C_5H_5W(CO)_4^+$ (116).

The acryloyl complex $C_5H_5W(CO)_3(COCH{=}CH_2)$ does not exhibit the molecular ion peak, but is decarbonylated in the spectrometer to the vinyl complex $C_5H_5W(CO)_3CH{=}CH_2$, and the ion $C_5H_5W(CO)_3CH{=}CH_2^+$ fragments by loss of CO and by loss of the vinyl group (114). The hydride $C_5H_5W(CO)_3H$ shows the usual stepwise loss of carbonyl groups, but the $[M-CO]^+$ ion also loses hydrogen and the series of ions $C_5H_5W(CO)_n^+$ ($n=0$–2) is observed, indicating that once one CO group is lost, cleavage of the hydrogen competes noticeably with cleavage of CO. Doubly charged ions are observed for $C_5H_5W(CO)_nH^{2+}$ ($n=0$–3) and $C_5H_5W(CO)_n^{2+}$ ($n=0$–2). Carrick and Glockling (40, 41) reported the spectra of some organogermanium derivatives of the type $C_5H_5M(CO)_3GeR_3$ ($M=Mo$, W; $R_3=Me_3$, Et_2H, Et_3, $n\text{-}Pr_3$) and found that they exhibit very strong molecular ion peaks and that most of the ion current (~70–80%) is carried by fragments containing a metal–metal bond. Considerable differences were observed in the spectra of the trimethylgermyl complexes C_5H_5M $(CO)_3GeMe_3$ ($M=Mo$, W). The molybdenum complex shows ions corresponding to loss of a methyl radical together with ions formed by loss of CO, while the tungsten complex shows loss of methyl and of CO_2 from the molecular ion. Loss of CO_2 only occurs in the molybdenum complex after elimination of two methyl groups. Fragmentation of the cyclopentadienyl group also is observed. For the ethyl and propyl complexes, loss of R gives the most abundant ion and loss of CO or CO_2 from M^+ gives ions of low abundance. In the spectra of dicarbonyl triphenylphosphine complexes $C_5H_5M(CO)_2PPh_3GeR_3$ (M = Mo, W; R_3 = Me$_3$, Et$_3$), fragments containing metal–metal bonds were of much lower abundance, and strong peaks were observed for ions containing the transition metal without germanium (41). The mass spectrum of $C_5H_5W(CO)_3GePh_3Pt(Ph_2PCH_2)_2$ showed the ions $GePh_3^+$ and $C_5H_5W(CO)_3^+$ and ions with a PtWGe isotopic pattern up to m/e 990, but in the region of the molecular ion (m/e 1230), only weak peaks could be detected and their isotopic pattern could not be distinguished (41).

The reaction of diphenylfulvene with $Cr(CO)_6$ gives a chromium tricarbonyl derivative in which the ligand is bonded to chromium via the five-membered ring, rather than by the six-membered ring. The mass spectrum of $Ph_2C_5H_4Cr(CO)_3$ shows no peaks due to the complex itself, but only peaks due to diphenylfulvene, and its fragmentation pattern is the same as the free ligand. Reduction of $Ph_2C_5H_4Cr(CO)_3$ with formic acid gave a green solid of stoichiometry $C_{21}H_{15}Cr(CO)_3$, and its mass spectrum shows no peaks due to the complex. The strongest peak occurs at m/e 232, 2 m.u. higher than diphenylfulvene itself (m/e 230). There are no peaks at higher mass, and fragmentation is similar to that of diphenyl-fulvene (47).

The observation of the molecular ion peak has been reported for the following cyclopentadienyl metal tricarbonyl derivatives: $C_5H_5Mo(CO)_3$ SiMe$_3$ (39), $C_5H_5M(CO)_3X$ [M = Mo, W; X = Cl, Br, I, SCN, Co(CO)$_4$] (139), and $C_5H_5W(CO)_3SnMe_3$ (38). $C_5H_5W(CO)_3SiMe_3$ shows at the highest peak the ion $[M - 14]^+$ (39), and loss of 14 m.u. has been observed in other trimethylsilyl complexes (39).

The cyclopentadienyl dicarbonyl nitrosyl complex $C_5H_5Mo(CO)_2NO$ (200) loses the carbonyl groups before cleavage of the nitrosyl group; also, cleavage of the cyclopentadienyl ring is observed, giving the ions C_3H_3Mo NO^+ and $C_3H_3Mo^+$. The spectrum was determined using a Bendix time-of-flight mass spectrometer, and the doubly charged ion $[C_5H_5Mo]^{2+}$ is very abundant. Cleavage of the carbonyl group is observed, giving the fragment ion MoC^+, a phenomenon previously observed only in the spectra of metal carbonyls (27).

The strongest ion in the spectra of the dicarbonyl cycloheptatrienyl complexes $C_5H_5M(CO)_2C_7H_7$ (M = Mo, W) corresponds to $C_5H_5MC_7H_7$, and no peak is observed for the $[M - CO]^+$ ion (114, 123). Loss of acetylene from the $(M - 2 CO)^+$ ion, as observed in the spectra of $C_5H_5MC_7H_7$ complexes (150), gave the bis(cyclopentadienyl) metal ion $(C_5H_5)_2M^+$.

The π-benzyl derivative $C_5H_5Mo(CO)_2CH_2Ph$ does not exhibit significant concentrations of the $(C_5H_5)_2Mo^+$ ion, indicating that elimination of acetylene from the benzyl ligand is much more difficult (114). However, the benzyl complex does lose ethylene, giving $C_5H_5MoC_5H_3^+$, a process not observed in the spectra of the cycloheptatrienyl complexes. The π-allyl complex $C_5H_5Mo(CO)_2$-π-C_3H_5 (114) first loses one carbonyl group, followed by the loss of 30 m.u., which corresponds to the loss of $CO + H_2$ (possibly as formaldehyde) to give the π-cyclopropenyl ion $C_5H_5MoC_3H_3^+$, the most abundant ion in the spectrum. Loss of a molecule of hydrogen from π-allyl complexes to give the π-cyclopropenyl ligand has also been observed for rhodium complexes (see Section VI,D). Further fragmentation of the ion gives $C_6H_6Mo^+$, $C_4H_3Mo^+$, and $C_2H_2Mo^+$.

King (118) has studied the spectra of some indenylmolybdenum carbonyl compounds and has observed that the indenyl ligand fragments by loss of acetylene in a similar manner to the fragmentation of the cyclopentadienyl ligand. The molecular ion of $C_9H_7Mo(CO)_3Me$ (V) exhibits stepwise loss of the CO groups, but after the loss of one CO, loss of methyl competes with loss of CO.

(V)

Also, loss of CO from $C_9H_7MoCOMe^+$ is accompanied by dehydrogenation to give $C_{10}H_8Mo^+$, the hydrogens coming from the methyl group, for $C_9H_7MoCO^+$ gives $C_9H_7Mo^+$ without loss of hydrogen. Both $C_{10}H_8Mo^+$ and $C_9H_7Mo^+$ fragment further by loss of acetylene. The π-allyl complex $C_9H_7Mo(CO)_2$-π-C_3H_5 loses one carbonyl group, followed by loss of $CO + H_2$ in a fragmentation path completely analogous to that for C_5H_5Mo $(CO)_2$-π-C_3H_5. Loss of the allyl ligand, giving $C_9H_7Mo^+$, was also observed. The mass spectrum of $C_9H_7Mo(CO)_2$-π-CH_2SMe is very similar to that of the cyclopentadienyl analog, with the stepwise loss of the CO groups followed by loss of methane giving the ion $C_{10}H_8SMo^+$, which is the most abundant metal-containing ion in the spectrum. The mass spectrum of $C_9H_7Mo(CO)_2I$ indicated that decomposition occurred in the spectrometer.

The peak at highest mass corresponded to $(C_9H_7)_2Mo_2I_4$, and another pyrolysis product identified in the spectrum was $(C_9H_7)_2Mo(CO)I$.

3. *Cyclopentadienyl Nitrosyl Compounds*

Some cyclopentadienyl-μ-mercapto-, μ-alkoxo-, and μ-dialkylamido-chromium nitrosyl complexes (VI)–(X) have been studied by Preston and Reed (*167*).

These compounds [with the exception of (X)] show systematic losses of the two NO groups, followed by simple losses of alkyl, alkyl(thio, oxy), and

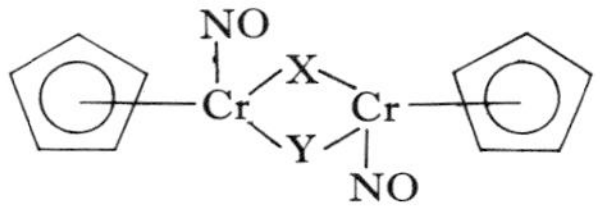

(VI)   X=Z=SMe
(VII)   X=Z=SPh
(VIII) X=SMe, Z=OH
(IX)   X=Z=OMe
(X)   X=Z=NMe$_2$

cyclopentadienyl radicals. However, in all cases there exists the ion m/e 182 which must be attributed to $(C_5H_5)_2Cr^+$, since further fragmentations occur to give m/e 117, m/e 65, and m/e 52. The formation of $(C_5H_5)_2Cr^+$ in, for example, (VI) is indicated by a metastable peak at m/e 111.1 for the transition $298 \rightarrow 182$.

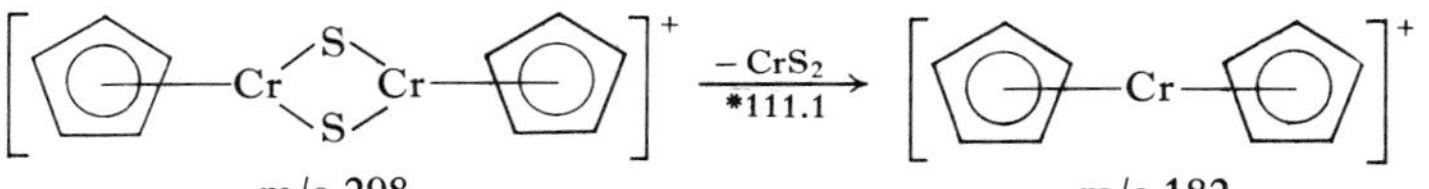

Such a fragmentation involves migration of a cyclopentadienyl group and the breaking of one Cr—Cr bond and two Cr—S bonds and could possibly occur via an intermediate such as (XI).

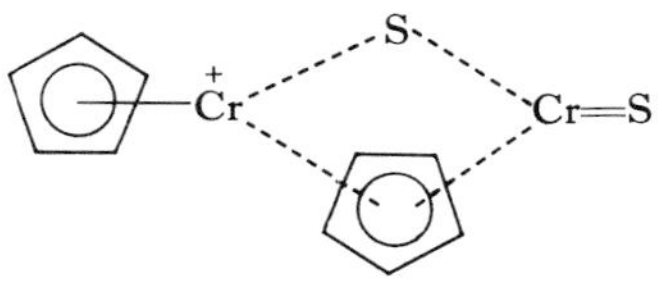

(XI)

The reaction of $C_5H_5Mo(CO)_2NO$ with $S_2C_2(CF_3)_2$ gives $[C_5H_5Mo(NO)S_2C_2(CF_3)_2]_2$, which shows the molecular ion and fragments for the successive loss of nitrosyl groups. The spectrum also shows the ions $[C_5H_5Mo(C_4F_6S_2)_2]^+$ and $[C_5H_5Mo(C_4F_6S_2)(C_4F_5S_2)]^+$ in large abundances. These could arise from fragmentation of the parent compound or from the species $C_5H_5Mo[S_2C_2(CF_3)_2]_2$, which is either an impurity present in the original sample or a product of vaporization in the spectrometer. Other ions in the mass spectrum of $[C_5H_5Mo(NO)S_2C_2(CF_3)_2]$ are $(C_5H_5)_2Mo_2(C_4F_6S_2)_3{}^+$ and fragmentation products formed by successive losses of fluorine. These must arise from a separate species, such as $(C_5H_5)_2Mo_2[S_2C_2(CF_3)_2]_3$, which is either present in the original sample or produced on vaporization (120).

Iodine reacts with $C_5H_5Mo(CO)_2NO$, giving the purple solid $[C_5H_5MoNOI_2]_2$, but its molecular ion peak is not observed, the peak at highest mass corresponding to the ion $[M-2 I]^+$. The presence of several ions containing two molybdenum atoms suggests that the molecule is dimeric, but the presence of much $[C_5H_5Mo(NO)I_2]^+$ and no $[C_5H_5Mo(NO)I_2]_2{}^+$ indicates that the binuclear complex easily converted to the mononuclear complex on vaporization (107).

4. Other Cyclopentadienyl Compounds

The mass spectrum of $C_5H_5CrC_7H_7$ (76, 117, 150) shows that the seven-membered ring is lost preferentially to the five-membered ring. Loss of acetylene from the molecular ion to give the bis(cyclopentadienyl) ion $(C_5H_5)_2Cr^+$ is also observed, but the ion $C_6H_6Cr^+$, formed by loss of a C_6H_6 from M^+, is of low abundance, whereas in the analogous vanadium complex $C_5H_5VC_7H_7$ this fragmentation path gave a very abundant ion $C_6H_6V^+$. The molybdenum complex $C_5H_5MoC_7H_7$ shows the molecular ion as the base peak and the fragmentation by loss of acetylene or C_6H_6 gives the ions $(C_5H_5)_2Mo^+$ and $C_6H_6Mo^+$. The ion $C_5H_5Mo^+$ is of low abundance (76).

The main peaks in the spectrum of $C_5H_5CrC_6H_6$ correspond to the ions M^+, $C_5H_5Cr^+$, and Cr^+, indicating that the six-membered ring is lost much more easily than the five-membered ring (150). An interesting feature is the observation of a metastable peak for the breakdown of the doubly charged molecular ion into two singly charged ions.

$$C_5H_5CrC_6H_6{}^{2+} \rightarrow C_5H_5Cr^+ + C_6H_6{}^+$$

$C_5H_5MoS_2C_2(CF_3)_2$ shows the molecular ion, and loss of fluorine occurs before cleavage of the cyclopentadienyl ring (120). Similar fragmentations were observed for $C_5H_5W[S_2C_2(CF_3)_2]_2$ (120).

B. Arene Complexes

The mass spectrum of $(C_6H_6)_2Cr$ exhibits strong peaks for the ions $(C_6H_6)_2Cr^+$, $C_6H_6Cr^+$, $C_6H_6^+$, and Cr^+ (54, 150, 165). Although very little fragmentation of the benzene ring occurs, the ions $C_4H_4Cr^+$, m/e 104 and $C_3H_3Cr^+$ are observed in low abundance, but the peak at m/e 104 also contains a contribution due to the doubly charged molecular ion. The base peak in one investigation (54) was benzene rather than the metal ion, but this is due to thermal decomposition of the compound on the probe of the spectrometer (166). The ionization potential of $(C_6H_6)_2Cr$ is 5.70 eV, which is 1 eV lower than the ionization potential of the free metal ion, indicating the high stability of the $(C_6H_6)_2Cr^+$ ion with respect to the neutral species (165). The extent of metalation of dibenzenechromium with TMEDA n-butyllithium has been determined by mass spectral analysis of the deuterolysis products (61). The mass spectra of benzenechromium tricarbonyl and several derivatives have been reported (see Table III), and it was observed for all these compounds that elimination of the carbonyl groups preceded any fragmentation of the aromatic ligand. However, in many cases the elimination of the carbonyl groups was not a stepwise process as observed for cyclopentadienylmetal carbonyl compounds, and usually two carbonyl groups were eliminated in one step from the molecular ion and this fragmentation path was confirmed in several spectra by observation of the appropriate metastable peaks. The $(M-2\ CO)^+$ ion then expels the third carbon monoxide molecule to give a strong peak for the $(M-3\ CO)^+$ ion. The mass spectrum of $C_6H_6Cr(CO)_3$ (118, 165) exhibits as the main peaks the ions $C_6H_6Cr(CO)_3^+$, $CrC_6H_6^+$, and Cr^+, while the ions $C_6H_6Cr(CO)_n^+$ ($n=1, 2$) are of low abundance, as is the ion $C_6H_6^+$. Thermal decomposition, giving as the main volatile products benzene and carbon monoxide, does not occur until 300° (166). No fragmentation of the benzene ring is observed while it is attached to the chromium. It was observed by Maoz et al. (136) that when the aromatic system bonded to chromium had an α-carbonyl substituent [(XII), (XIII)] migration of the group R to the metal occurred, giving the fragment ions Cr—R$^+$.

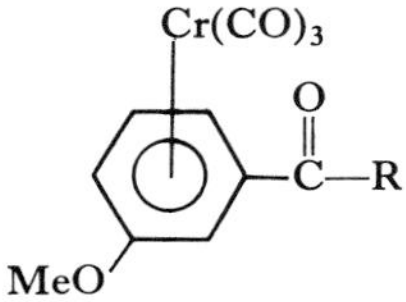

(XII) R = OMe

(XIII) R = OH

This rearrangement process has also been observed in α-carbonyl ferrocene and α-carbonyl cymantrene derivatives. An interesting fragment observed in the spectrum of the acid (XIII) was that corresponding to the molecular ion minus four carbonyl groups, i.e., as well as the usual elimination of the three carbonyl groups attached to the chromium, decarbonylation of the acid group also occurs. This is believed to be a one-step process from the molecular ion, and a metastable peak is observed for this transition.

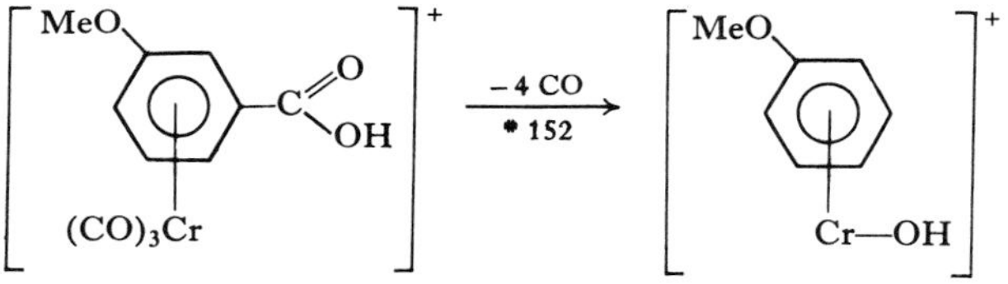

This type of fragmentation is not observed when the carbonyl group is not in an α-position to the metal-carrying π-moiety, and it is suggested that a possible mechanism for this decarbonylation process involves migration of R to the metal.

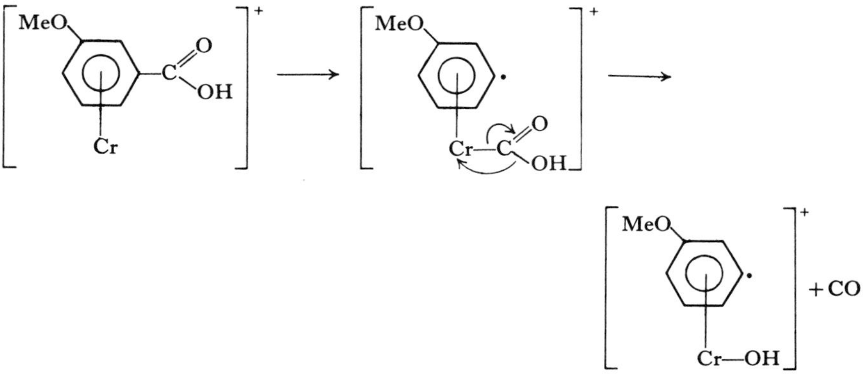

A similar phenomenon has been observed in the spectra of some diene–iron tricarbonyl compounds (136). The anisolechromium tricarbonyl complex shows loss of 2 H from the $[M - 3 \, CO]^+$ ion, giving $C_7H_6OCr^+$, which could have the tropone structure (XIV) while the analogous iso-propoxy complex loses propylene with the probable rearrangement to give

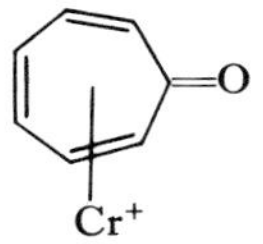

(XIV) 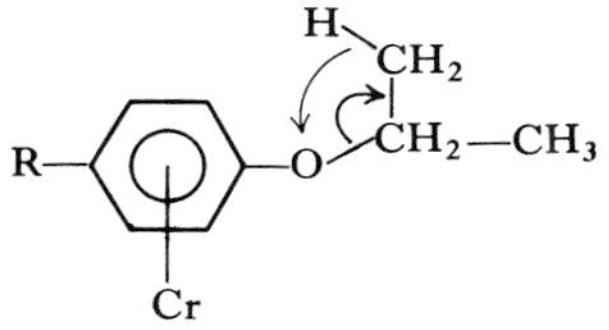   (XV)

phenol ions (XV). If R = Me, loss of hydrogen also occurs, with the formation of a hydroxytropylium ion (27). The carbinol derivative (XVI), after the usual loss of the three CO groups,

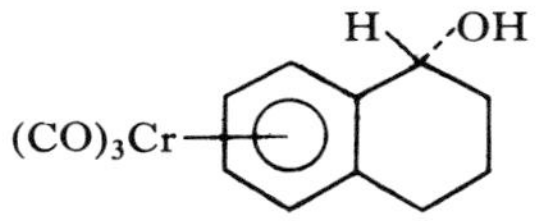

(XVI)

either loses two molecules of hydrogen, giving hydroxynaphthalene-chromium ion m/e 196, or eliminates one molecule of hydrogen followed by loss of water, or, alternatively, loss of water followed by elimination of hydrogen to give the fully aromatic system m/e 180 as a strong ion (58).

The molecular weights of the fused ring systems (XVII) and (XVIII) have been determined by mass spectrometry (55).

The hexamethylborazole complex $[B_3N_3(Me)_6]Cr(CO)_3$ exhibits as the main peaks the ions $[(B_3N_3(Me)_6]Cr(CO)_n]^+$ (n = 0, 1, 3) and $B_3N_3(Me)_6^+$ (168). The mass spectra of some bimetallic compounds (XIX)–(XXIV) have been measured (135) and the bimetal ions $[Cr—Cr]^+$ and $[Cr—Fe]^+$ observed. The elimination of the carbonyl groups is not predictable and does not usually occur in a stepwise manner, but rather the first three carbonyl groups are lost in one step, and the remaining three CO molecules are eliminated in a stepwise manner.

The mass spectra of three arenetungsten tricarbonyl complexes, arene-$W(CO)_3$ (arene = toluene, p-xylene, or mesitylene) show stepwise loss of

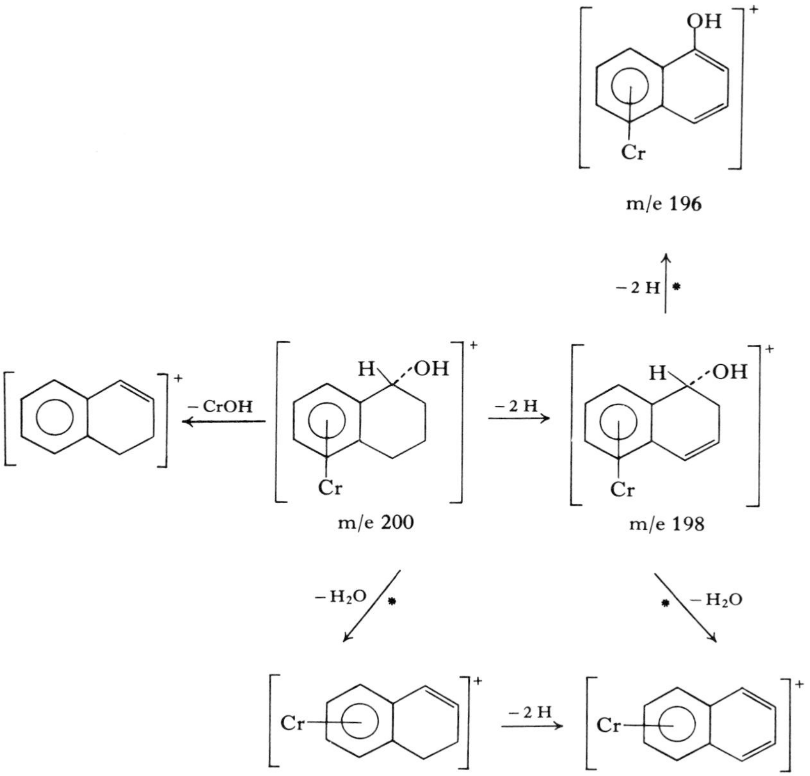
OH
Cr
m/e 196
− 2 H
H
OH
− CrOH
− 2 H
Cr
Cr
m/e 200
m/e 198
− H₂O
− H₂O
Cr
− 2 H
Cr
m/e 182
m/e 180

the three carbonyl groups giving the ions arene-W^+, which then lose H_2 fragments until an ion with fewer hydrogen atoms than carbon atoms is obtained. For toluene, *p*-xylene, and mesitylene, these are, respectively, $C_7H_6W^+$, $C_8H_6W^+$, and $C_9H_8W^+$, and these ions then fragment further by loss of acetylene (*119*).

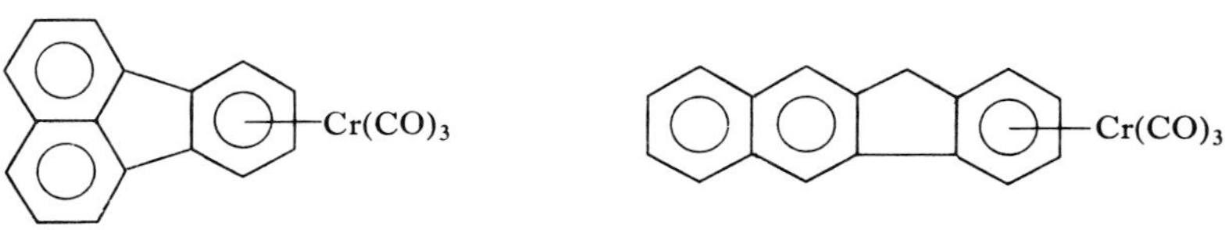
Cr(CO)₃
Cr(CO)₃

(XVII) (XVIII)

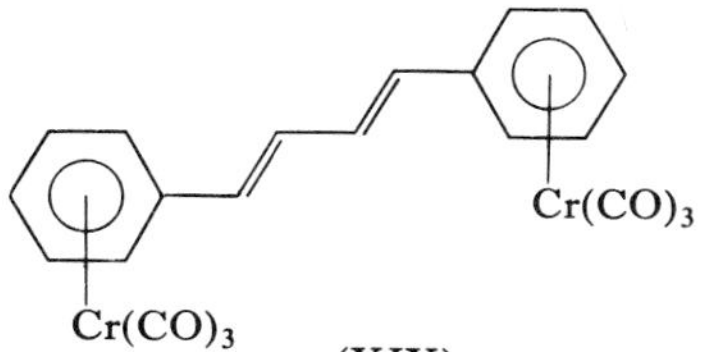

C. Olefin, Acetylene, and Allyl Complexes

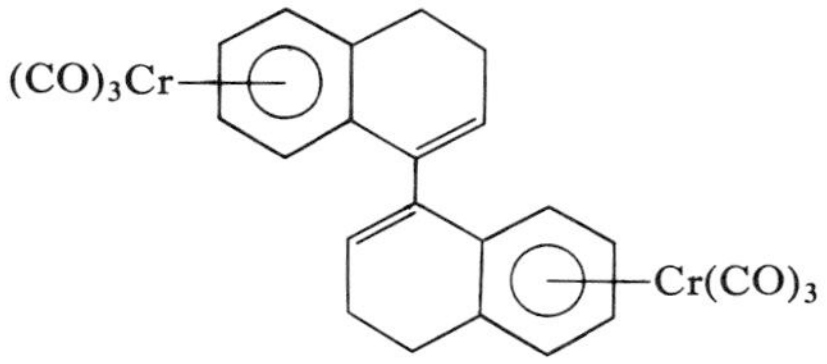

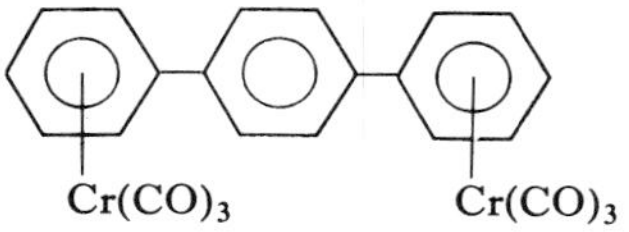

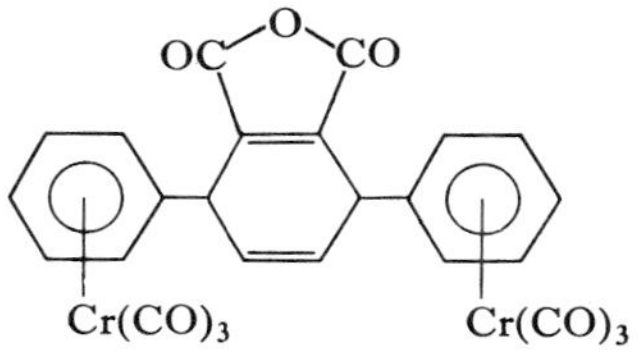

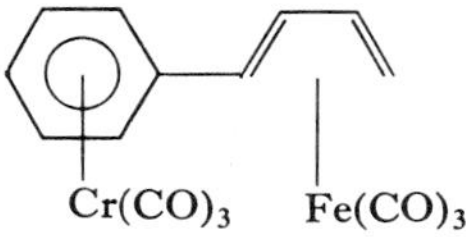

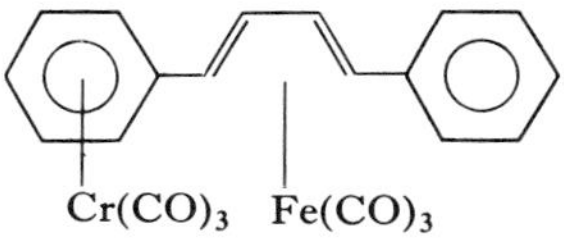

The composition of the tris(methyl vinyl ketone) complexes (MeCO
CH=CH$_2$)$_3$M (M = Mo, W) has been confirmed by mass spectrometry
(109, 119, 122, 123). The molecular ion loses one ligand to give the ion
(MeCOCH=CH$_2$)$_2$M$^+$. This ion in the molybdenum complex then loses
either ethylene or allene, giving the fragment ions (MeCOCH=CH$_2$)Mo
CH$_2$CO$^+$ and (MeCOCH=CH$_2$)MoCH$_2$$^+$, respectively. The ion MeCO
CH=CH$_2$MoO$^+$, containing a molybdenum–oxygen bond, is also observed.

The mass spectra of the cyclobutadiene and tetramethylcyclobutadiene
complexes (XXV) and (XXVI) show strong molecular ion peaks and the
major metal-containing fragments correspond to the successive loss of the
four CO groups (3). The hexamethyldewarbenzene complexes (XXVII)
show as the main peaks the series of ions [C$_6$Me$_6$M(CO)$_n$]$^+$ (n = 0–4) and
C$_6$Me$_6$$^+$ (66, 72). The 1,3-cyclohexadiene complexes (C$_6$H$_8$)$_2$M(CO)$_2$
(M = Mo, W) both show the molecular ion peak, and fragmentation of this
ion is by loss of one CO group followed by loss of H$_2$ giving C$_6$H$_8$MCO

$C_6H_6^+$. The most abundant metal-containing fragment for both complexes is $(C_6H_6)_2M^+$, showing that the cyclohexadiene ligand very readily loses two hydrogen atoms to give the more stable dibenzene-metal ion. The stability of the dibenzene-metal ions is further illustrated by the high abundance of the doubly charged species $(C_6H_6)_2M^{2+}$ (118, 123).

The mass spectrum of cycloheptatrienetungsten tricarbonyl C_7H_8W $(CO)_3$ (119) is very similar to that of the toluene analog, the only differences being that the cycloheptatriene complex exhibited the ions $C_6H_6^+$ and $C_4H_4^+$, which are not present in significant quantities in the spectrum of

M(CO)₄	M(CO)₄	M(CO)₄
(XXV)	(XXVI)	(XXVII)
M = Mo, W	M = Cr, Mo, W	M = Cr, Mo, W

the toluene complex. A possible explanation for this may be that in cycloheptatriene, but not in toluene, each of six adjacent carbon atoms has a hydrogen bonded to it. Hence, a benzene ion may be formed without a hydrogen shift, while for toluene, formation of C_6H_6 requires both bond cleavage and hydrogen transfer. The mass spectrum of the cycloheptatriene-chromium complex $C_7H_8Cr(CO)_3$ differs from its tungsten analog by the absence of dipositive ions and by the absence of many of the peaks between $C_7H_8M^+$ and M^+ observed in the spectrum of the tungsten complex. The greater abundance of dipositive ions in tungsten complexes, a $5d$ transition metal as compared with analogous complexes of chromium, a $3d$ transition metal, has been observed in many other cases.

The spectra of the norbornadiene complexes $C_7H_8Cr(CO)_4$ (27) and $C_7H_8W(CO)_4$ (119) show the expected stepwise loss of the four CO groups to give $C_7H_8M^+$, which then loses acetylene. The chromium complex also shows loss of CH_2 from the molecular ion to give $C_6H_6Cr(CO)_4^+$, a fragmentation not observed for the tungsten complex. The chromium complex also exhibits the series of ions $C_5H_5Cr(CO)_n^{2+}$ ($n=0$–3), which are not observed as their singly charged counterparts. $C_7H_8W(CO)_4$ loses C_5H_6 from the ions $C_7H_8W(CO)_n^+$ ($n=0$–2), giving the ions C_2H_2W $(CO)_n^+$.

The cycloocta-1,5-diene complexes $C_8H_{12}M(CO)_4$ (M = Mo, W) show the normal stepwise loss of CO, but also loss of hydrogen competes with loss of CO. Both spectra show ions $C_8H_{10}M(CO)_n^+$ ($n = 0$–2), and the ions $C_8H_8M^+$, $C_8H_6M^+$ are also very abundant. These ions then undergo the usual fragmentation by loss of acetylene (118). Similarly, the 1,3,5-cyclo-octatriene complex $C_8H_{10}W(CO)_3$ shows the ready loss of H_2 as well as of CO, and it is possible that the facile elimination of H_2 from π-bonded ligands may be indicative of the presence of an adjacent methylene group (119).

The mass spectrum of the cyclooctatetraene complex $C_8H_8Cr(CO)_3$, which has the structure (XXVIII), shows the typical stepwise loss of the

Cr(CO)₃

(XXVIII)

carbonyl groups, but an interesting feature is the presence of ion $C_6H_6Cr(CO)_3^+$, which is not observed in the spectrum of $C_8H_8Fe(CO)_3$ (109, 119). The molecular ion of the dicyclopentadiene complex $C_{10}H_{12}W(CO)_4$ undergoes stepwise loss of two CO groups, giving $C_{10}H_{12}W(CO)_2^+$ (119, 123). The next loss of CO is accompanied by loss of H_2, giving $C_{10}H_{10}WCO^+$, which then loses CO to give $C_{10}H_{10}W^+$. This last ion then fragments by loss of acetylene. The ions $C_5H_6W(CO)_n^+$ ($n = 0$–3) are also observed and arise by cleavage of the carbon–carbon bond between the two cyclo-pentadiene rings.

The mass spectrum of $(\pi\text{-}C_3H_5)_3Cr$ has been recorded and shows a much weaker molecular ion than the bis(π-allyl) complexes of nickel, palladium, and platinum. The major fragmentation is loss of an allyl radical to give as the dominant metal-containing species, the ion $(C_3H_5)_2Cr^+$. The fragments $C_3H_5Cr^+$ and Cr^+ were also observed (12). The mass spectrum of cyclo-heptatrienylchromium-cyclohepta-1,3-diene $C_7H_7CrC_7H_{10}$ shows the molecular peak.

Loss of the cycloheptadiene ring occurs preferentially, giving as the base peak $C_7H_7Cr^+$, and no peak is observed for $C_7H_{10}Cr^+$ (75). The mass spectra of the hexafluorobutyne complexes $[(CF_3)_2C_2]_3MNCMe$ (M = MoW), show that the acetonitrile ligand is readily eliminated from the molecular ion, and that transfer of fluorine atoms from the ligand to the metal occurs.

The hexafluorobutyne ligand is eliminated as tetrafluorobutatriene C_4F_4 with a two-step transfer of two fluorine atoms (*116, 123*).

Fischer and co-workers have prepared several novel carbene derivatives of chromium pentacarbonyl and have confirmed the molecular formulas by mass spectrometry (*5, 6, 73, 74, 77, 128, 147*). The spectra show the expected stepwise loss of the five carbonyl groups giving the ions [Ligand $Cr(CO)_n]^+$ ($n = 0$–5). The compounds for which spectral data have been reported are given in Table III. The selenium derivative (XXIX) shows

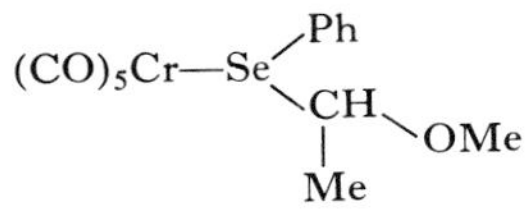

$$(CO)_5Cr\!-\!Se\!\begin{array}{l} \diagup Ph \\ \diagdown CH\diagdown OMe \\ \ \ | \\ \ \ Me \end{array}$$

(XXIX)

fragment ions $[CrSePhCH(OMe)Me]^+$ and $CrSePh^+$, which provide good evidence for a chromium–selenium bond (*74*).

V

GROUP VIIB METALS

A. Manganese

1. Cyclopentadienyl Compounds

The fragmentation pattern of $(C_5H_5)_2Mn$ (*81, 148*) is similar to that of the ionic compound $(C_5H_5)_2Mg$, and shows as the base peak the fragment ion $C_5H_5Mn^+$, whereas covalent bis(cyclopentadienyl) compounds $(C_5H_5)_2$ M (M = V, Cr, Fe, Co, Ni) have the molecular ion as the base peak of the spectrum. The total ion current carried by the ion $(C_5H_5)_2Mn^+$ is 18.5%, while for $(C_5H_5)_2M$ (M = Fe, Co) the total ion current of the molecular ion is 60%. It is concluded that the bonding between the cyclopentadienyl rings and manganese is mainly ionic, and an estimation of the metal–ring bond strengths shows that the metal–ring bond energy for the manganese compound is substantially less than for bis(cyclopentadienyl) compounds of the other first-row transition metals. The mass spectrum of benzene-(cyclopentadienyl)manganese shows as the main ions, $C_6H_6(C_5H_5)Mn^+$, $C_5H_5Mn^+$, and Mn^+, and the principal decomposition paths are shown below (*54, 150*).

Metastable peaks are observed for the breakdown of the doubly charged molecular ion into two singly charged species as shown. Examination of the metal-containing ions shows that (a) the species C_2HMn^+ is the fourth most abundant; (b) the peak for $C_6H_6Mn^+$ is small, while the peak for $C_6H_6MnH^+$ is four times larger. The criterion for stability of manganese-containing ions appears to be association of a formally dipositive d^5 manganese ion with a species which can form a relatively stable anion, and

$$C_5H_5MnC_6H_6^+ \xrightarrow{*\ 72.7} C_5H_5Mn^+ + C_6H_6$$

$$C_5H_5Mn^+ \xrightarrow{*\ 25.2} C_5H_5 + Mn^+$$

$$C_5H_5MnC_6H_6^{2+} \xrightarrow{*} C_5H_5Mn^+ + C_6H_6^+$$

$C_5H_5Mn^+$ and C_2HMn^+ ions can be considered as containing cyclopentadienide and acetylide ions, respectively. The ion $C_6H_6Mn^+$ does not satisfy this requirement and is therefore in low abundance. The very low abundance of $C_6H_6Mn^+$, even at 70 eV, is explained by the very rapid dissociation to Mn^+ and benzene. At an ionizing energy below 20 eV no peak for $C_6H_6Mn^+$ is observed at all, but the ion $C_6H_6MnH^+$ is observed and its appearance potential, 12 eV, is comparable to that of $C_5H_5Mn^+$. The formation of this ion is explained in terms of a concerted mechanism for a transfer of hydrogen which accounts for the low appearance potential, since one bond is formed as another is broken. However, no metastable peak was observed to confirm this rearrangement process.

2. Cyclopentadienyl Nitrosyl Compounds

The monomeric nature of $C_5H_5Mn(NO)S_2C_2(CF_3)_2$ has been confirmed by mass spectrometry, and loss of the nitrosyl group and of fluorine is observed (120). $[C_5H_5Mn(NO)_2]_n$ is not volatile and its mass spectrum shows as the major ions $(C_5H_5)_2MnH^+$ and $C_5H_xMn^+$ ($x = 3$–5), indicative of drastic decomposition on attempted vaporization (107). $(C_5H_5)_3Mn_3(NO)_4$ has been reported as showing a very weak molecular ion peak (60a).

3. Cymantrene Derivatives

The parent compound $C_5H_5Mn(CO)_3$ has been studied by Winters and Kiser (199). Successive loss of carbonyl groups occurs to give as the base peak of the spectrum $C_5H_5Mn^+$, and very little cleavage of the cyclopentadienyl ring is observed before removal of all the carbonyl groups.

The only metal carbonyl fragment observed, MnCO$^+$, is of very low abundance (0.2% R.A.). An estimation of the bond strengths for the ions [C$_5$H$_5$—Mn]$^+$ and [C$_3$H$_3$—Mn]$^+$ indicates that the bond strength for the cyclopentadienylmanganese ion should be negative, and this is confirmed by the very low abundance of this ion. From the measurement of appearance potentials, assuming fragmentation is according to the equation

$$e + C_5H_5Mn(CO)_3 \rightarrow Mn^+ + C_3H_3 + C_2H_2 + 3\ CO + 2\ e$$

and that neutral fragments are formed in their ground states, the heat of formation of C$_5$H$_5$Mn(CO)$_3$ has been calculated as 67 kcal/mole. The negative ion mass spectrum shows peaks for the ions C$_5$H$_y$Mn(CO)$_x^-$ ($y = 3$–5, $x = 0$–3), the most abundant being when $x = 3$. Although an ion-pair production process is postulated [Eqs. (4) and (5)], because of the poor resolution in measuring the exact mass numbers it is not possible to determine if resonance electron capture [Eq. (6)] does occur.

$$C_5H_5Mn(CO)_3 + e \rightarrow C_5H_4Mn(CO)_3^- + H^+ + e \qquad (4)$$

$$C_5H_5Mn(CO)_3 + e \rightarrow C_5H_3Mn(CO)_3^- + H_2^+ + e \qquad (5)$$

$$C_5H_5Mn(CO)_3 + e \rightarrow C_5H_5Mn(CO)_3^- \qquad (6)$$

The mass spectra of a variety of monosubstituted cymantrenes have been studied and the fragmentation patterns discussed in detail (37, 136, 194). The presence of the metal atom in many of the fragment ions seems to play an important role in determining the frágmentation path, and intramolecular rearrangements, involving the metal atom, have been observed (37). The spectra show the expected loss of the three carbonyl groups, usually before fragmentation in the side chain of the cyclopentadienyl ring occurs.

The carbonyl groups are eliminated in two steps, the first being the simultaneous elimination of two CO groups, followed by loss of the third CO group, giving the [M − 3 CO]$^+$ ion, which is often the base peak of the spectrum. Cleavage of the side chain, R, gives C$_5$H$_4$Mn$^+$ in low abundances, but further fragmentation of the cyclopentadienyl ring gives the fragments C$_2$HMn$^+$ and C$_3$H$_2$Mn$^+$ in significant abundances. It is suggested that the ion C$_3$H$_2$Mn$^+$ has the structure of a cyclopropenyl group σ-bonded to manganese. Rearrangement processes involving the migration of species such as H, Me, OH, or OMe to the manganese atom are observed from aldehydes, ketones, alcohols, acids, or methyl esters, but ethyl esters show a peak for MnOH$^+$, and no ion corresponding to MnOEt$^+$, possibly because this ion is very unstable and eliminates the neutral fragment

ethylene with the formation of MnOH⁺. When the side chain, R, contains
two or more carbons, the formation of ions with bicyclic structures occurs
[see Eqs. (7)], leading ultimately to the ion $C_7H_5^+$ for which the bicyclic
structure (XXX) is preferred to the monocyclic structure (XXXI). A
similar bicyclic structure (XXXII) has been considered for the ion $C_7H_7^+$
which is observed in the spectra of cymantrenes having a saturated side
chain [e.g., in Eqs. (8)] although the tropylium structure may seem more
likely, in ions containing no metals atoms.

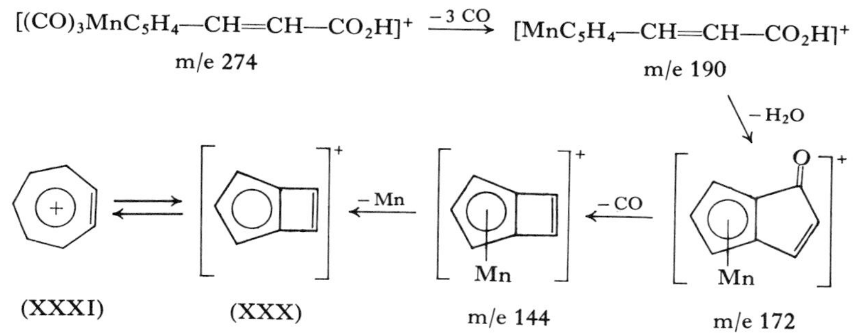

$$[(CO)_3MnC_5H_4\!-\!CH\!=\!CH\!-\!CO_2H]^+ \xrightarrow{-3\ CO} [MnC_5H_4\!-\!CH\!=\!CH\!-\!CO_2H]^+$$

m/e 274   m/e 190

(XXXI)   (XXX)   m/e 144   m/e 172   (7)

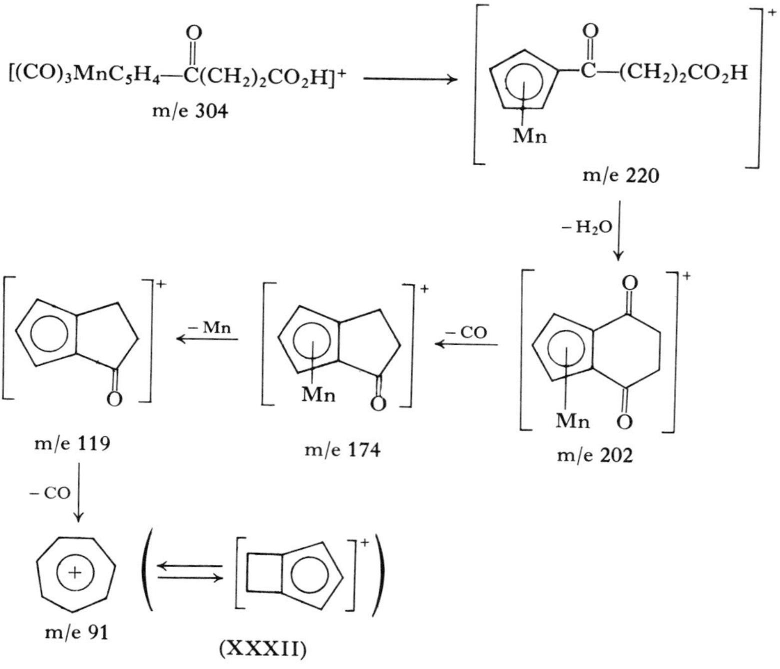

$$[(CO)_3MnC_5H_4\!-\!\overset{\overset{\displaystyle O}{\|}}{C}(CH_2)_2CO_2H]^+ \longrightarrow$$

m/e 304

m/e 220

m/e 119   m/e 174   m/e 202   (8)

m/e 91

(XXXII)

The mass spectra of several 1,2-disubstituted cymantrenes have been investigated and in general they exhibit much stronger molecular ion peaks and little fragmentation occurs after the cleavage of the carbonyl

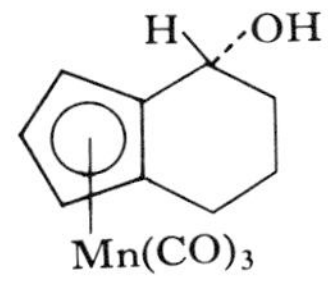

(XXXIII)

groups (58, 133). The alcohol (XXXIII) shows loss of the carbonyl groups, giving the $(M-3\ CO)^+$ ion which then eliminates water, giving m/e 172, followed by elimination of a molecule of hydrogen to give

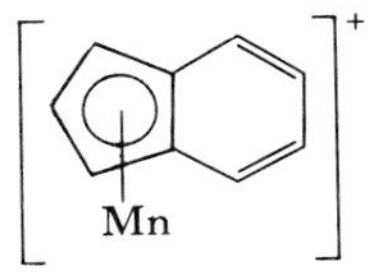

m/e 170

Both the m/e 172 and m/e 170 fragments lose the metal atom, giving $C_9H_9^+$ and $C_9H_7^+$, respectively. Migration of hydroxyl to the metal also occurs, giving $MnOH^+$. The mass spectra of some bimetallic and tri-metallic compounds have been measured and bimetallic fragments $(Mn—Mn)^+$ and $(Mn—Fe)^+$ are observed (136).

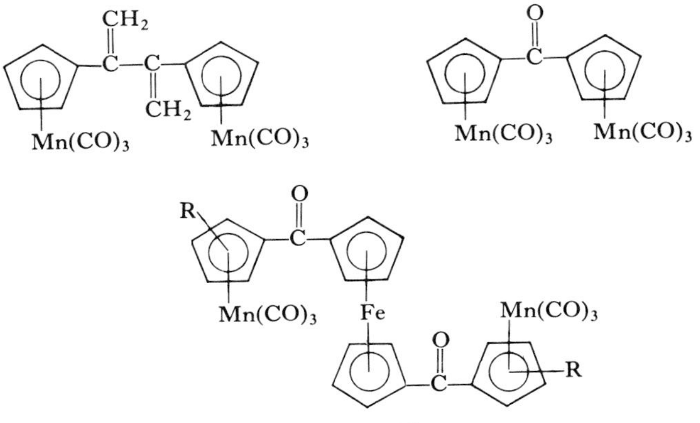

R = H, Me

Loss of carbonyl groups does not occur in a stepwise manner as usually observed, but three carbonyl groups are eliminated in one step, and for the trimetallic compounds even the remaining three carbonyl groups are eliminated in one step. This suggested pattern is supported by the presence of the appropriate metastable peaks in all the spectra.

In the course of preparing $[C_{10}H_8Mn(CO)_3]_2$ Bird and Churchill (22) isolated azulenedimanganese hexacarbonyl $C_{10}H_8Mn_2(CO)_6$. Its mass spectrum shows the molecular ion and fragments corresponding to stepwise loss of the six carbonyl groups.

The mass spectra of benzene(hexamethylbenzene)manganese(I) salts $[C_6H_6MnMe_6C_6]^+$ X^- ($X = PF_6$, BPh_4) show only peaks for benzene and hexamethylbenzene. Reduction of $[C_6H_6MnMe_6C_6]PF_6$ with $LiAlH_4$ gives $C_6H_7MnMe_6C_6$. The spectrum of this complex shows that benzene is eliminated from the molecular ion preferentially to the hexamethylbenzene ligand. The base peak of the spectrum is $[M-C_6H_6]^+$, and hydrogen is transferred to the rest of remaining fragment ion. Migration of the hydrogen to manganese occurs as shown by the appearance of MnH^+ m/e 56, but the predominant rearrangement is of hydrogen to the hexamethylbenzene, giving a strong peak for the hexamethylcyclohexadienyl ion $[Me_6C_6H]^+$, m/e 163 (69).

The mass spectrum of $C_5H_5Mn(CO)_2CNH$ (78) shows the molecular ion peak, and this fragments by loss of two CO groups, giving C_5H_5Mn CNH^+. Further fragmentation gives the ions $C_5H_5Mn^+$, $C_3H_2Mn^+$, $MnCNH^+$, $MnCN^+$, and MnC_2H^+.

B. Technetium and Rhenium

Although the mass spectra of $Re_2(CO)_{10}$ (131, 190), $ReMn(CO)_{10}$ (97, 190), and hydrido (99, 187), halogeno (57), sulfido (57), seleno (1), and fluorocarbon (25, 46) derivatives of rhenium carbonyl have been reported, only a few π-bonded derivatives of rhenium and one technetium compound have been investigated by mass spectrometry. The spectrum of $(C_5H_5)_2ReH$ shows a strong molecular ion peak. The other major ions are $(C_5H_5)_2Re^+$ and $C_8H_xRe^+$, and the abundance of Re^+ is very low (65, 81, 148). Metastable peaks indicate that fragmentation of the cyclopentadienyl rings arises from the fragment ion $(C_5H_5)_2Re^+$ and not from the molecular ion. The corresponding technetium compound $(C_5H_5)_2TeH$ also shows a strong molecular ion peak and an equally strong peak for the $(M-H)^+$ ion. A comparison of the intensities of the molecular ion and the $(M-H)^+$ ion indicates that the metal–hydrogen bond is weaker in the technetium compound (65, 148).

$$[(C_5H_5)_2TeH^+]:[(C_5H_5)_2Te^+] = 100.0:106.0$$

$$[(C_5H_5)_2ReH^+]:[(C_5H_5)_2Re^+] = 100.0:69.9$$

The mass spectrum of benzene(cyclopentadienyl)rhenium shows a very strong molecular ion peak. Very little fragmentation to give $C_5H_5Re^+$ was observed (71). This is in contrast to the spectrum of benzene(cyclopentadienyl)manganese, which shows as the base peak $C_5H_5Mn^+$ (54).

King (114) has compared the spectrum of $C_5H_5Re(CO)_3$ with that obtained by Winters and Kiser (199) for $C_5H_5Mn(CO)_3$. The rhenium complex shows the expected stepwise loss of the carbonyl groups but shows many more ions between $C_5H_5M^+$ and M^+ than are observed in the spectrum of the manganese complex. These ions include $C_5H_3Re^+$, $C_3H_3Re^+$, C_3HRe^+, C_2HRe^+, and CRe^+. The presence of significant quantities of these ions is probably a result of the stronger bonds formed between carbon and a heavy third-row (5d) transition metal, rhenium, as compared with the lighter third-row (3d) transition metal manganese. The tendency of the 5d elements to form doubly charged ions is also observed, and the rhenium complex exhibits the ions $C_5H_5Re(CO)_n^{2+}$ $(n=0-3)$, which are not present in the spectrum of the manganese analog.

Cycloocta-1,3-diene or cycloocta-1,5-diene react with $Re_2(CO)_{10}$ to give $C_8H_9Re(CO)_3$, which has been shown by single-crystal analysis to be a

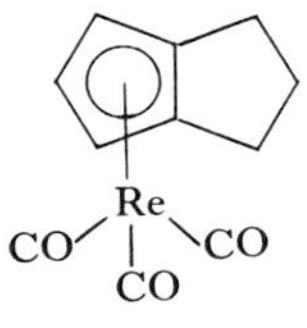

(XXXIV)

trihydropentalene derivative. Its mass spectrum shows an abundant molecular ion peak and loss of one carbonyl group occurs, followed by elimination of a molecule of hydrogen, giving $[C_8H_7Re(CO)_2]^+$; successive elimination of the remaining two carbonyl groups occurs from this ion. Doubly charged ions of all the species in this fragmentation sequence are detected in high abundance, indicative of high stability of these ions. Further fragmentation ultimately gives Re^+ and hydrocarbon fragments of low abundance, the charge remaining mainly on the metal atom (102).

VI

GROUP VIII METALS

A. Iron

Of all the metallocene and related compounds that have been examined by mass spectrometry, by far the greatest number are the compounds in which iron is the central atom, and of the iron compounds studied, a high proportion are compounds derived from ferrocene. In many cases mass spectrometry has been used only to determine the molecular weight by recording the molecular ion, with perhaps a few other major peaks being listed. Recently there have been several detailed studies on the fragmentation patterns of ferrocenes (58, 64, 134, 172–174), aided in some cases by deuterium labeling, but the structures of many of the fragment ions are still not known with certainty. Ferrocene and its derivatives will be discussed according to the degree of substitution of the cyclopentadienyl rings, and the remaining iron compounds will be discussed according to the classification: cyclopentadienyliron carbonyls, olefin-iron carbonyls, and fluorocarbon-iron carbonyls.

1. Ferrocene

At a low ionizing voltage of 8 eV ferrocene exhibits only the molecular ion (81, 189). At higher ionizing energies fragmentation occurs, and the major ions observed are $(C_5H_5)_2Fe^+$, $C_5H_5Fe^+$, and Fe^+, although the molecular ion is still the base peak of the spectrum (54, 81, 141, 181), and this accounts for 60% of the total ion current. At an ionizing energy of 70 eV some fragmentation of the cyclopentadienyl ring does occur, and an interesting fragment is $C_9H_7Fe^+$, that is $[M - CH_3]^+$ and the mechanism proposed is shown on p. 242.

The main cleavage of the cyclopentadienyl ring is, however, by elimination of a molecule of acetylene from $C_5H_5Fe^+$, giving the cyclopropenyl ion $C_3H_3Fe^+$, and another noticeable fragment is C_2HFe^+, which can be considered to contain the acetylide ion (148). Peaks corresponding to elimination of the iron atom from the molecular ion have also been observed, at m/e 128 ($C_{10}H_8^+$), m/e 129 ($C_{10}H_9^+$), and m/e 130 ($C_{10}H_{10}^+$), and it is possible that the m/e 128 fragment could be the fulvene cation or some rearranged product thereof (135, 148). An investigation of the mass spectrum of ferrocene at an ionizing voltage of 12 to 20 eV and a relatively high

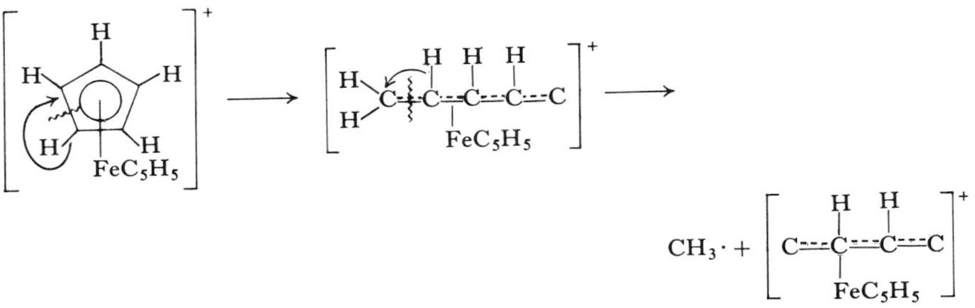

pressure of 10^{-5} torr (181) shows that under these conditions ion–molecule reactions occur to give associated molecules in small abundances. The ions observed at masses higher than the molecular ion were $Fe_2(C_5H_5)$ $C_3H_3^+$, $Fe_2(C_5H_5)_2^+$, $Fe_2(C_5H_5)_2C_3H_3^+$, and $Fe_2(C_5H_5)_3^+$, and "triple-decker sandwich" structures were proposed for these ions.

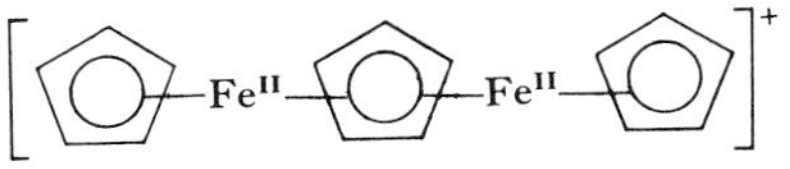

It has recently been shown that decomposition of ferrocene does not occur in the spectrometer until 900°–1000° C (165). Therefore it is reasonable to believe that the associated products observed by Schumacher and Taubenest are indeed the products of ion–molecule reactions.

When a mixture of ferrocene and nickelocene was investigated, peaks due to $Fe_2(C_5H_5)_3^+$, $FeNi(C_5H_5)_3^+$, and $Ni_2(C_5H_5)_3^+$ were observed (181). The mass spectrum of bis(pentamethylcyclopentadienyl)iron [C_5Me_5]$_2$Fe shows a very strong molecular ion peak, ions corresponding to the loss of one and two methyl groups, and the doubly charged molecular ion, as the only ions above m/e 160 (121).

2. Monosubstituted Ferrocenes

Low-voltage mass spectrometry has been used to identify ferrocene and its derivatives, for by using an ionizing voltage of 8 eV only the molecular ion peaks are observed (44, 189). Reed and Tabrizi studied the mass spectra of ten monosubstituted ferrocenes (*170*) and found that the base peak was the molecular ion and that there was no evidence for the loss of iron. The fragmentations which were observed were in accord with the cleavage of usually weak bonds in the ions. No further details were given.

Following the paper by Reed and Tabrizi, Mandelbaum and Cais (*135*) reported the mass spectra of six monosubstituted ferrocenes having a carbonyl group in the α-position to the cyclopentadienyl ring. In addition

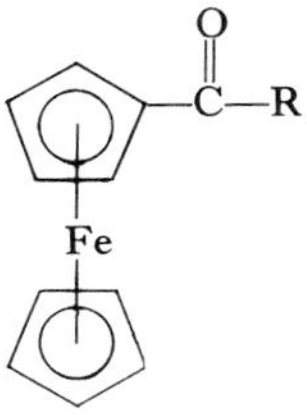

(XXXV) R = Me	(XXXIX) R = OD
(XXXVI) R = Ph	(XL) R = OMe
(XXXVII) R = *p*-MeOC₆H₄	(XLI) R = NHMe
(XXXVIII) R = OH	

to the expected cleavage on both sides of the carbonyl group, cleavage of the bonds between the metal and the two cyclopentadienyl rings was observed.

However, the more interesting aspects of the fragmentation of these compounds were the formation of rearranged ions in which the group R

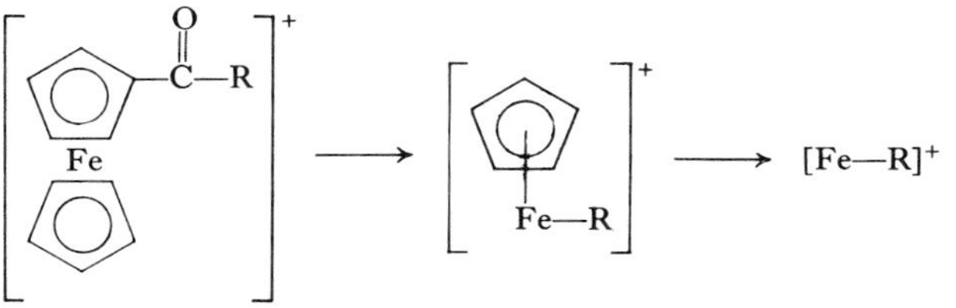

was associated with the metal atom and also the elimination of the iron atom from the $[M - COR]^+$ fragment ion. This gives rise to the ion $(C_5H_5—C_5H_4)^+$ m/e 129, which loses hydrogen to give $[C_5H_4C_5H_4]^+$, which could be the fulvalene cation or some rearranged product thereof. It is noteworthy that the fragment $[C_5H_5FeR]^+$ was the base peak for the carboxylic acid (R = OH).

Mass spectrometry has been used as a means of identification of various ferrocene derivatives, as well as in the study of specific problems such as determination of configuration and the interesting loss of formaldehyde in some methyl esters.

First of all, the compounds for which mass spectral data have been used just for identification will be listed, and a more detailed description will be given of papers dealing with specific mass spectral problems.

The reaction of aminoferrocene with nitrosobenzene gives phenylazo-ferrocene as the main product (56%) as well as the previously unknown phenylazoxyferrocene in 20% yield. The mass spectrum of the azoxy complex shows the molecular ion peak and fragments $[C_5H_5FeC_5H_4N_2O]^+$, $[C_5H_5FeC_5H_4N_2]^+$, $[C_5H_5FeC_5H_4N]^+$, and $PhNO^+$. No fragment corresponding to $[C_5H_5FeC_5H_4NO]^+$ m/e 215 was observed (154).

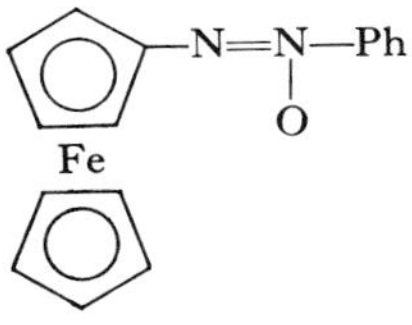

Schlögl has used mass spectrometry to confirm the molecular weights of ferrocene derivatives such as

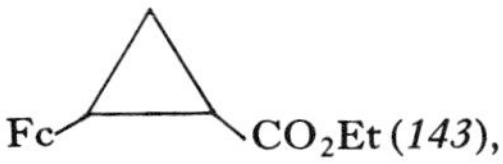

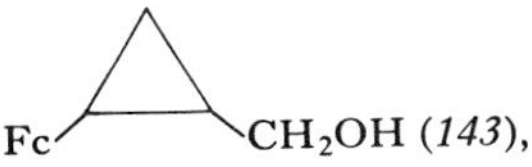

$Fc-(C{=}C)_4Ph$ (179),   and $Fc-CH(OH)CH_2-CH{=}CH_2$ (143)

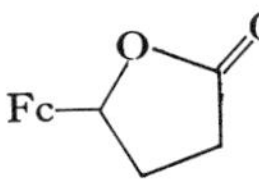

$(Fc = C_5H_5FeC_5H_4)$

The mass spectrum of dimethylaminomethylferrocene exhibits as the base peak the molecular ion, as well as a significant peak for loss of one hydrogen, a feature commonly observed in the mass spectra of simple amines (186). The tertiary alcohol, diphenylhydroxymethylferrocene, shows as the base peak an ion at m/e 230 for which the structure (XLII) is proposed or a rearrangement product thereof. The molecular ion gives a peak of 11.2% relative abundance, and the other very strong peak in the spectrum is for the phenyl cation, $C_6H_5^+$ (186).

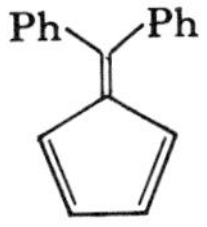

m/e 230

(XLII)

Egger has investigated the mass spectra of various primary, secondary, and tertiary carbinols of ferrocene and attempted to correlate fragmentations with stereochemistry (58). The important fragmentations are loss of water, loss of the unsubstituted ring, cleavage of the substituted ring leaving $C_5H_5Fe^+$, or migration of the hydroxyl group to the iron atom giving the fragment ion $[C_5H_5FeOH]^+$ m/e 138. The monosubstituted compounds will be discussed in this section, the others in the relevant sections following.

For the series of primary alcohols (XLIII)–(XLVI)

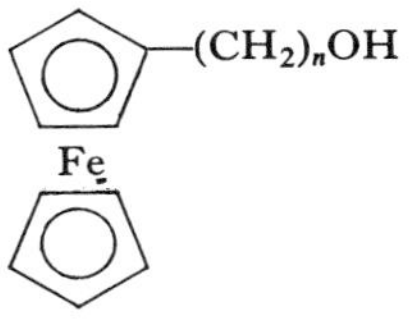

(XLIII) $n = 1$
(XLIV) $n = 2$
(LXV) $n = 3$
(XLVI) $n = 4$

the migration of the hydroxyl group to the iron decreases relative to cleavage, giving $C_5H_5Fe^+$ m/e 121, as n increases from 1 to 4.

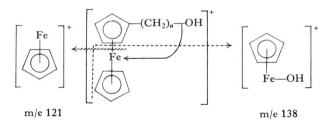

m/e 121   m/e 138

Ratio of Intensities, peaks 138 *to* 121

(XLIII)   10
(XLIV)   0.30
(XLV)   0.077
(XLVI)   0.03

The loss of water is also observed from the molecular ion, and deuteration studies indicate that the hydrogen comes from the cyclopentadienyl ring for (XLIII). The dominating fragmentation of the hydroxyferrocene (XLIV) is loss of CH_2OH, giving the ion $C_5H_5FeC_5H_4CH_2^+$ m/e 199. The ion $[C_5H_5FeOH]^+$ gives rise to the fragment ion $FeOH^+$ m/e 73 by loss of the cyclopentadienyl ring.

The secondary alcohols investigated were (XLVII) and (XLVIII). The former loses water more easily than the corresponding primary alcohol (XLIII), and deuteration studies show that the water is eliminated by a 1,2-mechanism. Fragment ions for loss of C_5H_5 and $C_5H_5 + H_2O$ are also observed. However, the second most abundant peak (the base peak being the molecular ion) is $C_5H_5FeOH^+$.

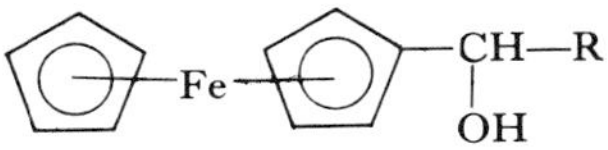

(XLVII) R = Me
(XLVIII) R = Ph

The mass spectrum of ferrocenylphenylcarbinol (XLVIII) does not exhibit loss of water from the molecular ion; the main fragmention is loss of the unsubstituted cyclopentadienyl ring, giving m/e 227, which then loses H_2O followed by loss of iron atom. Migration of the hydroxyl group occurs, giving m/e 138 and also $[M-138]^+$, and the mass spectrum of the deuterated compound (XLIX) shows that loss of water from the $[M - C_5H_5]^+$ involves "participation of the aromatic ring hydrogen atoms." Two isomeric tertiary alcohols, (L) and (LI), were investigated by Egger, who found that (L) loses water more readily than (LI) and that the ratios for the peaks 138–121 are 0.4 and 0.2 for (L) and (LI), respectively.

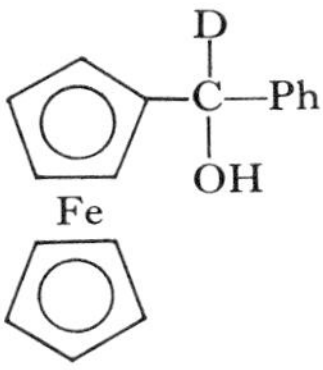

(XLIX) 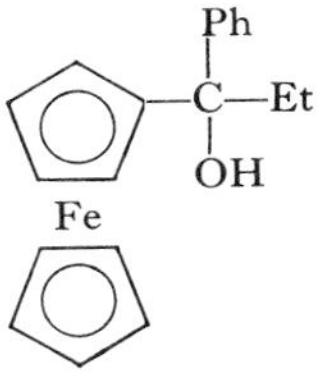   (L) 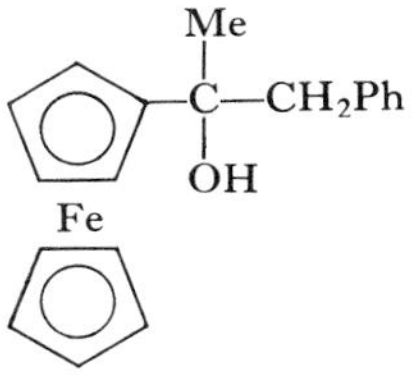   (LI)

Roberts *et al.* have observed the unusual fragmentation of loss of formaldehyde in methyl esters of ortho-substituted ferrocenyl-benzenes (*172*). This novel fragmentation of loss of aldehyde has also been observed by Lupin *et al.* for methyl, ethyl, and *n*-propyl esters of ortho-substituted ferrocenyl benzenes as well as for methyl esters of alkenyl ferrocenes (*134*). The "ortho" effect, in a rearrangement depicting a six-membered transition state, has been proposed by Bursey *et al.* to explain this fragmentation, but it is more likely that the metal atom plays an important role, possibly with the migration of the methoxyl group to the metal (*134*).

Bursey *et al.* studied methyl *o*- and *m*-ferrocenyl benzoates and found that for the ortho isomer, fragmentation of the unsubstituted ring was followed by loss of 30 m.u. (metastable peaks observed). Loss of 30 m.u. is attributed to loss of formaldehyde with transfer of hydrogen to the charge-retaining fragment, giving m/e 225, which then loses CO and Fe, giving m/e 141. Transfer of methoxyl group to iron, followed by loss of CH_2O, would leave the hydrogen on the metal, but no evidence for a Fe—H bond could be obtained. The mass spectrum of the methyl *m*-ferrocenylbenzoate does not show loss of 30 m.u. but remarkably shows a metastable peak for simultaneous loss of methoxyl and cyclopentadienyl iron from the molecular ion. Otherwise fragmentation typical of ferrocenes is found. The ethyl ester of *o*-ferrocenylbenzoic acid shows transfer of hydrogen rather than methyl from the $[M - C_5H_5]^+$ ion with the expulsion of acet-aldehyde, and the fragmentation pattern is similar to that of the methyl ester, except that the expected loss of ethylene from the molecular ion and the $[M - C_5H_5]^+$ ion is observed as well as loss of the ethoxy radical from the M^+ ion. The *m*-ethyl ester shows loss of ethylene from the molecular ion and the resulting fragment ion, m/e 306, undergoes the unusual frag-mentation observed for the *meta*-methyl ester, a simultaneous loss of hydroxyl, iron, and cyclopentadienyl, $[C_5H_5FeOH]^+$, a process which has been observed for some of the alcohols described by Egger (*58*).

Loss of aldehyde was also observed in the mass spectra of *o*-ferrocenyl phenylacetate esters. Thus methyl *o*-ferrocenylphenyl acetate loses the cyclopentadienyl radical followed by CH_2O, but this elimination is not so prevalent as in the spectrum of the benzoate ester. The molecular ion also loses a methoxyl radical followed by loss of cyclopentadiene, C_5H_6, and then loss of iron, giving m/e 153. The meta and para isomers of the phenyl-acetate esters show no simultaneous loss of cyclopentadienyl iron and alkoxy fragments and the fragmentation is quite straightforward. Loss of the methoxycarbonyl group gives a stable ion at m/e 275, as evidenced by a doubly charged species at m/e 137.5. It is possible that ring expansion has occurred to give a tropylium ion. The mass spectra of *o*, *m*, and *p*-ferro-cenylacetophenones were studied, but the ortho isomer did not show any evidence of a rearrangement, and the mass spectra are interpretable by simple cleavage processes. The most notable difference between the ortho isomer and the meta and para isomers is the presence of an $[M - MeCO]^+$ ion in the spectrum of the ortho isomer and the absence of $[M - C_5H_5]^+$ and $[M - C_5H_5 - MeCO]^+$ ions for the meta and para isomers (*172*).

A low-voltage study of the trideuteriomethyl-*o*-ferrocenyl benzoate showed no isotope effect and it is suggested that a σ- or π-bond is formed between the ferrocene unit and the transferred hydrogen before the rate-determining step, which may be pictured as the cleavage of the C—O bond. However, there is a definite isotope effect in the ferrocenylphenyl acetate rearrangement which is relatively constant over a wide range of voltage. The data seem to be consistent with the C–D bond rupture as the rate-determining step but does not preclude complex formation prior to this step nor participation of the C—O bond rupture also. Transfer of the methoxy group to the iron atom is not considered and the rearrangement is discussed in terms of the "ortho effect" in analogy to aromatic systems, without fully taking into account the effect of the metal.

Roberts *et al.* have also investigated the mass spectra of the ferrocenyl esters (LII) and (LIII) when *n* has the values 0 to 5 (*173*). The spectra of

$$\text{Fe}\overset{\displaystyle\bigcirc}{\underset{\displaystyle\bigcirc}{}}\!\!\!-(CH_2)_nCO_2R$$

(LII) R = Me
(LIII) R = Et

the six methyl esters are in general quite similar, with the major fragments corresponding to fragmentation of the ferrocene moiety and to cleavage of the ester function. Typical fragmentations of aliphatic esters, giving an $[M-31]^+$ ion and an m/e 74 ion, are suppressed by location of the charge on the ferrocene portion of the molecule. In each case the molecule ion loses cyclopentadienyl radical, but only when $n=0$, 2, or 3 does loss of CH_2O occur as the sequential step, and metastable peaks are present to confirm the eliminations. Another interesting feature observed in several of the spectra is the loss of CH_2O from fragments which have previously lost the carbonyl group. Thus for (LII) ($n=0$), loss of C_5H_4CO occurs with the migration of the methoxyl group to the metal, giving the fragment ion (LIV) as proposed by Mandelbaum and Cais (135). This ion then eliminates CH_2O.

The trideuteriomethyl ester, $C_5H_5FeC_5H_4COOCD_3$ fragments such that the m/e 155 ion $[C_5H_5FeOCD_3]^+$ loses CH_2O, CHDO, and CD_2O in the ratio 4:7:7. Also half the intensity of the ion m/e 121 $C_5H_5Fe^+$ is shifted to m/e 122. These results indicate that partial randomization of the methoxy hydrogens and the cyclopentadienyl hydrogens occurs in at least some of the ions of structure (LIV). A similar sequence is observed for methyl ferrocenylacetate (LII) ($n=1$), but the trideuteriomethyl ester shows that elimination of formaldehyde occurs before equilibration of the methoxy and ring hydrogens occurs.

(LIV)

The spectrum of methyl β-ferrocenylpropionate shows a similar transfer of methoxy group, the loss of cyclopentadienyl being followed by loss of $CH_2=C=O$, migration of the methoxy group, and then elimination of formaldehyde. Transfer of methoxy is not observed in the mass spectra of the other homologous methyl esters. Other interesting features in these spectra are the occurrence of fragments containing C_6H_6 and C_7H_7 units such as $MeOFeC_6H_6^+$ m/e 165, $FeC_7H_7CO^+$ m/e 175, and $FeC_7H_7^+$ m/e 147.

The mass spectra of the ethyl esters are dominated by the elimination of ethylene from the molecular ion, and the subsequent fragmentation paths are similar to that of the corresponding carboxylic acid. Only the ethyl esters of β-ferrocenylpropionate and γ-ferrocenylbutyric acid ($n=2, 3$) show loss of C_5H_5 followed by loss of acetaldehyde, and numerous ions in the spectra seem to contain the C_6H_6 and C_7H_7 ligands.

No isotope effect was observed when $n=2$, but was observed when $n=3$ for the methyl esters (LII). These results agree with the absence and presence of isotope effects for the methyl o-ferrocenylbenzoate and methyl o-ferrocenylphenylacetate. Thus when there are two carbons between the ferrocene unit and the ester function, no isotope effect is observed, and the hydrogen is transferred to the metal before the rate-determining step. When three carbons intervene between the ferrocene unit and the ester, the isotope effect is observed, suggesting that the C—D bond rupture is the rate-determining step. The ion $C_5H_5FeOMe^+$ formed from methyl ferrocenylcarboxylate shows by labeling studies considerable scrambling of the cyclopentadienyl and methyl hydrogens. The ion of the same composition from methyl ferrocenylacetate loses CH_2O without scrambling, so it is certain that ions written as $MeOFeC_5H_5^+$ have different properties and possibly different structures, depending on whether they are formed from the ferrocenecarboxylate ester or the ferrocenylacetate ester (173).

The mass spectra of some vinylic ferrocenes have been measured (134) and all show a strong molecular ion peak, and cleavage of the unsubstituted ring usually occurs before fragmentation of the vinyl substituent.

(LV)   R = CO_2H
(LVI) R = COMe
(LVII) R = cis-CO_2Me
(LVIII) R = trans-CO_2Me
(LIX) R = $(CH_2)_2OOCCH_3$

(LX) X = Cl
(LXI) X = Br

However, when the unsaturated part of the functional group R is not in conjugation with the cyclopentadienyl ring, as in the case of (LIX), cleavage of the side chain is observed before loss of the unsubstituted ring. The spectra of the two isomeric acrylic esters (LVII) and (LVIII) are very similar, the only significant difference being that the trans isomer shows an $[M-OMe]^+$ fragment which is not observed in the spectrum of the cis isomer. The main fragmentation for both these esters is, after the loss of C_5H_5, elimination of formaldehyde; a similar rearrangement process was observed in the cymantrene derivative $(CO)_3MnC_5H_4-CH=CH-CO_2$ Me (37). Fragmentations of the $[M-C_5H_5]^+$ ions occur in which the iron atom is still attached to the organic moiety. Bicyclic structures can be envisaged for these ions and it is possible that the bicyclic form is still retained even after the cleavage of the metal atom. The fragmentation patterns of the two dihalovinylic compounds (LX) and (LXI) are markedly different from those for the vinylic compounds discussed above. The molecular ion peak is observed in both cases, but no cleavage of the unsubstituted ring is observed, rather, elimination of X_2 gives the ion $C_{12}H_{10}Fe^+$ m/e 210.

The base peak for both spectra corresponds to $C_{12}H_9^+$ m/e 153, but as no metastable peaks are observed, the fragmentation path leading to this ion is not clear. Three routes are possible, namely, simultaneous loss of HCl and FeCl from M^+, loss of FeH from ion m/e 210, or loss of the Fe from ion m/e 210 giving m/e 154, which then expels a hydrogen atom. The other major fragments observed in the spectrum correspond to $C_7H_5^+$ and $C_5H_3^+$ which could be formed from cleavage of the m/e 153 ion. The predominant fragmentation of the ferrocenylacetylene carboxylic acid is elimination of carbon dioxide followed by either loss oᶜ acetylene or FeH (both processes accompanied by metastable peaks), and the base peak corresponds to $C_{12}H_8^+$ m/e 152 (*134*).

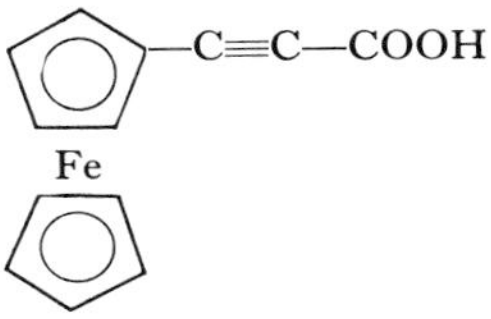

Recently attention has been drawn to the peaks at masses corresponding to $C_6H_6Fe^+$ and $C_7H_7Fe^+$ in the mass spectra of many substituted ferrocenes, and particularly in the case of the acrylic ester (LVII) or (LVIII), and it was suggested that the latter ion may have the tropylium structure, but the evidence is not conclusive and does not rule out the bicyclic structure discussed above (*174*).

3. Polysubstituted Ferrocenes

a. *Disubstituted Ferrocenes.* Although the molecular ion peaks were not observed for the isomeric 1,1'- and 1,2-disubstituted ferrocenes (LXII) and (LXIII), analysis of the fragmentation patterns enabled the compounds to be distinguished (*186*).

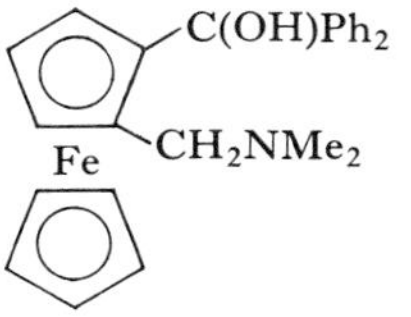

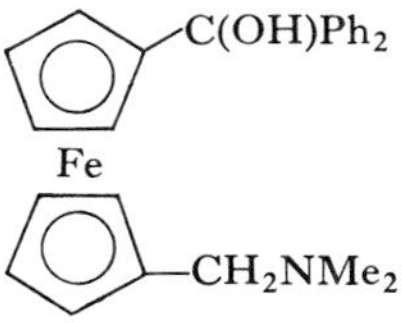

Thus the 1,1'-disubstituted compound shows as the base peak an ion at m/e 230 attributed to diphenylfulvene (LXIV), and formation of such an ion from the 1,2-disubstituted compound would involve the energetically less favorable transfer of hydrogen. The 1,2-disubstituted compound showed a strong peak at m/e 242 for $[C_5H_5FeC_5H_3CH_2NMe_2]^+$ which was of low abundance in the spectrum of (LXIII). Elimination of $C(OH)Ph_2$ from an otherwise unsubstituted ring such as in (LXV) is not a favored process.

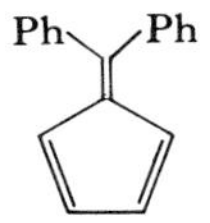

LXIV

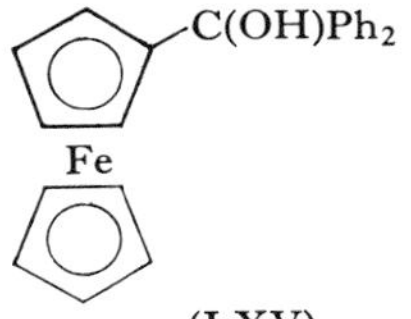

(LXV)

A considerable number of· 1,2- and 1,1′-disubstituted ferrocenes have been investigated by Egger (58) and Egger and Falk (59), and a good correlation between mass spectra and stereochemistry has been found. The homonuclear disubstituted compounds (LXVIa), (LXVIb), and (LXVII) are thermally very sensitive, and their spectra differed only in the intensities of some of the fragments.

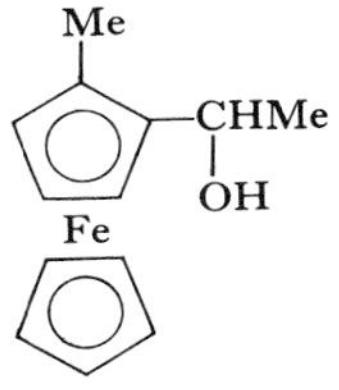

(LXVI)   (LXVII)
(a) ψ-endo
(b) ψ-exo

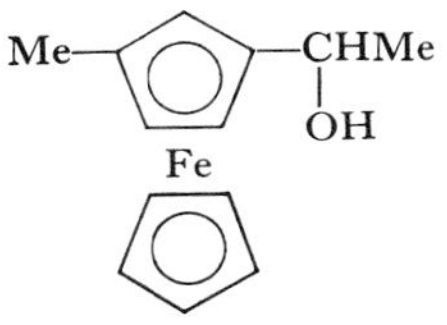

The general fragmentation path resulted in the ions $[M - H_2O]^+$, $[M - C_5H_5]^+$, $[M - H_2O - C_5H_5]^+$ and migration of the hydroxyl group to the iron atom with cleavage of C_5H_5FeOH, giving the $[M - 138]^+$. Loss of methyl radical only occurs from this $[M - 138]^+$ fragment. Egger also investigated the mass spectra of several cyclic secondary alcohols of varying stereochemistry (LXVIII)–(LXX) and found that fragmentation of the endo isomers was much more pronounced than fragmentation of the exo isomers.

Thus for the endo isomers, the highest peak after the molecular ion is $[M - C_5H_5FeOH]^+$, but this peak is very weak in the spectra of the exo isomers. Elimination of water from (LXVIIIa) and (LXIXa) is followed by loss of hydrogen and aromatization of the six-membered ring. The ratio of the peaks m/e 138/121 for the endo isomers is approximately ten

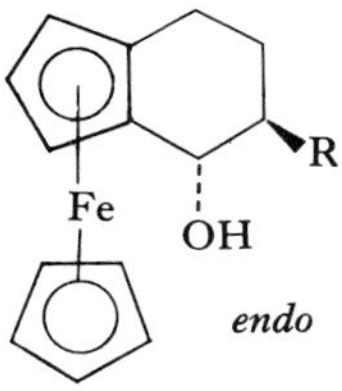

(LXVIII)   (LXIX)   (LXX)
(a) R = H   (a) R = H
(b) R = Ph   (b) R = Ph

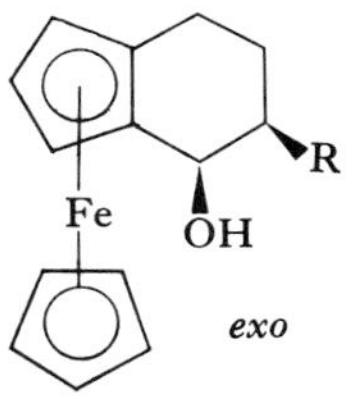

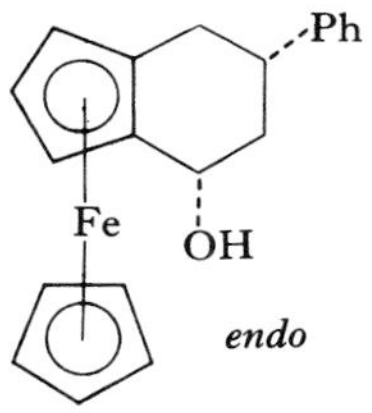

times that for the exo isomers. The diol compounds (LXXI) and (LXXII) show loss of two molecules of water, but the endo/endo isomer shows only a weak peak for loss of one molecule of water.

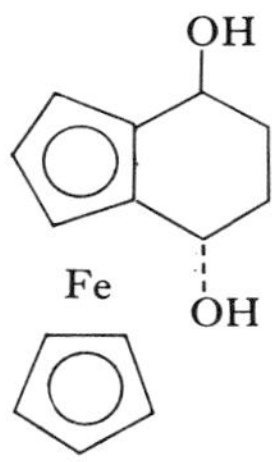

exo/endo   endo/endo
(LXXI)   (LXXII)

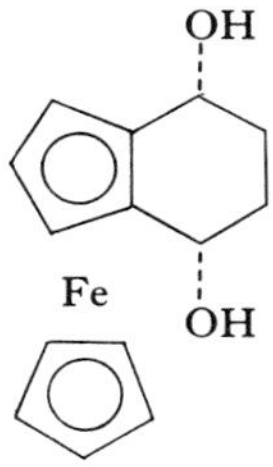

In both cases the main fragmentation corresponds to the loss of iron, the unsubstituted ring and both hydroxyl groups giving the fragment ion $C_9H_9^+$ m/e 117, which eliminates a molecule of hydrogen giving m/e 115 (58). The 1,1'-diols (LXXIIIa) and (LXIIIb) eliminate water with the formation of the ether fragment (LXXXIV).

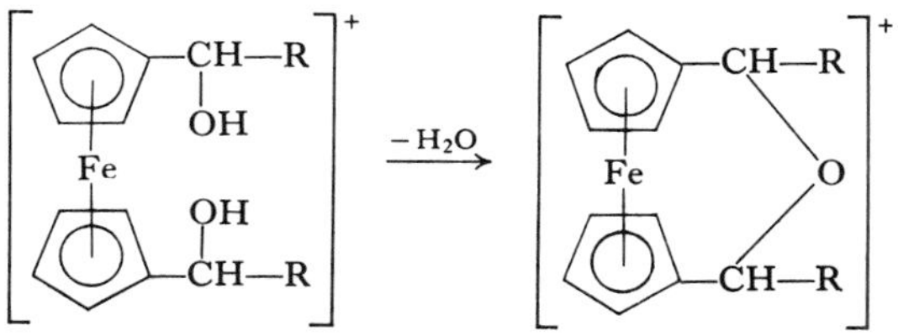

(LXXIII)   (LXXIV)
(a) R = H
(b) R = Ph

The base peak for the primary alcohol (LXXIIa) occurs at m/e 78, $C_6H_6^+$. The corresponding ion $[M - 78]^+$ is also observed, and the latter ion loses water, giving m/e 150. The spectra of both diols show a peak at m/e 121 due to $C_5H_5Fe^+$, indicating that rearrangement with hydrogen transfer has occurred, and so the criterion that the peak at m/e 121 can be used to differentiate between homo- and heteronuclear-substituted ferrocenes must be treated with caution. The bridged system (LXXV) exhibits a very strong molecular ion peak and the main fragmentation is elimination of C_6H_6, giving the fragment m/e 164 which then eliminates water.

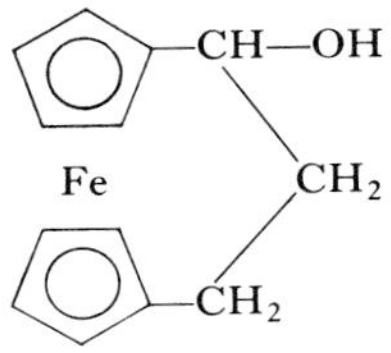

(LXXV)

Egger and Falk (59) have reported the partial mass spectra of the hydroxy ketones (LXXVIa, b, and c).

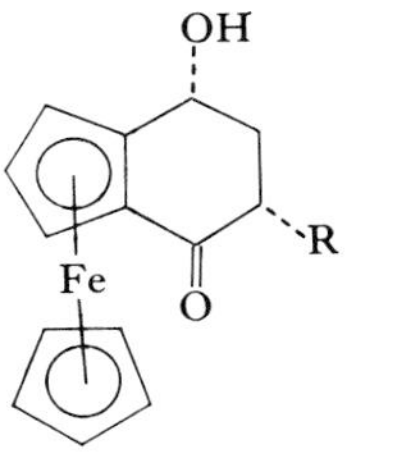

(LXXVI)
(a) R = H
(b) R = Ph
(c) R = C_6H_{11}

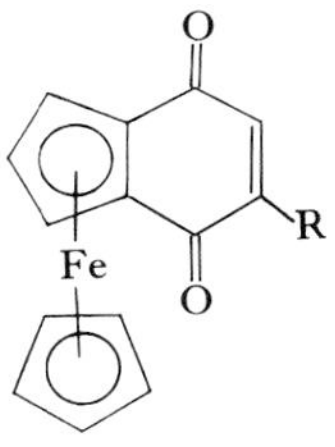

(LXXVII)
(a) R = H
(b) R = Ph

The greatest fragmentation occurs when R=H, and loss of water is observed from both the molecular ion and the $[M - C_5H_5]^+$ ion. The other hydroxy ketones show as the main fragmentation path, loss of the unsubstituted cyclopentadienyl ring. The diketones (LXXVIIa) and (LXXVIIb) exhibit very strong molecular ion peaks, and all other peaks are below 10% of this peak, which is consistent with the structure of these molecules. The diketones (LXXVIII) show strong molecular ion peaks. Also, the fragment ions corresponding to the elimination of the CO groups are of significant abundance. The diol LXXIX has a strong molecular ion peak and the major fragment ions are $[M - 2\ H_2O]^+$ and $[M - C_5H_5Fe - 2\ OH]^+$ (58).

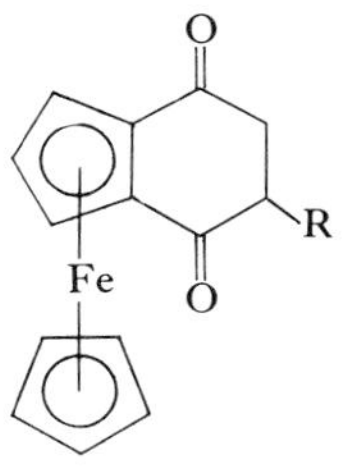

(LXXVIII)
(a) R = H
(b) R = Ph

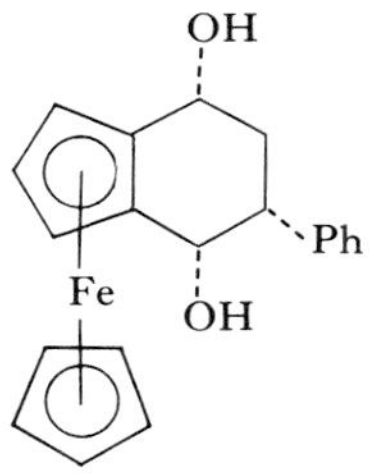

(LXXIX)

The spectra of various disubstituted ferrocenes were examined by Clancy and Spilners (44) at low ionizing voltage, and they observed only the molecular ions. They have also used mass spectrometry to identify products from the reaction of lithioferrocene with benzyl chloride; one of the products was dibenzylferrocene (189). The mass spectra of [5]-ferroceno-phanes show very intense molecular ion peaks and fragmentation is limited (9, 10). The ketone derivatives (LXXXI) and (LXXXII) fragment by elimination of carbonyl groups first.

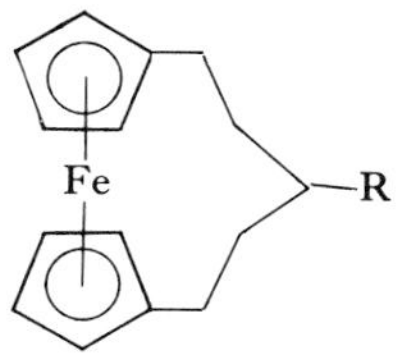

(LXXX)
R = H, Ph

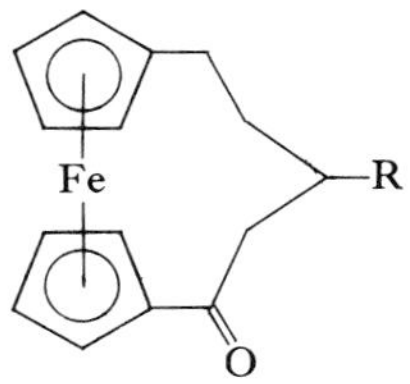

(LXXXI)
R = H, Ph

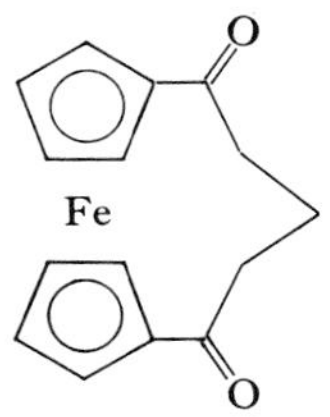

(LXXXII)

As a result of the observation that a peak corresponding to $C_7H_7Fe^+$ occurs in the spectra of many ferrocenes containing at least a two-carbon side chain, Roberts et al. (174) have examined the nature of this ion. They used as the parent compound 1,1′-divinylferrocene, for which the principal ions are M^+, $(M - C_2H_2)^+$, $C_7H_7Fe^+$, $C_5H_5Fe^+$, $C_7H_7^+$, $C_5H_5^+$, and Fe^+. The ion $C_7H_7^+$ loses acetylene to give $C_5H_5^+$ (metastable peak observed) and from the spectrum of the α,α-dideuteriodivinylferrocene it appears that most of the $C_5H_5^+$ is formed from the tropylium ion $C_7H_7^+$. However, the variation of the deuterium distribution in the $C_5H_5^+$ fragments with ionizing voltage indicates that there is more than one path leading to $C_5H_5^+$. The more important question as to the form of the C_7H_7 ligand in the ion $C_7H_7Fe^+$ is not so clear. $C_7H_7Fe^+$ loses acetylene, giving the ion $C_5H_5Fe^+$ m/e 121, and an analysis of the deuterium distribution of this fragment ion in the spectrum of the dideuterio compound indicates that there may be a certain contribution due to the tropylium form.

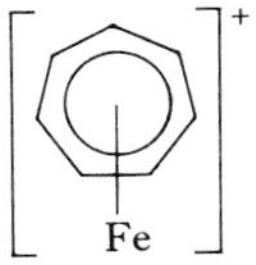

This is by no means the only structure contributing to the $C_7H_7Fe^+$; others such as the bicyclic form suggested previously may exist, and, on cleavage of iron from this moiety, rearrangement to the seven-membered ring could occur.

b. *Trisubstituted Ferrocenes.* The only trisubstituted ferrocenes for which mass spectral data have been reported are (LXXXIII) (*186*), the isomeric methyl esters of 1,1′-dimethylferrocene carboxylic acid (LXXXIVa) and

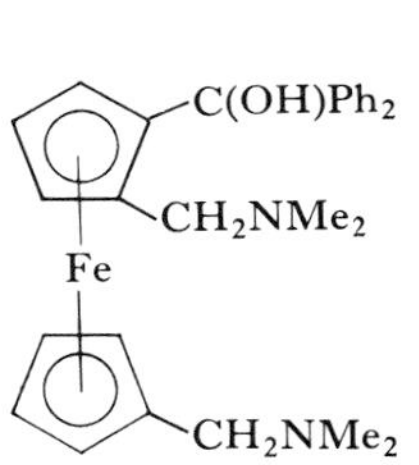

(LXXXIII)

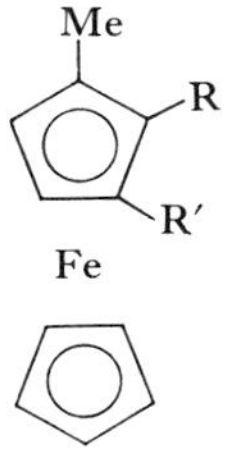

(LXXXIV)
(a) R = H, R′ = CO$_2$Me
(b) R = CO$_2$Me, R = H

(LXXXIVb) (*64*), and triethylferrocene (*189*). The 1,2′,1′-trisubstituted complex (LXXXIII) does not exhibit a molecular ion peak, but has a strong peak at m/e 242 corresponding to the loss of both carbinol groups and one hydrogen, and a peak at m/e 230 corresponding to the diphenyl-fulvene cation which was also observed in the spectrum of $C_5H_5FeC_5H_4$ $C(C_6H_5)_2OH$ (LXV). For (LXXXIVa) and (LXXXIVb) and triethyl-ferrocene only the observation of the molecular ion peaks has been reported.

c. *Tetrasubstituted Ferrocenes.* Egger (*58, 60*) has reported the spectra of the 1,2,1'2'-bicyclic systems (LXXXV) and (LXXXVI) and found that elimination of water is much more pronounced for the exo/endo isomers than the endo/endo forms.

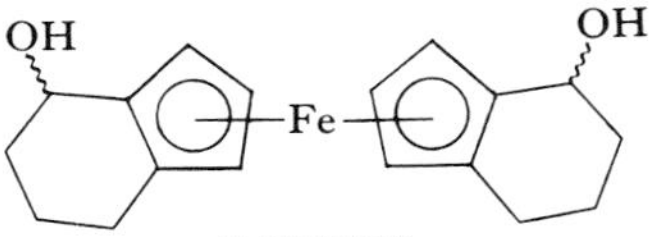

(LXXXV)
(a) *endo/endo*
(b) *exo/endo*

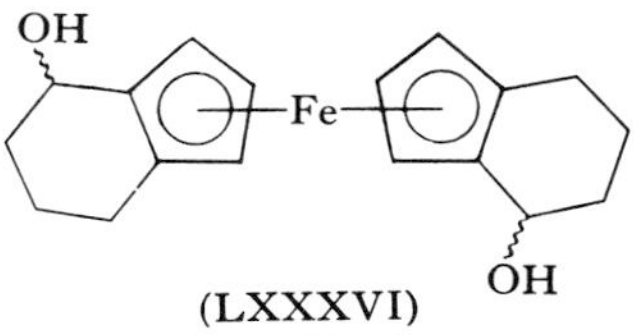

(LXXXVI)
(a) *endo/endo*
(b) *exo/endo*

The endo/endo isomers show a very strong peak at m/e 118 corresponding to (LXXXVII), and this peak is very weak in the spectra of the exo/endo forms.

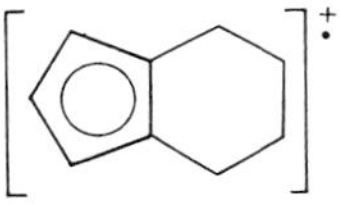

(LXXXVII)
m/e 118

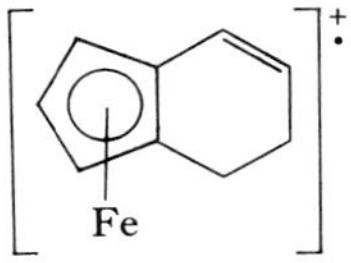

m/e 172

Another important cleavage is the loss of one of the organic ligands followed by loss of water, giving the ion m/e 172 which is an abundant fragment in the spectra of all four compounds (*60*). The iron derivative of hydropentalene (LXXXVIII) exhibits a very strong molecular ion peak and fairly intense peaks for $C_8H_6Fe^+$ and $C_8H_6^+$ which could possibly involve the pentalene cation (*103*). The mass spectrum of the bis(trimethylene)-bridged complex (LXXXIX) has been reported by Cordes and Rinehart (*48*) to show a very strong molecular ion peak and also a strong doubly charged species M^{2+}. Fragmentations involve loss of methyl, acetylene, ethylene, and propene from the molecular ion, followed by further degradation of the resulting ions.

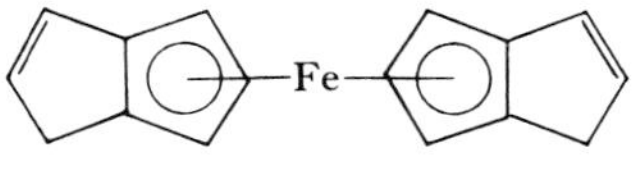

(LXXXVIII)

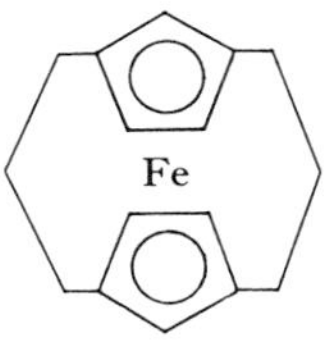

(LXXXIX)

4. Polyferrocenes

Mass spectrometry has proved extremely useful in determining the composition of various polyferrocene products such as those formed by trimerization of ferrocenylacetylenes (180), reactions of ferrocenyl acetylenes with metal carbonyls (175), oxidative coupling of ferrocenyl polyacetylenes (179), and lithiation of ferrocene (186). The 1,12-dimethyl [1,1]ferrocenophane (XC) shows a very strong molecular ion and fragment $[M—Me]^+$ $[M—2\ Me]^+$ and the doubly charged species M^{2+}, $[M-Me]^{2+}$ and $[M-2\ Me]^{2+}$ (197).

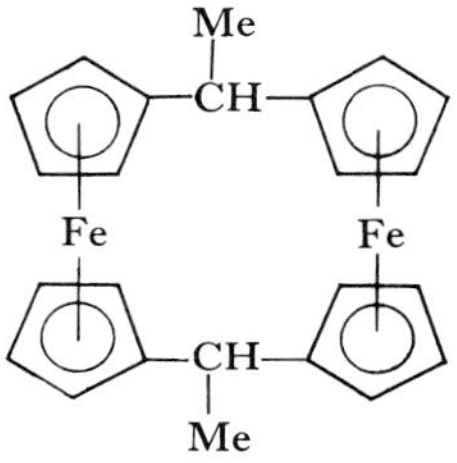

(XC)

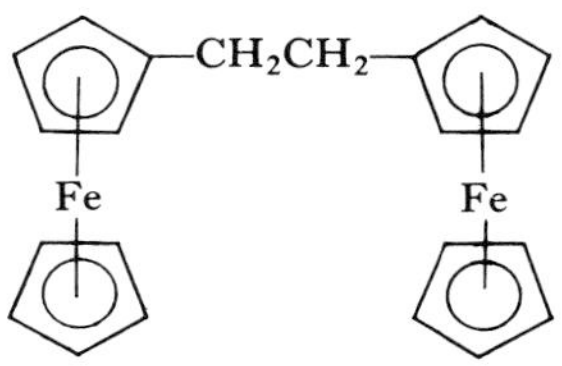

(XCI)

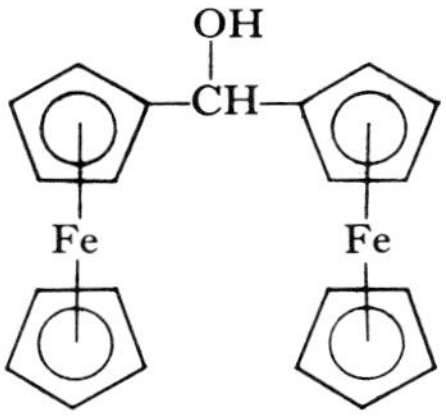

(XCII)

Diferrocenyl alkanes such as (XCI) show only a moderately strong molecular ion peak and alkane cleavage is a pronounced fragmentation. An unusual fragmentation was observed in which cleavage of the alkane was accompanied by transfer of an iron atom, giving a fragment $[C_{11}H_{11}Fe_2]^+$ (48). The carbinol (XCII) shows as the main fragmentation, loss of C_5H_5FeOH, giving as the base peak the $[M-138]^+$ ion (57). Diferrocenyl-mercury and polyferrocenylmercury show only peaks for $C_{10}H_{10}Fe^+$, Hg, and Fe^+; no peaks were observed for $C_{10}H_9Fe^+$ and $C_{10}H_8Fe^+$ (169). The ferrocene derivative of benzopentalene has been shown by mass spectrometry to have the dimeric structure (XCIII) (35).

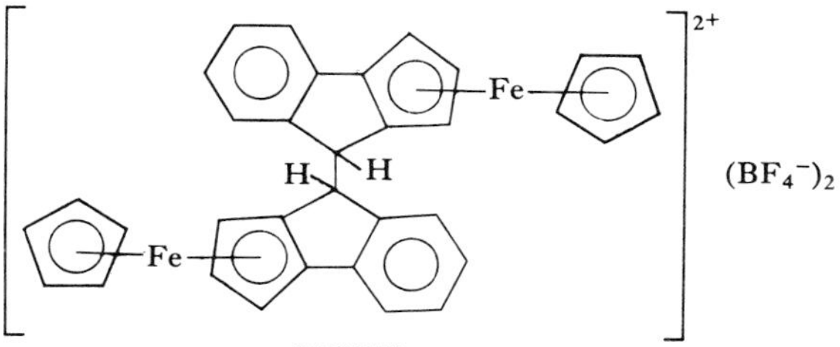

(XCIII)

5. Cyclopentadienyliron Carbonyls

The mass spectrum of the carbonyl-bridged complex $[C_5H_5Fe(CO)_2]_2$ shows stepwise loss of carbon monoxide to give the fragment ion $(C_5H_5)_2$ Fe_2^+ (*131, 182*). At high electron energy (70 eV) this ion is the base peak of the spectrum, but at a lower electron energy (20 eV) (*182*) the base peak is $(C_5H_5)_2Fe^+$, formed by migration of a cyclopentadienyl ring and elimination of an iron atom (metastable peak observed). An alternative fragmentation of the molecular ion (XCIV) by symmetrical cleavage gives C_5H_5 $Fe(CO)_2^+$, which can eliminate carbon monoxide giving $C_5H_5FeCO^+$.

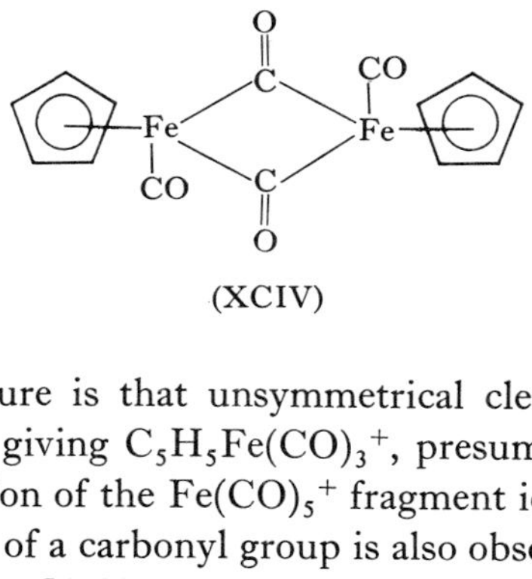

(XCIV)

An interesting feature is that unsymmetrical cleavage of the dimeric molecule also occurs, giving $C_5H_5Fe(CO)_3^+$, presumably in an analogous manner to the formation of the $Fe(CO)_5^+$ fragment ion in the spectrum of $Fe_3(CO)_{12}$. Migration of a carbonyl group is also observed at 20 eV, giving in low abundance the ion $[C_5H_5Fe(CO)_4]^+$. While at 70 eV no fragmentation of the cyclopentadienyl ring is observed, at 20 eV $C_5H_5Fe^+$ eliminates acetylene to give the cyclopropenyliron ion $C_3H_3Fe^+$ (metastable peak observed). Schumacher and Taubenest studied the mass spectrum under high pressure (10^{-5} torr) and observed ions such as $(C_5H_5)_4Fe_4(CO)_4^+$, $(C_5H_5)_3Fe_3(CO)_4^+$, and $(C_5H_5)_3Fe_2(CO)_4^+$ formed by ion–molecule reactions. Fragmentation of the associated ion $(C_5H_5)_3Fe_2(CO)_4^+$ occurred according to the process shown below, and the resulting ion showed step-wise loss of CO to give the ions $(C_5H_5)_2CH_2Fe_2(CO)_n^+$ $(n=0-3)$.

$$(C_5H_5)_3Fe_2(CO)_4^+ \rightarrow (C_5H_5)_2CH_2Fe_2(CO)_4^+ + C_4H_3$$

King has prepared the tetrameric cyclopentadienyliron carbonyl $[C_5H_5FeCO]_4$ by refluxing $[C_5H_5Fe(CO)_2]_2$ in xylene for 12 days, and its mass spectrum confirms the tetrameric composition (104). The two most intense peaks are for $C_5H_5Fe^+$ and $(C_5H_5)_2Fe^+$, the latter being formed by migration of a cyclopentadienyl ring, but this might be a thermal process rather than an electron-induced rearrangement.

Schumacher and Taubenest (182) have also studied the spectrum of the mononuclear complex $C_5H_5Fe(CO)_2Br$ and find that the carbonyl groups are eliminated preferentially to the bromine and the base peak is the ion $[C_5H_5FeBr]^+$. Fragmentation of the cyclopentadienyl ring occurs and the ions $C_3H_3FeBr^+$ and $C_3H_3Fe^+$ are observed. Although the spectrum was run under similar conditions (20 eV, 10^{-5} torr) as cyclopentadienyliron dicarbonyl dimer (XCIV), no ion–molecule reactions were observed.

Two sulfido-bridged complexes, (XCV) and (XCVI), have been examined by Preston and Reed (167), and the expected loss of carbonyl and alkyl (or aryl) groups is observed.

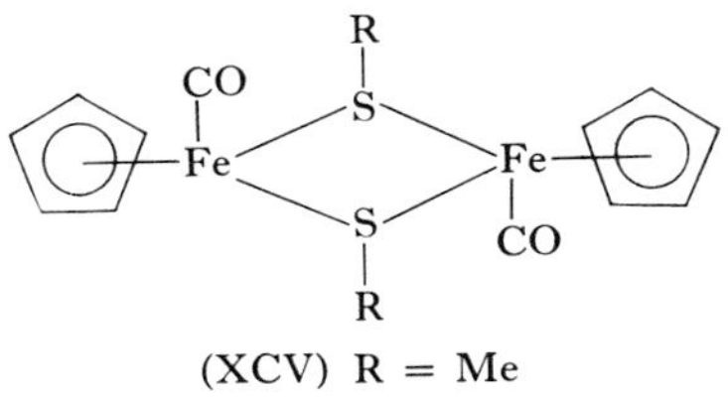

(XCV) R = Me
(XCVI) R = Ph

However, the $[M-2\ CO]^+$ ion for (XCV) ion also dissociates by loss of MeSH followed by loss of MeS. The structure of the $[M-2\ CO-MeSH]^+$ ion possibly involves a cyclopentadienyl group simultaneously σ-bonded to one iron and π-bonded to the second iron atom, in an analogous manner to some of the structures suggested for fragment ions of $[C_5H_5ZrCl]_2O$ (171). An ion corresponding to $(C_5H_5)_2Fe$ is also observed in the spectra of both (XCV) and (XCVI) but is thought to occur by a thermal process.

$$(XCV) \xrightarrow{-2\ CO}$$

m/e 336 → m/e 288

Lewis *et al.* (*131*) have measured spectra of some tin and mercury derivatives of cyclopentadienyliron carbonyl (XCVII)–(XCIX).

$[C_5H_5Fe(CO)_2]SnPh_3$ $[C_5H_5Fe(CO)_2]_2SnCl_2$ $[C_5H_5Fe(CO)_2]_2Hg$

(XCVII)   (XCVIII)   (XCIX)

The triphenyltin complex (XCVII) shows a weak molecular ion and a weak $[M - CO]^+$ fragment, and the major concentration of metal-containing ions contains no carbonyl groups. Transfer both of phenyl groups to iron and of cyclopentadienyl to tin occurs and metastable peaks indicate that $PhSnFe(C_5H_5)^+$ is the precursor of both $C_5H_5Sn^+$ and $C_6H_5Fe^+$.

The chlorotin complex (XCVIII), involving three metal atoms, in addition to giving ions resulting from fission of one of the tin–iron bonds, e.g., $[ClSnFe(CO)_2C_5H_5]^+$, yields mainly ions which still contain three metal atoms, e.g., $[Cl_2SnFe_2(CO)_3(C_5H_5)_2]^+$. As expected, preferential loss of carbonyl groups occurs, and transfer of a cyclopentadienyl group from iron to tin is observed; however, the corresponding transfer of chlorine from tin to iron was not observed. The spectrum of the mercury complex (XCIX) is very similar in type to that of the chlorotin complex (XCVIII), but no migration of any group from one metal to another was observed.

Bruce (*24, 26*) has examined the mass spectrum of the π-bonded benzyl complex $C_5H_5Fe(CO)_2CH_2Ph$ which exhibits weak peaks for M^+ and $[M - CO]^+$, and the base peak is the $[M - 2\ CO]^+$ ion. There is a strong peak at m/e 91, $C_7H_7^+$, which is believed to be the tropylium cation formed by rearrangement of the benzyl cation, and by analogy it is suggested that a σ–π rearrangement occurs in the $[M - 2\ CO]^+$ fragmention, which would then be the π-cyclopentadienyliron-π-tropylium ion $C_5H_5FeC_7H_7^+$. The fact that no peak is observed for $C_7H_7Fe^+$ is taken as an indication that fragmentation of the $C_5H_5FeC_7H_7^+$ ion occurs by fission of the weaker Fe—C_7H_7 bond. Recently, Hawthorne *et al.* (*88*) have suggested that the intermediate formed in the σ–π rearrangement of benzyl metal complexes may have the π-allylic structure. However, only with the aid of detailed deuterium substitution investigations on these compounds can the structure of the σ–π intermediates be determined.

King (114) has measured the mass spectra of some cyclopentadienyliron dicarbonyl compounds $C_5H_5Fe(CO)_2R$ where $R = COMe$, COPh, Ph, COCH=CHPh, and CH_2OCOMe. The general features of stepwise loss of the carbonyl groups and fragment ions formed by loss of acetylene from the cyclopentadienyl group were observed. Also all the spectra exhibited the ion $(C_5H_5)_2Fe^+$ formed by thermal decomposition processes. The mass spectrum of the acetyl complex exhibited sufficient metastable peaks for a detailed fragmentation scheme to be elucidated. Stepwise loss of the three carbonyl groups was followed by loss of H_2 from $CH_3FeC_5H_5^+$, giving $C_6H_6Fe^+$, which is probably the benzene-iron cation. This loses the C_6H_6 fragment to give Fe^+. An alternative fragmentation path is by cleavage of the methyl group from the molecular ion followed by stepwise elimination of the three carbonyl groups. The mass spectra of the benzoyl and phenyl complexes were very similar and the molecular ion of the benzoyl complex was not observed. It appears that the benzoyl complex is decarbonylated in the spectrometer. An interesting feature of the spectrum of the phenyl derivative is the expulsion of the iron atom from the $[M-2\,CO]^+$ ion, giving $C_6H_5C_5H_5^+$, a process related to that observed in substituted ferrocenes (135). The spectrum of $C_5H_5Fe(CO)_2COCH=CHPh$ exhibits the molecular ion, and fragmentation is similar to that discussed for the acetyl complex, namely stepwise loss of the three carbonyl groups. The spectrum contains peaks for ions with C_6H_5 as well as C_8H_7 substituents, indicating that expulsion of acetylene occurs from ions containing the PhCH=CH group. The methyl ester complex fragments by two different pathways: first, by loss of the two CO groups, the $[M-2\,CO]^+$ then losing ketene with migration of the methoxyl group to the iron atom; or, alternatively, by cleavage of the methoxyl group from the molecular ion, followed by loss of ketene giving $C_5H_5Fe(CO)_2^+$.

The mass spectrum of $C_5H_5Fe(CO)_2C\equiv CPh$ shows the molecular ion and fragments including $FeC\equiv CPh^+$, and the elimination of the iron atom from the molecular ion giving $C_5H_5C\equiv CPh^+$ is confirmed by a metastable peak at m/e 124 (85).

Bruce (24, 26) has examined the spectra of the fluoroaromatic compounds $C_5H_5Fe(CO)_2CH_2C_6H_5$ and $C_5H_5Fe(CO)_2CH_2C_6F_4(CO)_2FeC_5H_5$. The molecular ion of the pentafluorobenzyl complex shows stepwise loss of CO and also loss of fluorine, giving $[M-F]^+$ which then loses CO. The binuclear complex exhibits a very weak molecular ion peak, but the $[M-4\ CO]^+$ fragment ion is relatively intense, and this latter ion loses one metal atom, giving $C_5H_5FeCH_2C_6F_4C_5H_5{}^+$. A strong peak for $(C_5H_5)_2Fe^+$ is observed in the spectrum of the binuclear complex, probably formed by thermal decomposition. Both compounds show ions such as FeF^+ and $C_5H_5FeF^+$ in which rearrangement with transfer of fluorine to the metal has occurred, but corresponding ions containing carbonyl groups are absent.

The spectra of some mononuclear fluorocarbon derivatives of cyclopentadienyliron dicarbonyl have been studied by King (112, 116). The spectrum of $C_5H_5Fe(CO)_2COC_3F_7$ exhibits the molecular ion peak and the fragmentation is similar to other arylcyclopentadienyl iron dicarbonyl complexes. The spectra of fluoroaryl complexes $C_5H_5Fe(CO)_2R$ $(R=C_6F_5,$ $p\text{-}C_6F_4H$, $3,4\text{-}C_6H_2F_3$, $p\text{-}CF_3{-}C_6F_4)$ are very similar, and the main features are stepwise losses of the two neutral HF fragments. Transfer of fluorine from the aromatic ring to iron was also observed, giving the ions $C_5H_5FeF^+$ and FeF^+. Metastable peaks were observed for elimination of the neutral fragment FeF_2 from both the $[M-2\ CO]^+$ and $[M-2\ CO-HF]^+$ ions.

The organosulfur compounds $C_5H_5Fe(CO)_2SMe$ and $C_5H_5Fe(CO)_2$ SOCPh show stepwise loss of the carbonyl groups, and the methyl sulfide complex then loses hydrogen, giving $C_6H_6SFe^+$. No loss of methyl group was observed, although this is a predominant process for complexes having bridging MeS groups and indicates a stronger bonding between the metal and sulfur groups in the bridged complexes (115).

The molecular ion of $[C_5H_5FeP(OPh)_3]_2$ was not observed, as it was beyond the range of the instrument used, but it was observed that the triphenylphosphite ligands were eliminated before the cyclopentadienyl group (156).

The π-pyrrolyl complex $C_5H_5FeC_4H_4N$ shows four fragmentation paths of the molecular ion, all of which are confirmed by observation of the appropriate metastable peaks. Thus M^+ fragments either by loss of C_2H_2, HCN, C_2H_2N, or the entire C_4H_4N unit. Further fragmentations are as expected for cyclopentadienyl compounds (117).

Bis(indenyl)iron $(C_9H_7)_2Fe$ shows a strong molecular ion peak and loses one indenyl ligand, giving $C_9H_7Fe^+$, which then loses the iron atom giving $C_9H_7{}^+$ (118). $C_5H_5FeC_9H_7$ loses the indenyl ligand much more readily than the cyclopentadienyl ligand from the molecular ion and while fragmentation of $C_5H_5Fe^+$ gives Fe^+, fragmentation of $C_9H_7Fe^+$ gives $C_9H_7^+$ (117). Hydrogenation of $C_5H_5FeC_9H_7$ gives $C_5H_5FeC_9H_{11}$ and the mass spectrum of this complex is more complicated than the indenyl complex because dehydrogenation processes can occur. $C_5H_5FeC_9H_{11}^+$ can fragment either by loss of a methyl group, a process which must involve a hydrogen shift, or by elimination of ethylene, or by elimination of an allyl radical giving the stable cation $C_5H_5FeC_6H_6^+$, or by elimination of C_5H_6 and H_2 in one step (118).

The composition of the partially hydrogenated complex $C_5H_5FeC_9H_9$ was confirmed by its mass spectrum. The molecular ion undergoes both dehydrogenation, giving $C_5H_5FeC_9H_8^+$, which is the most abundant metal-containing ion, and acetylene elimination to give $C_5H_5FeC_7H_7^+$. The latter ion then loses a second molecule of acetylene to give C_5H_5Fe $C_5H_5^+$. The relative abundances of $C_9H_8Fe^+$ and $C_5H_5Fe^+$ are nearly equal, suggesting that the C_9H_8 ligand is as strongly bonded to iron as the π-C_5H_5 ligand (117).

The molecular weights of the following compounds have been determined by mass spectrometry: $C_5H_5Fe(CO)_2HgX$ [X = Cl, Br, I, SCN, Co(CO)$_4$] (139); $C_5H_5Fe(CO)_2SC_6F_5$ (45); $C_5H_5Fe(CO)_2SC_6H_5$ (45); C_5H_5Fe $(CO)_2NC_4F_5$ (46); $C_5H_5Fe(CO)_2C_4F_6$ (84); $C_5H_5Fe(CO)_2CH{=}CHCF_3$ (31); and $C_5H_5Fe(CO)_2CH{=}CHC_6F_5$ (31).

6. Olefin and Acetylene Complexes

A series of 14 monoolefin–iron tetracarbonyl compounds has been reported by Koerner von Gustorf et al. (86).

The molecular ion peak was observed in all cases except when the olefin was maleic anhydride (Ca) and dimethyl maleate (Cc), in which case the peak at highest mass corresponded to $[LFe(CO)_3]^+$. In all cases stepwise loss of the CO groups was observed. The series of ions $Fe(CO)_n^+$ ($n = 1$–4) was also present. When the olefin contained two halogens a series of ions $[(L{-}X)Fe(CO)_n]^+$ ($n = 0$–4) was also observed. In most of the spectra there was a peak at m/e 82, corresponding to $C_2H_2Fe^+$, which may be formed by elimination of HX (where X = Cl, Br, CN, OEt, Ph) or by elimination of two CO_2Me groups, and in three of the spectra (Cl), (Cm), and (Cn) the ion $C_2H_2Fe(CO)^+$ is also found. An interesting fragmentation for dimethyl fumarate and dimethyl maleate complexes is the loss of CO followed by loss of formaldehyde, whereas the free ligands show the normal cleavage of esters giving an $[M - OMe]^+$. It is possible that in these complexes, migration of the methoxyl group to the iron atom occurs, followed by loss of the carbonyl group and then loss of the formaldehyde, as has been observed in methyl ester derivatives of ferrocene and cymantrene. As would be expected, the vinyl halide complexes show very strong peaks for FeX_2^+

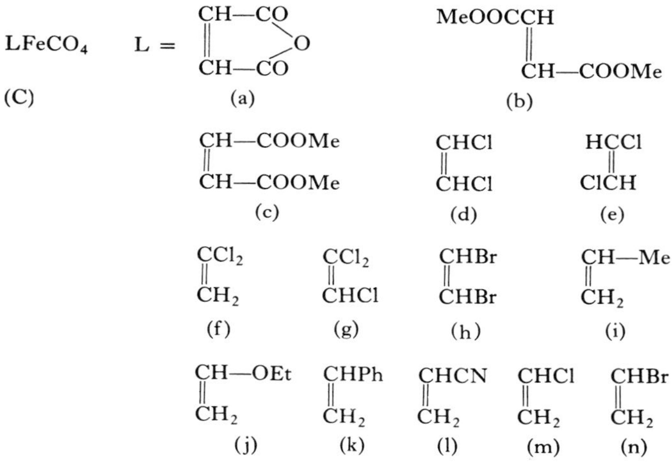

$LFeCO_4$ $L =$

(C)

CH—CO ‖ O CH—CO (a)		MeOOCCH ‖ CH—COOMe (b)
CH—COOMe ‖ CH—COOMe (c)	CHCl ‖ CHCl (d)	HCCl ‖ ClCH (e)
CCl$_2$ ‖ CH$_2$ (f)	CCl$_2$ ‖ CHCl (g)	CHBr ‖ CHBr (h)

CH—Me
‖
CH$_2$
(i)

CH—OEt   CHPh   CHCN   CHCl   CHBr
‖   ‖   ‖   ‖   ‖
CH$_2$ CH$_2$ CH$_2$ CH$_2$ CH$_2$
(j)   (k)   (l)   (m)   (n)

and FeX^+. There are notable differences between the fragmentation patterns of the free ligands and the fragmentations of the ligands when attached to the $Fe(CO)_4$ unit. For example, the base peak in the spectrum of vinyl chloride is the ion $C_2H_3^+$, but when the olefin is complexed to $Fe(CO)_4$ the main ion in the spectrum corresponds to $C_2H_3Cl^+$. The effect of the different donor abilities of the ligands and hence the stabilization of the positive charge influences the relative abundances of the $LFe(CO)_x^+$ ions and can be expressed by the relationship

$$F = \frac{LFe(CO)_3^+ + LFe(CO)_2^+}{LFe(CO)^+ + LFe^+}$$

The change in donor abilities of the ligands also affects the carbonyl stretching frequency $\nu(C{\equiv}O)$, and a good correlation is obtained between the F value and $\nu(C{\equiv}O)$.

The mass spectra of several substituted butadieneiron tricarbonyl compounds [(CI)–(CVII)] have been reported (136), and the migration of the carbonyl substituent, R^1, to the metal atom giving the fragment $[FeR^1]^+$ was observed in all cases.

$$\left[\begin{array}{c} R^2\!\!-\!\!\diagdown\!\!/\!\!\diagdown\!\!-\!\!\overset{O}{\underset{R^1}{C}} \\ \underset{CO \quad CO \quad CO}{\overset{|}{Fe}} \end{array} \right]^{+} \longrightarrow [Fe\!-\!R^1]^{+}$$

(CI)	R^1 = OH	R^2 = Me
(CII)	R^1 = OD	R^2 = Me
(CIII)	R^1 = NH_2	R^2 = Me
(CIV)	R^1 = ND_2	R^2 = Me
(CV)	R^1 = OH	R^2 = CO_2H
(CVI)	R^1 = OMe	R^2 = CO_2Me

The compounds all exhibit a molecular ion peak and also peaks for stepwise loss of three carbonyl groups. Surprisingly, it was observed that decarbonylation of the ligand occurred as well giving an $[M-4\,CO]^+$ fragment ion. Such decarbonylation does not occur in the mass spectra of the free ligands or when the carbonyl-containing group COR^1 is not attached directly to the metal carrying π-organic moiety, and it is suggested that the mechanism involves transfer of the group R^1 to the metal.

The mass spectrum of cyclobutadieneiron tricarbonyl shows strong peaks for the molecular ion $C_4H_4Fe(CO)_3{}^+$ and for fragments due to stepwise loss of carbonyl groups (3, 62, 196). The absence of peaks corresponding to loss of C_2H_2 units is taken as strong evidence that the ligand does have the cyclobutadiene structure rather than the complex having the conceivable bis(acetylene)iron tricarbonyl structure (62). Tyerman et al. (196) have used kinetic mass spectrometry to follow the formation of free cyclobutadiene formed by flash photolysis of $C_4H_4Fe(CO)_3$. The mass spectrum of tetramethylcyclobutadieneiron tricarbonyl also shows the molecular ion and stepwise loss of the CO groups (3). Cyclohexa-1,3-dieneiron tricarbonyl exhibits a weak molecular ion peak. Stepwise loss of two CO groups occurs, and besides the ion $C_6H_8FeCO^+$ the fragment C_6H_6FeCO was observed. Also no ion corresponding to cleavage of the three CO groups was detected but, rather, an intense peak for $C_6H_6Fe^+$ was observed (200.) A series of substituted cyclohexa-1,3-dieneiron tricarbonyl compounds

has been studied by Haas and Wilson (87), who found that although loss of a molecule of hydrogen did not occur readily from the molecular ion, it did occur when the number of carbonyl groups was 0–2, and especially from the $[M - 3\ CO]^+$ ion. Very few of the uncomplexed dienes have been examined, but those that have show that elimination of a single hydrogen atom, or of a radical is much more probable, while the diene iron carbonyls decompose by elimination of a hydrogen molecule to odd-electron fragments. After elimination of one or more carbonyl groups the metal atom is electron-deficient and subsequent loss of H_2 converts the organic moiety from a 4 π- to a 6 π-electron system which can better stabilize the electron configuration of the iron atom, so providing a driving force for the elimination of a molecule of hydrogen. If the substituent group in the 5-position, R^1, contains a β-carbonyl group, e.g., the dimedone derivative (CVII), then dehydrogenation does not compete with decarbonylation, and all three CO groups are eliminated before any ligand decomposition takes place.

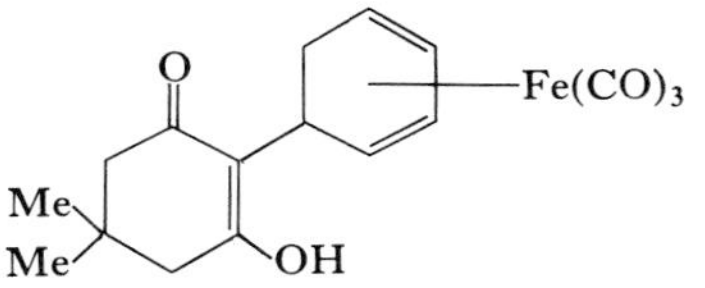

(CVII)

This observation is consistent with the idea that in this case, the oxygen of the carbonyl group of the ligand can approach the iron atom sufficiently closely to coordinate to an electron-deficient iron atom, and so the driving force for elimination of a molecule of hydrogen is not so strong.

The mass spectra of some iron tricarbonyl derivatives of β-ionone and related compounds (CVIII)–(CX) show the usual stepwise loss of CO groups from the molecular ion, but further analysis of the spectra was not attempted because of the complex nature of the ligands (*36*). Vitamin A

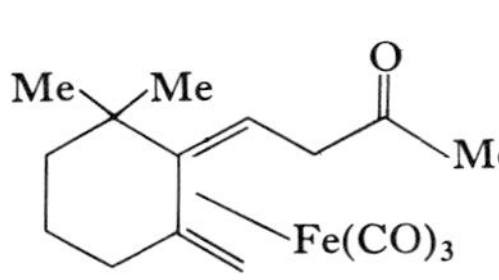

(CVIII)

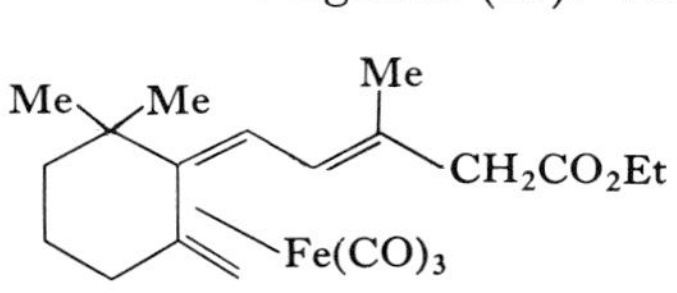

(CIX)

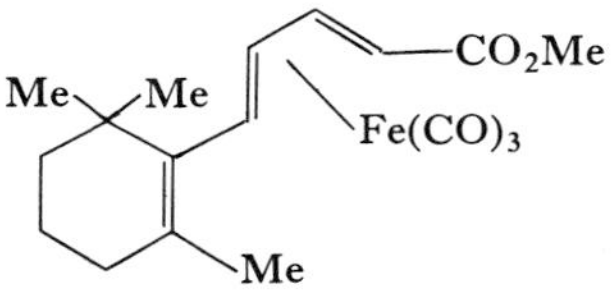

(CX)

acetate–iron tricarbonyl exhibits the peaks M^+ and $[M-3\,CO]^+$ at m/e 468 and m/e 384, respectively (20). The following compounds 1,4-di-methoxycyclohexa-1,3-dieneiron tricarbonyl, cyclohepta-4,6-diene-1,3-dioneiron tricarbonyl, [5-(2-cyclohepta-4,6-diene-1,3-dionato)cyclohexa-1,3-diene]diiron hexacarbonyl, and (2-methoxy[5-^{2}H$_1$]cyclohexa-1,3-diene)iron tricarbonyl exhibit the molecular ion peak but 5-[2-(5,5-dimethyl-cyclohexane-1,3-dionato)]cyclohexa-1,3-dieneiron tricarbonyl shows as its highest peak the $[M-CO]^+$ ion (21). The spectrum of γ-pyroneiron tricarbonyl exhibits peaks for the ions

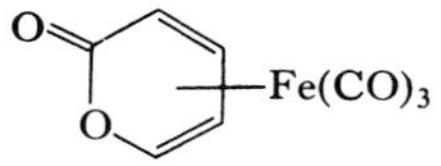

$[C_5H_4O_2Fe(CO)_n]^+$ $(n=0\text{–}3)$, $C_4H_4OFe^+$ and $C_5H_4O_2^+$ (176). The mononuclear complex diacetyldianil iron tricarbonyl (PhCN=C(Me)—C(Me)=NPh)Fe(CO)$_3$ shows a series of peaks for the ions [diacetyldianil Fe(CO)$_n$]$^+$ $(n=0\text{–}3)$ (158).

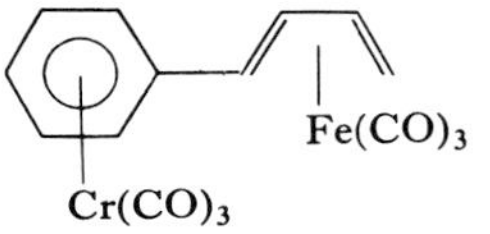

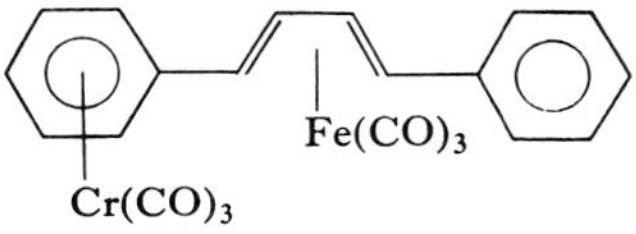

The bimetallic complexes (CXI) and (CXII) show the bimetallic ion $[Fe\text{—}Cr]^+$ (136). Loss of the CO groups is not stepwise, for the species $LCrFe(CO)_5$ is not detected, but peaks for the ions $LCrFe(CO)_n$ $(n=0\text{–}4)$ are observed.

The mass spectrum of norbornadiene-7-oneiron tricarbonyl shows the molecular ion, the fragment ions $[M-(CO)_n]^+$ $(n=1\text{–}4)$, and a peak corresponding to $C_6H_6^+$. If the sample is heated to 200°C, only the ion $C_6H_6^+$ is observed (130). The norbornadiene complex $C_7H_8Fe_2(CO)_4$ (SMe)$_2$ shows loss of the four carbonyl groups followed by loss of the two methyl groups, giving the most abundant metal-containing peak $C_7H_8Fe_2S_2^+$. The norbornadiene ligand then undergoes degradation with the expulsion of acetylene, giving $C_5H_6Fe_2S_2^+$. This ion loses a hydrogen to give $C_5H_5Fe_2S_2^+$, which may contain the cyclopentadienyl group (115).

In the mass spectrum of the cycloheptadienol complex (CXIII), dehydrogenation, dehydration, and elimination of acetylene occur and compete with stepwise loss of carbonyl groups from the molecular ion (119).

Fe(CO)₃
(CXIII)

The mass spectrum of cyclooctatetraeneiron tricarbonyl shows stepwise loss of the three carbonyl groups from the molecular ion, followed by elimination of acetylene giving $C_6H_6Fe^+$, and further breakdown gives Fe^+ (119). The mass spectra of two substituted cyclooctatetraene complexes (see Table VIII) show the molecular ion and stepwise loss of the carbonyl groups (83).

The reaction of $Fe_3(CO)_{12}$ with 1,4-dibromobutyne and zinc gives a butatriene complex originally formulated as $C_4H_4Fe_2(CO)_5$ (151), but mass spectrometry has shown that the compound contains six carbonyl groups, namely, $C_4H_4Fe_2(CO)_6$, and ions for the successive loss of six CO groups are observed (111, 152, 158). The spectrum of 1,4-dimethylbutatrienediiron hexacarbonyl has also been measured, and both these complexes show a strong peak for Fe_2^+, indicative of a relatively strong iron–iron bond (158). The spectrum of the hexapentaeneiron carbonyl complex $C_6H_4Fe_3(CO)_{7-8}$ failed to show the molecular ion, but did exhibit a peak at m/e 168 assigned to Fe_3^+. No strong peak was observed at m/e 112 (Fe_2^+) (158). $Fe_3(CO)_{12}$ and acenaphthalene give a red-violet solid originally formulated as the diiron hexacarbonyl derivative $C_{12}H_8Fe_2(CO)_6$ (124), but the mass spectrum shows as the ion of highest mass $C_{12}H_8Fe_2(CO)_5$, indicating that the complex is diiron pentacarbonyl (111). The composition of the deep red azulene complex prepared by Burton, Pratt, and Wilkinson (34) from azulene and $Fe(CO)_5$ has been confirmed by mass spectrometry by King (111), who also found that by altering the reaction conditions slightly the dark red complex $[C_{10}H_8Fe(CO)_2]_2$ could be obtained and this was characterized by its mass spectrum. A volatile yellow solid obtained by reaction of allene dimer with $Fe_3(CO)_{12}$ has been shown by mass spectrometry to be a dicarbonyl complex $C_{12}H_{16}Fe(CO)_2$ (111, 119). The formation of diphenylvinylidenediiron octacarbonyl deduced by X-ray analysis is in agreement with the mass spectrum (*145*). The molecular ion is observed together with fragment ions corresponding to successive loss of eight CO groups, while peaks at m/e 178, 179, and 180 are indicative of diphenylvinyl-idene, Ph_2C_2, Ph_2C=CH, and Ph_2C=CH_2, respectively. A strong peak is observed also for tetraphenylbutatriene, while a trimer peak at m/e 534 could raise by attack of diphenylvinylidene on the central bond of the cumulene system to give (CXIV).

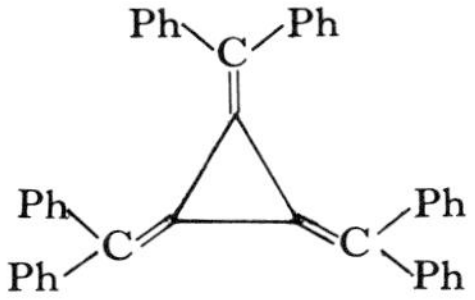

(CXIV)

The peak at highest mass occurs at m/e 636, and together with fragments corresponding to loss of six CO groups indicates the presence of tetraphenyl-butatrienediiron hexacarbonyl. Allene reacts with $Fe_2(CO)_9$ to give the allylic complex (CXV).

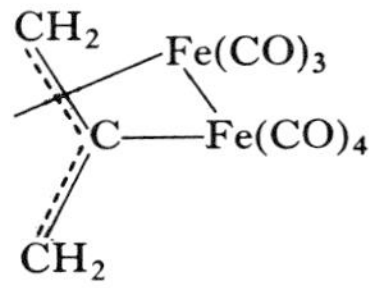

(CXV) 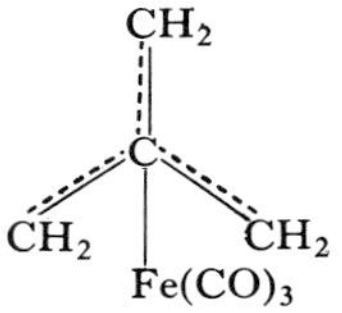   (CXVI)

Its mass spectrum shows the molecular ion and peaks corresponding to successive loss of seven carbonyl groups plus an intense peak at m/e 112, suggesting the presence of an iron–iron bond (*13*).

3-Chloro(2-chloromethyl)propene reacts with $Fe_2(CO)_9$ to give tri-methyleneiron tricarbonyl (CXVI) which exhibits strong peaks for the molecular ion and fragments due to successive losses of the carbonyl groups (*63*). In the spectrum of the partially fluorinated bicyclo[2.2.2]octatriene complex $C_{14}H_{14}F_6Fe(CO)_3$, loss of fluorine from the molecular ion com-petes with loss of the carbonyl groups (*116*).

Octafluorocyclohexa-1,3-dieneiron tricarbonyl exhibits the series of ions $C_6F_8Fe(CO)_n^+$ ($n=0$–3), but cyclohexadieneiron fragments are of low abundance. The major ions are $C_6F_6^+$ and $Fe(CO)_n^+$ ($n=0$–3), indicating that the bonding between the perfluorohexadienyl ligand and iron is much weaker than for hexa-1,3-dieneiron tricarbonyl (91). The mass spectrum of perfluorotetramethyleneiron tetracarbonyl $C_4F_8Fe(CO)_4$ exhibits the molecular ion peak and stepwise loss of the carbonyl groups. Although no loss of fluorine is observed from iron-containing fragments, the base peak of the spectrum is $C_4F_6^+$ (91).

B. Ruthenium and Osmium

The mass spectrum of ruthenocene $(C_5H_5)_2Ru$ (81, 148) shows a very strong molecular ion peak, and the yields of $C_5H_5Ru^+$ and Ru^+ are lower than the analogous ions in the spectrum of ferrocene. Also the fragment ion corresponding to loss of a C_2 unit is much more abundant for ruthenium than for iron. The mass spectrum of $(C_5H_5)_2Os$ shows that cleavage of C_5H_5 unit is further reduced, and again the base peak is the molecular ion. The second most abundant ion is the fragment caused by cleavage of a C_2 unit from the molecular ion, giving $[C_8H_xO_5]^+$ (148).

One of the products of the reaction between cyclooctatetraene and $Ru_3(CO)_{12}$ is $C_8H_8Ru(CO)_3$, and its mass spectrum is consistent with the assigned formulation (29, 50). C_8H_8 reacts with $Ru_3(CO)_{12}$ in refluxing octane to give $(C_8H_8)_2Ru_3(CO)_4$. Its mass spectrum shows the molecular ion and stepwise loss of carbonyl groups by the appearance of both singly charged ions $[(C_8H_8)_2Ru_3(CO)_n]^+$ ($n=0-4$) and doubly charged ions $(C_8H_8)_2Ru_3(CO)_n]^{2+}$ ($n=0-3$) (29).

Cyclobutadieneruthenium tricarbonyl shows a strong molecular ion peak and prominent fragments corresponding to the successive loss of CO groups (3). Similarly, tetraphenylbutadieneruthenium tricarbonyl shows the ions $[Ph_4C_4Ru(CO)_n]^+$ ($n=0-3$) (183). p,p'-Dichlorodiphenylacetylene reacts with $Ru_3(CO)_{12}$ to give a dark red compound $C_{24}H_{16}Ru_3(CO)_8$, and its mass spectrum shows the molecular ion and peaks corresponding to successive loss of eight carbonyl groups followed by loss of four chlorine atoms (183). When $Ru_3(CO)_{12}$ is refluxed with an arene (arene $= MeC_6H_5$, $Me_2C_6H_4$, $Me_3C_6H_3$) purple compounds of composition $Ru_6C(CO)_{14}$ arene are obtained, and the composition is confirmed by the mass spectra which show the molecular ion peaks followed by loss of 14 carbonyl groups, giving $[Ru_6C(arene)]^+$, which breaks down ultimately to Ru_6C^+. The corresponding doubly charged ions are also present in high abundances (100).

4,6,8-Trimethylazulene reacts with $Ru_3(CO)_{12}$ to give $Ru_6(CO)_{17}C$ and $Me_3C_{10}H_5Ru_4(CO)_9$, which were identified by mass spectrometry (*43*). Tetracyclone reacts with $Ru_3(CO)_{12}$ to give the cyclopentadienone complex (CXVII), which exhibits the molecular ion and fragments corresponding to the loss of four carbonyl groups (*30*).

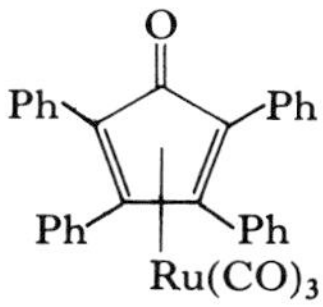

(CXVII) 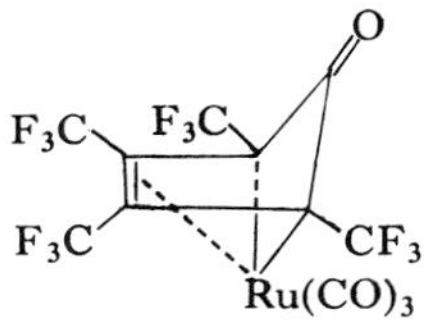   (CXVIII)

Hexafluorobut-2-yne and $Ru_3(CO)_{12}$ give the cyclopentadienone complex (CXVIII), and the molecular weight was determined by mass spectrometry (*30*).

C. Cobalt

The mass spectrum of $(C_5H_5)_2Co$ was first investigated by Friedman *et al.* (*81*), who found that the main ions observed were $(C_5H_5)_2Co^+$, $C_5H_5Co^+$, and Co^+. The molecular ion accounted for 60% of the total ion current, indicating that the ion $(C_5H_5)_2Co^+$ is a very stable species. Further evidence as to the high stability of the molecular ion relative to the neutral species comes from the measurement of its appearance potential (*81, 148, 165*), which is 1.7 eV lower than the ionization potential of the cobalt atom. A reinvestigation of the spectrum by Müller and D'Or (*148*) has shown that fragmentation of the cyclopentadienyl ring by expulsion of a molecule of acetylene and the formation of cyclopropenyl metal fragments does occur to a limited extent. The fragmentation scheme [Eqs. (9)] accounts for the formation of the fragment ions observed and is supported by the presence of metastable peaks for many of the fragmentation processes. Although the most metastable peaks were observed for $(C_5H_5)_2Co$ the same breakdown pattern probably applies to all the bis(cyclopentadienyl) compounds of the first-row transition metals.

Winters and Kiser (*199*) have examined the mass spectrum of C_5H_5 $Co(CO)_2$ and found that elimination of the carbonyl groups occurs much more readily than loss of cyclopentadienyl and the base peak of the spectrum

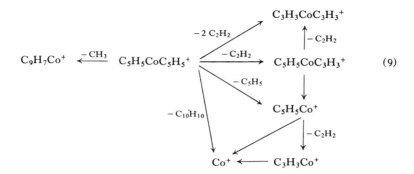

$$e + C_5H_5Co(CO)_2 \rightarrow Co^+ + C_3H_3 + C_2H_2 + 2\,CO + 2\,e$$

is for the ion $C_5H_5Co^+$. Fragmentation of the cyclopentadienyl ring was also observed, giving the cyclopropenylcobalt cation $C_3H_3Co^+$ (28% R.A.). From measurement of appearance potentials of the various fragment ions, the heat of formation of $C_5H_5Co(CO)_2$, assuming the process shown below, was calculated as 15 kcal/mole (see Section VII).

$$e + C_5H_5Co(CO)_2 \rightarrow Co^+ + C_3H_3 + C_2H_2 + 2\,CO + 2\,e$$

Winters and Kiser also recorded the negative ion mass spectrum and observed the ions $C_5H_yCo(CO)_2^-$ and $C_5H_yCoCO^-$ ($y = 3$–5) which are formed by either electron capture or ion-pair production processes.

$$AB + e \rightarrow A^- + B$$
$$AB + e \rightarrow A^- + B^+ + e$$

The formation of $C_5H_5Co^-$ was observed at low ionizing energy, and is probably formed by a dissociative electron capture process.

$$C_5H_5Co(CO)_2 + e \rightarrow C_5H_5Co^- + 2\,CO$$

The observation of $C_5H_yCo(CO)_2^-$ at high ionizing energies (70 eV) indicates that the formation processes must involve ion-pair production, such as the two possible modes illustrated below.

$$C_5H_5Co(CO)_2 + e \rightarrow C_5H_4Co(CO)_2^- + H^+ + e$$
$$C_5H_5Co(CO)_2 + e \rightarrow C_5H_3Co(CO)_2^- + H_2^+ + e$$

An investigation by Pignataro and Lossing (*166*) on the thermal decomposition of organometallic compounds in the ion source of the mass spectrometer indicates that at temperatures of 250°–300° C, $C_5H_5Co(CO)_2$ decomposes to $(C_5H_5)_2Co$, giving rise to ions $(C_5H_5)_2Co^+$ (m/e 189), $C_5H_6^+$, and $C_5H_5^+$; the m/e 189 peak showed a maximum at 350° C. King (*104*) has also observed the phenomenon of thermal decomposition in the ion source, for the spectrum of the trimer $(C_5H_5CoCO)_3$ corresponds to the unknown tetramer $(C_5H_5)_4Co_4(CO)_2$ (M^+ at m/e 552).

The main peaks in the mass spectrum of $C_5H_5CoS_2C_2(CF_3)_2$ are M^+, $C_5H_5CoS_2^+$, and $C_5H_5Co^+$ as well as ions derived from the ligand itself. Loss of hexafluorobutyne from M^+ to give $C_5H_5CoS_2^+$ is confirmed by observation of the appropriate metastable peak, and cleavage of the dithietene ligand occurs before loss of the cyclopentadienyl ring (*120*).

The most abundant ion in the mass spectrum of the partially fluorinated bicyclo[2.2.2]octatriene complex $C_5H_5CoC_{14}H_{14}F_6$ is the molecular ion, which fragments either by loss of a methyl group or by elimination of a butyne bridge. The elimination of a neutral C_5H_5CoF fragment from the molecular ion and from the fragment ion $C_5H_5CoC_{10}H_8F_6$ was observed (*112, 116*).

(*tert*-BuN)$_2$S reacts with $C_5H_5Co(CO)_2$ to give the dark green complex $(C_5H_5Co)_2(tert\text{-BuN})_2CO$, and its mass spectrum shows the molecular ion and binuclear fragments such as $(C_5H_5Co)_2(NR)CON^+$ ($R = C_4H_9$ or C_3H_6), $(C_5H_5Co)_2N^+$, $(C_5H_5Co)_2C^+$, and $(C_5H_5Co)_2^+$ (*160*). The base peak is $(C_5H_5)_2Co^+$. There are several peaks which point to the presence of an N—CO—N group and the overall fragmentation pattern indicates a reluctance to release the carbonyl group, suggesting that this is not a metal carbonyl complex. An X-ray analysis shows that the carbonyl group is part of an alkyl urea moiety in which the nitrogen atoms bridge the two cobalt atoms (*160*).

The molecular weight of $C_5H_5CoCO(GeCl_3)_2$ was confirmed by exact mass measurement of the molecular ion (*129*). The mass spectra of the complexes $C_5H_5CoN_4R_2$ (CXIX) both show the molecular ion peak but fragmentation paths are considerably different. The molecular ion peak of the methyl-substituted complex is strong and an $[M+1]$ is also observed. The M^+ ion breaks down by loss of either an N_2Me fragment or by an N_4Me_2 group, giving $C_5H_5Co^+$ as the base peak of the spectrum. The remaining ions in the spectrum are formed by cleavage of the $C_5H_5Co^+$ ion.

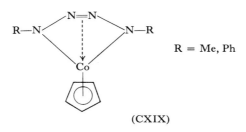

R = Me, Ph

(CXIX)

The molecular ion of $C_5H_5CoN_4Ph_2$ is weak, and the fragment ions $[M-N_2]^+$, $[M-N_3H]^+$, $[M-2\,N_2]^+$, and $[M-2\,N_2-H]^+$ are also observed in low abundances; the base peak of the spectrum corresponds to the biphenyl cation $[C_6H_5-C_6H_5]^+$ and the ion $C_5H_5Co^+$ is observed in relatively low abundance (*153*).

Cyclobutadiene(cyclopentadienyl)cobalt has been prepared by Amiet and Pettit (*4*) from dichlorocyclobutene and sodium tetracarbonyl cobaltate, while Rosenblum and North (*177*) prepared it from photo-α-pyrene and $C_5H_5Co(CO)_2$. The mass spectrum shows that fragmentation of the cyclo-butadiene ring occurs before loss of the cyclopentadienyl ring, and the major ions in the spectrum correspond to M^+ m/e 176, $[M-C_2H_2]^+$ m/e 150, and $[M-C_4H_4]^+$ m/e 124 (*4, 177*). Metastable peaks have been reported by Rosenblum and North for the transitions $176 \to 150$, $150 \to 124$, and $124 \to 98$, indicating that cleavage of the cyclobutadiene ring does not occur, but rather the four-membered ring eliminates acetylene. The high stability of the complex is illustrated by the fact that the molecular ion accounts for almost 40% of the total ion current. The observation of the molecular ion for a series of substituted cyclobutadiene(cyclopenta-dienyl)cobalt complexes has been reported (*90*) (see Table XI).

The reaction of 3,3,3-trifluoropropyne with $Co_2(CO)_8$ at $60°\,C$ gives $Co_2(CO)_6(CF_3C{\equiv}CH)$, but its mass spectrum does not show the molecular ion, the peak at highest mass corresponding to $[Co_2(CO)_5(CF_3C{\equiv}CH)]^+$, possibly because of thermal instability of the complex *in vacuo* (*56*). When the reaction is carried out at $110°\,C$, the product is $Co_2(CO)_4(CF_3C{\equiv}CH)_3$, which does exhibit a molecular ion peak. This fragments by successive loss of carbonyl groups, giving $Co_2(CF_3C\equiv CH)_3^+$, which then loses the metal atoms to leave $[CF_3C\equiv CH]^+$ (56).

D. Rhodium and Iridium

The mass spectrum of $(C_5H_5RhCO)_3$ shows the molecular ion and fragment ions corresponding to successive loss of carbonyl groups as well as peaks which correspond approximately to the expulsion of a single carbon atom together with a variable number of hydrogens (146). A peak was also observed at m/e 728, of greater intensity than the molecular ion corresponding to $(C_5H_5)_4Rh_4(CO)_2^+$, which probably arises by thermal decomposition of the trimer on the probe, as was found for the cobalt trimer (104). One of the products of the reaction of $RhCl_3$ with C_5H_5MgBr is $(C_5H_5)_4Rh_3H$, and its composition has been confirmed by mass spectrometry and X-ray crystallography (68, 70). The molecular ion is observed at m/e 570 and the main fragmentation gives peaks at 504, 502, and 500 associated with the ion $(C_5H_5)_3Rh_3^+$; at m/e 233, $(C_5H_5)_2Rh^+$; and at m/e 168, $C_5H_5Rh^+$. In addition, a number of small peaks are observed in the region m/e 309–335 which may be associated with Rh_3^+ residues. The spectrum of $C_5H_5Rh_2$ $S_2C_2(CF_3)_2$ has been briefly reported and the main peaks are M^+, C_5H_5 RhS_2^+, and $C_5H_5Rh^+$ (120). The corresponding iridium complex $C_5H_5IrS_2C_2(CF_3)_2$ shows strong peaks for M^+ and $C_5H_5IrS_2^+$ (120). The formation of this latter ion, by elimination of hexafluorobutyne from the molecular ion, was confirmed for both the rhodium and iridium compounds by observation of the appropriate metastable peaks. Loss of fluorine was also observed. The mass spectrum of the bis(ethylene) complex C_5H_5Rh $(C_2H_4)_2$ shows stepwise loss of the ethylene groups from the molecular ion in an analogous manner to loss of carbonyl groups in cyclopentadienyl metal carbonyls, and the base peak of the spectrum corresponds to the ion $C_5H_5Rh^+$ (117). Metal-free fragments observed include $C_4H_9^+$ and $C_4H_7^+$, indicating that coupling of the ethylene ligands occurs together with hydrogenation or dehydrogenation, but the fragments may be formed by thermal processes rather than by electron-induced processes.

The mass spectra of the cycloocta-1,5-diene complexes $C_5H_5MC_8H_{12}$ (M = Rh, Ir) both show the molecular ion as the base peak. The rhodium complex exhibits several metastable peaks and a detailed fragmentation scheme has been proposed. The molecular ion fragments by several alternative pathways in which the cyclooctadiene ligand is cleaved completely or suffers only partial fragmentation to give the ions $C_5H_5RhC_8H_{10}^+$, $C_5H_5RhC_6H_6^+$ (loss of C_2H_6), or $C_5H_5RhC_4H_6^+$ (loss of C_4H_6). Further fragmentations involving the elimination of neutral fragments such as C_6H_6, C_5H_6, C_4H_6, C_2H_2, or H lead to Rh^+. The cyclopropenyl ion $C_3H_3Rh^+$ is formed by loss of acetylene from $C_5H_5Rh^+$. The iridium complex shows a greater tendency to form dipositive ions. Ions involving fragmentation of the ligands are of lower abundance because of the greater strength of iridium–carbon bonds as compared with analogous rhodium–carbon bonds (117).

The cycloocta-1,5-diene rhodium derivative of hydropentalene (CXX) has as its base peak the molecular ion and also an equally intense peak corresponding to $C_8H_7Rh^+$, while that attributable to $C_8H_6Rh^+$ is only slightly weaker. However, the peak at m/e 102, for free pentalene, is very weak (103).

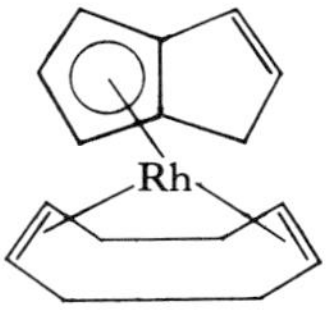

(CXX)

The mass spectra of the tris(π-allyl) complexes of rhodium and iridium have briefly been mentioned. $(C_3H_5)_3Rh$ shows a strong M^+, and peaks due to $(C_3H_5)_2Rh^+$, $C_3H_5Rh^+$, and Rh^+ are observed (11). $(C_3H_5)_2Ir$ shows the molecular ion with the peaks at m/e 314 and m/e 316 corresponding to ^{191}Ir and ^{193}Ir (42). The spectra of two π-allylic chloro-bridged complexes

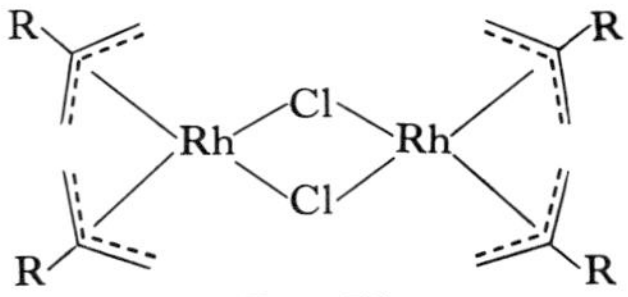

R = H
R = Me

have been studied, and it was found that the fragmentation patterns were considerably different in that the π-allyl complex (R = H) showed far more fragments containing two rhodium atoms than were present in the spectrum of the 2-methylallyl complex (132). An interesting feature in these spectra was the loss of two hydrogens from species containing only allylic groups bonded to the metal with the formation of cyclopropenyl structures, as illustrated below.

It was also suggested that a change of valency of the central metal atom from Rh(III) to Rh(I) is an important factor in determining the fragmentation path. The effect of changes in valency has previously been discussed by Shannon and Swan and by Reid *et al.* for complexes of iron, gold, and zirconium (*171, 184*).

E. Nickel, Palladium, and Platinum

1. Cyclopentadienyl Compounds

The mass spectrum of $(C_5H_5)_2Ni$ was first investigated by Friedman *et al.* (*81*), who observed three strong peaks for $(C_5H_5)_2Ni^+$, $C_5H_5Ni^+$, and Ni^+; the molecular ion is the base peak of the spectrum. The total ion current of $(C_5H_5)_2Ni^+$ is 46%, compared with the total ion current of the molecular ion of ferrocene of 60%, indicating a weakening of the metal–ring bond for the nickel complex. This suggestion is supported by thermochemical evidence which indicates a weakening of 24 kcal/mole of the metal–ring bonds in $(C_5H_5)_2Ni$ compared with $(C_5H_5)_2Fe$ (*164*). However, weakening of the metal–ring bond is not sufficient to account for the increased yield of $C_5H_5Ni^+$ ions compared with the abundance of $C_5H_5Fe^+$, and the greater stability of the $C_5H_5Ni^+$ must also be considered. A reinvestigation of the spectrum by Müller and D'Or (*148*) has shown that besides the three main peaks observed by Friedman *et al.*, fragmentation of the cyclopentadienyl ring occurs to give an abundant peak for $C_3H_3Ni^+$. The spectrum of $(C_5H_5)_2Ni$ has also been investigated by Schumacher and Taubenest (*181*) at low ionizing energies (20 eV), relatively high pressure (2×10^{-5} Torr), and low temperature (probe temperature approximately $25°C$), and they found that under these conditions ion–molecule reactions occurred to give associated species such as $(C_5H_5)_3Ni_2^+$, $(C_5H_5)_2C_3H_3Ni_2^+$, and $(C_5H_5)C_3H_2Ni_2^+$. A "triple-decker sandwich" structure was suggested for these associates. It has been shown that thermal decomposition of nickelocene occurs at $700°–800°\,C$ (166), indicating these are indeed ion–molecule reactions and not thermal decomposition products.

The mass spectrum of $[C_5H_5NiCO]_2$ has been measured by Schumacher and Taubenest under conditions similar to those for $(C_5H_5)_2Ni$ (182). Loss of the carbonyl groups occurs readily, and the main ions are $(C_5H_5)_2Ni_2^+$, $(C_5H_5)_2Ni^+$, and $C_5H_5Ni^+$; fragmentation of the cyclopentadienyl ring also is observed. Associated species such as $(C_5H_5)_4Ni_4(CO)_2^+$, $(C_5H_5)_4Ni_4^+$, $(C_5H_5)_3Ni_3(CO)_2^+$, and $(C_5H_5)_3Ni_2(CO)_2^+$ are formed by ion–molecule reactions. An interesting feature of this spectrum is migration of the cyclopentadienyl group from one nickel atom to the other, giving the ion $(C_5H_5)_2Ni^+$.

$$(C_5H_5)_2Ni_2^+ \xrightarrow{\;\bullet\;144\;} (C_5H_5)_2Ni^+ + Ni$$

The loss of hydrogen has been observed from the ions $(C_5H_5)_2Ni_2^+$ (182) and $(C_5H_5)_2Ni^+$ (148, 182), and the resulting species could have the cyclopentadienyl either π- or σ-bonded to the nickel.

The spectrum of $(C_5H_5Ni)_3(CO)_2$ exhibits a weak molecular ion peak, and the base peak is $(C_5H_5)_2Ni^+$ (182). The ion $(C_5H_5)_4Ni_3(CO)_2^+$, formed by an ion–molecule reaction, was also observed.

The molecular weight of $(C_5H_5)Ni(CO)C_2F_5$ was determined by mass spectrometry (140). $C_5H_5NiS_2C_2(CF_3)_2$ exhibits a strong molecular ion and also strong peaks for $C_5H_5Ni^+$, $C_4F_5S_2^+$, and $C_4F_6^+$ (120). The mass spectra of the binuclear mercapto-bridged complexes $[C_5H_5NiSR]_2$ ($R=$ Me, Ph) have been investigated and a peak corresponding to $(C_5H_5)_2Ni^+$ observed, but it is believed that migration of the cyclopentadienyl group occurs by a thermal process, rather than by an electron-bombardment-induced process (167). The normal fragmentation path of these compounds is by loss of RSH followed by elimination of RS. $(C_5H_5)_2Ni$ or $[C_5H_5NiCO]_2$ reacts with $(N\text{-}tert\text{-Bu})_2S$ to give $(C_5H_5Ni)_3N\text{-}tert\text{-Bu}$, which exhibits a strong molecular ion peak as well as the fragment ions $(C_5H_5Ni)_3C_3H_6N^+$ and $(C_5H_5Ni)_3N^+$ (160). The spectrum of C_5H_5NiNO shows that the nitrosyl group is eliminated before the cyclopentadienyl group, giving as the base peak $C_5H_5Ni^+$ (79). This observation indicates that the ligand with higher ionization potential is eliminated first.

The mass spectrum of $C_5H_5PdC_{10}H_{12}OMe$, (CXXI), shows that the cyclopentadienyl–palladium bond is not very strong and the molecular ion readily loses C_5H_6 to give the ion $C_{10}H_{11}OMePd^+$ (117).

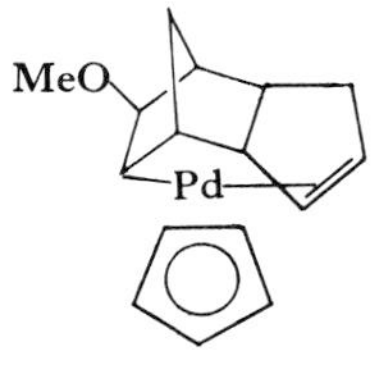

(CXXI)

The most abundant palladium-containing ion is $C_5H_5PdOMe^+$, which is probably formed by elimination of $C_{10}H_{12}$ with concurrent migration of the methoxyl group to palladium. The only other palladium-containing ions observed are $C_5H_6Pd^+$, $C_5H_5Pd^+$, and $C_3H_3Pd^+$, and metal-free ions are very abundant.

The molecular ion of $C_5H_5PtMe_3$ fragments by either loss of the cyclopentadienyl ring, giving $PtMe_3{}^+$, or by loss of methyl groups, giving $C_5H_5PtMe^+$ and $C_5H_5Pt^+$. Dehydrogenation of the methyl groups also occurs, giving the ions $C_7H_9Pt^+$, $C_7H_7Pt^+$, $C_6H_7Pt^+$, and $C_6H_5Pt^+$ (117). The hydropentalenyl complex (CXXII) exhibits as the base peak of the spectrum, the protonated pentalene ion $C_8H_7{}^+$, indicating that the molecule fragments easily (103). The molecular ion peak has been observed for bis(cyclopentadienyl)platinum (69a).

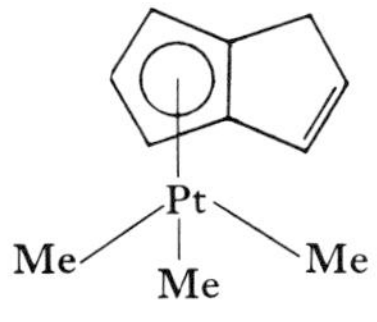

(CXXII)

2. Olefinic and Acetylenic Compounds

Bis(cycloocta-1,5-diene)nickel, $(C_8H_{12})_2Ni$, shows the molecular ion, and fragment ions $C_8H_{12}Ni^+$, $C_4H_6Ni^+$, and Ni^+ (149). The platinum analog, $(C_8H_{12})_2Pt$, shows as the principal ions $(C_8H_{12})_2Pt^+$, $C_8H_{12}Pt^+$, and $C_8H_{12}^+$, but the spectrum is more complicated than that of the nickel complex, and fragments due to loss of C_2 and C_3 fragments are observed (*149*). The diene-palladium compounds for which mass spectral data have been reported, $C_6H_{10}PdCl_2$ (*132*), $C_7H_8PdCl_2$ (*132*), and $(C_6H_5)_4C_4PdCl_2$ (*15*), do not show any metal-containing ions, but only ions attributable to LCl_2^+, LCl^+, and L^+, which are probably formed by thermal decomposition processes.

The composition of the acetylene complex $[(CF_3)C{\equiv}C(CF_3)]_3Ni_4(CO)_3$ was established by mass spectrometry. It shows a strong molecular ion, fragment ions $[M-(CO)_n]^+$ ($n=0$–2), and also fragmentation of the fluorinated ligand (*125*).

One of the products of the reaction of tetramethylcyclobutadienenickel chloride dimer and $Fe_3(CO)_{12}$ has the composition $(Me)_8C_8FeNi(CO)_3$. It exhibits the molecular ion peak, and successive loss of the carbonyl groups occurs to give $(Me)_8C_8FeNi^+$. Further breakdown is not clear, but it appears that the organic ligand is composed of one $(Me)_8C_8$ unit rather than two discrete $(Me)_4C_4$ units (*32*).

The reaction of o-diiodobenzene with $Ni(CO)_4$ gives a black compound of composition $[NiC_7H_4I_2O]_2$ which is thought to be a benzyne derivative (*82*).

Its mass spectrum does not show the molecular ion, the ion of highest mass observed corresponding to $C_6H_4NiI_2^+$. Other ions observed include $C_6H_5I_2^+$, $C_{12}H_8^+$, $C_6H_5CO^+$, I_2^+, CO^+, and C_6H_5, but no nickel-containing ions other than $C_6H_4NiI_2^+$ are reported.

3. Allylic Compounds

The mass spectra of the isoleptic complexes $(C_3H_5)_2M$ ($M=Ni$, Pd, Pt) have been measured. $(C_3H_5)_2Ni$ shows a molecular ion with the expected isotopic distribution (*49, 198*). A more detailed study of π-allyl metal complexes (*12*) has shown that in the field-ionization mass spectra, where the initial ionization is not accompanied by electronic excitation, little fragmentation occurs, and the molecular ion only was observed for

$(C_3H_5)_2M$ $(M = Ni, Pt)$. Under electron-bombardment conditions, the spectra are dominated by hydrocarbon fragments, but for $(C_3H_5)_2M$ the molecular ions and peaks formed by successive loss of allyl radicals are also strong. The predominant fragmentation of $(C_3H_5)_2Pt$ was elimination of propylene from the molecular ion, while for $(C_3H_5)_2Ni$, a peak at $[M - 28]^+$ corresponds to elimination of a molecule of ethylene (12). The molecular ion has been observed for $(2\text{-methylallyl})_2Ni$ as well as the fragment ions $C_4H_8Ni^+$, C_4H_7Ni, and $C_8H_{13}^+$ (49).

The mass spectrum of $(\pi\text{-}C_3H_5PdCl)_2$ shows loss of both chlorine and allyl radicals from the molecular ion, but no ion $C_3H_5PdCl^+$ corresponding to half the dimerized molecule was observed, and metal–metal interaction is proposed to account for the abundance of fragments containing the Pd_2Cl unit. The base peak of the spectrum corresponds to $C_3H_5Pd^+$, which can be formed from many of the other ions (132). The mass spectrum of the methoxyallyl complex (CXXIII)

$$Me-\!\!\!<\!\!\!\!\!\underset{MeOCH_2}{\overset{}{\Big\langle}}\!\!\!-Pd\!\!<\!\!\!\overset{Cl}{\underset{Cl}{\diagup}}\!\!\!>\!\!Pd-\!\!\!\underset{}{\overset{CH_2OMe}{\Big\rangle}}\!\!\!-Me$$

(CXXIII)

shows only three very weak clusters of peaks for palladium-containing ions corresponding to M^+, $[M - HCl]^+$, and $[M - HCl - C_6H_{11}OCl]^+$, and the majority of the ions observed are derived from the organic ligand. The high abundance of L^+ and $[L - H]^+$ could be due to thermal decomposition. The spectra of several other methoxylallylpalladium complexes did not show any metal-containing ions, but have as the base peak of the spectrum, the ion $[L - H]^+$ (132).

The mass spectrum of $C_5H_5Ni(1\text{-methylallyl})$ shows the molecular ion and the fragment ions $C_5H_6Ni^+$, $C_5H_5Ni^+$, $C_4H_6Ni^+$, and Ni^+ (140). The compound $C_5H_5NiC_{13}H_{17}$ has the structure (CXXIV) and its mass spectrum (117) shows that the cyclobutenyl ligand is bonded to the nickel more strongly than the cyclopentadienyl ligand, and the cyclopentadienyl ligand is lost from M^+ as C_5H_6, leaving $C_{13}H_{16}Ni^+$.

This latter ion breaks down by loss of two methyl groups, the second of which can also be eliminated as CH_4, giving $C_{12}H_{12}Ni^+$ which fragments further by loss of acetylene. The nickel-free ion of highest mass is $C_{13}H_{16}^+$, which arises by elimination of C_5H_6Ni from the molecular ion. The molecular ion of $C_5H_5Pd(\pi\text{-}C_3H_5)$ fragments either by the loss of allene, giving $C_5H_6Pd^+$, or by loss of the C_5H_5 radical, giving $C_3H_5Pd^+$; the ion $C_5H_5Pd^+$, formed by simple cleavage of the allyl group, was not observed (117). The only other palladium-containing ion in the spectrum is $C_3H_3Pd^+$, which

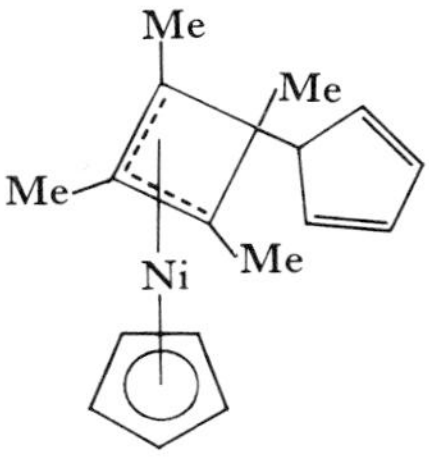

(CXXIV)

probably has the cyclopropenyl structure and could be formed by elimination of H_2 from the ion $C_3H_5Pd^+$, although this process was not observed in the spectrum of $(\pi\text{-}C_3H_5PdCl)_2$ (132).

The σ-allyl complex $(PPh_3)_2Pt(\sigma\text{-}C_3H_5)_2$ exhibits a fragmentation due to the triphenylphosphine as well as the ions $(C_3H_5)_2Pt^+$, $(C_3H_5)Pt^+$, and $C_3H_5^+$. The fragment ions $C_3H_4Pt^+$ and Pt^+ were observed in lower abundances (12)

VII

APPEARANCE POTENTIALS, BOND DISSOCIATION ENERGIES, AND HEATS OF FORMATION

A. Appearance Potentials

The appearance potentials for molecular ions (ionization potentials) and for fragment ions formed in the mass spectra of metallocenes and related compounds are listed in Table XIII. These appearance potentials have been used to calculate bond dissociation energies and heats of formation of organometallic compounds, but the results obtained must be treated cautiously because the appearance potentials of fragment ions include excess energy due to excited species. The values obtained for the heats of formation are best considered as upper limits, rather than precise determinations. The extent to which energy due to excited states can contribute to the appearance potentials is illustrated by the metal ions M^+ ($M = V$, Cr, Mn, Fe, Co, Ni, Ru), whose appearance potentials are approximately twice the value of the ionization potentials obtained by spectroscopic methods.

The values obtained for the ionization potentials of organometallic molecules are much lower than the ionization potentials of the ligands (Table XIV) and are much closer to the ionization potentials of the central metal atom (Table XV), indicating that ionization subsequent to electron impact involves an electron associated with the metal atom.

In several cases shown in Table XV, the ionization potential of the molecule is 1–1.5 eV lower than that of the metal atom, suggesting a high stability of the molecular ion with respect to neutral species, and this is more noticeable with the second- and third-row transition metals such as molybdenum, rhenium, tungsten, and osmium.

The appearance in the mass spectra of organometallic compounds of ions of species formed by thermal decomposition processes has already been discussed and recently it has been found by Müller and Göser (150) that such fragment ions can be observed in appearance potential measurements. Thus in the spectra of $C_5H_5MC_7H_7$ ($M = V$, Cr), $M(C_6H_6)_2$ ($M = V$, Cr), and $C_5H_5CrC_6H_6$, the ionization curves of the fragment ions $C_6H_6M^+$, $C_5H_5M^+$, and M^+ showed two or three stages of ionization. For example, the fragment $C_6H_6Cr^+$ in the spectrum of $(C_6H_6)_2Cr$ has appearance potentials at 8.8 and 6.4 eV. The higher value corresponds to the process

$$(C_6H_6)_2Cr \rightarrow (C_6H_6)_2Cr^+ \rightarrow C_6H_6Cr^+ + C_6H_6$$

The value of 6.4 eV is in between the ionization potentials of the complex and the free metal atom and is probably due to a C_6H_6Cr fragment formed by thermal decomposition in the spectrometer.

B. Bond Dissociation Energies

Bond dissociation energies have been calculated from appearance potentials according to the equation

$$A.P.[C_5H_5M^+] \geqslant D[C_5H_5M - C_5H_5] + I.P.[(C_5H_5)_2M]$$

but no correction has been made for excess kinetic energy of the fragment ions and the neutral fragments lost, so that the values obtained cannot be considered as very reliable and give the upper limits only.

The dissociation energies for the loss of the second cyclopentadienyl radical from $(C_5H_5)_2M$

$$C_5H_5M^+ \rightarrow M^+ + C_5H_5$$

compounds have been calculated from the appearance potentials of the fragment ions (148), and although the results are subject to the reservations discussed above, it is apparent that the energy required to cleave the second cyclopentadienyl group is much less than that required to cleave the first C_5H_5 radical. Müller and Göser (150) have assumed that the appearance potentials of $C_5H_5M^+$ and $C_6H_6M^+$ fragment ions formed by thermal decomposition processes give a value for the ionization potentials of the radicals C_5H_5M and C_6H_6M and have used these values to calculate the dissociation energies for cleavage of C_7H_7 and C_6H_6 ligands in some vanadium and chromium complexes.

A different approach by Winters and Kiser (199) involves the calculation of heats of formation of fragment ions (discussed below). Using the appropriate thermochemical values, ΔH (reaction) gives the bond strength $D[R-M^+]$ for the reaction

$$RM^+ \rightarrow R + M^+$$

and they have calculated the dissociation energies for the fragment ions $C_5H_5M^+$ ($M = V$, Mn, Co) and $C_3H_3M^+$ ($M = V$, Co). The bond dissociation energies of the cyclopropenylmetal ions are 20–30 kcal/mole less than the dissociation energies for cyclopentadienylmetal ions for vanadium and cobalt, and as $D[C_5H_5 - Mn^+]$ is only 2 kcal/mole it would be expected that $D[C_3H_3 - Mn^+]$ would have a negative value. This is supported by the very low abundance of $C_3H_3Mn^+$ in the spectrum of $C_5H_5Mn(CO)_3$, whereas large relative abundances of $C_3H_3V^+$ and $C_3H_3Co^+$ are observed in the spectra of the vanadium and cobalt compounds. The bond dissociation energies reported so far are shown in Table XVI. The values of the dissociation energies of a cyclopentadienyl radical from $(C_5H_5)_2M^+$ ions give an indication of the relative strength of bonding in the various complexes, and a big difference is observed between the energy required to cleave a cyclopentadienyl radical from $(C_5H_5)_2Mn$ and from bis(cyclopentadienyl) complexes of other transition metals. The value obtained for the manganese complex is much closer to that for the ionic compound

$(C_5H_5)_2Mg$, rather than the covalently bonded compounds of vanadium, chromium, iron, and nickel, indicating a greater degree of ionic bonding in $(C_5H_5)_2Mn$.

C. Heats of Formation

Müller and D'Or (*148*) and Müller and Göser (*150*) have estimated the heats for formation of some metallocenes for the process

$$M(g) + Ar(g) + Ar'(g) \rightarrow ArMAr'(g)$$

$$(Ar = C_5H_5, C_6H_6, \text{ or } C_7H_7)$$

from the differences between the appearance potential of the metal ion formed from the complex and the ionization potential of the free metal atom. The values obtained are given in Table XVII; also shown are Skinner's selected heats of formation obtained from heat of combustion and heat of sublimation data (*185*). The values obtained from appearance potential measurements are rather higher than those obtained thermochemically, possibly because the metal ion M^+ is formed with excess kinetic energy.

A more sophisticated approach has been used by Winters and Kiser (*199*) to calculate the heats of formation of the cyclopentadienylmetal carbonyls $C_5H_5V(CO)_4$, $C_5H_5Mn(CO)_3$, and $C_5H_5Co(CO)_2$ as well as the heats of formation of some of the fragment ions. The heats of formation were calculated for the process

$$C_5H_5M(CO)_x \rightarrow M^+ + C_3H_3 + C_2H_2 + x\,CO$$

assuming that the neutral fragments are formed in their ground states and the values obtained for $C_5H_5V(CO)_4$, $C_5H_5M(CO)_3$, and $C_5H_5\,Co(CO)_2$ are, respectively, -139 kcal/mole, -67 kcal/mole, and -15 kcal/mole. The heats of formation of the fragment ions are summarized in Table XVIII.

Recently, Müller and Göser (*150*) have observed metastable peaks for the decomposition of doubly charged molecular ions of $(C_6H_6)_2V$ and $(C_6H_6)_2Cr$ into two singly charged species according to the following process:

$$(C_6H_6)_2M^{2+} \rightarrow C_6H_6M^+ + C_6H_6^+$$

From the width of the metastable peaks they calculated the total kinetic energy evolved in the metastable decay, which is the total kinetic energy of the two fragment ions. The values obtained were for $(C_6H_6)_2V$, 2.3 eV and for $(C_6H_6)_2Cr$, 3.0 eV. Assuming that the kinetic energy is exclusively due to electrostatic repulsion in the doubly charged ions, the distance between the charges can be calculated directly and the values of 4.9 and 6.3 Å are obtained for the vanadium and chromium complexes, respectively. The Cr—C distance in gaseous $(C_6H_6)_2Cr$ is 2.15 Å, and the value obtained for the charge separation in $(C_6H_6)_2Cr^{2+}$ is more than twice this distance, indicating that the positive charges are localized on the rings.

ACKNOWLEDGMENT

We wish to thank all our colleagues who have kindly sent us unpublished data for inclusion in this review.

VIII

TABULAR SUMMARY OF MASS SPECTRAL DATA FOR TRANSITION-METAL ORGANOMETALLIC COMPOUNDS SINCE 1967

TABLE I

COMPLEXES OF TITANIUM, ZIRCONIUM, AND HAFNIUM

Compound	Section	References
$C_5H_5Ti(OEt)_3$	II	155
$C_5H_5Ti(OEt)_2Cl$	II	155
$C_5H_5Ti(OEt)Cl_2$	II	155
$C_5H_5TiCl_3$	II	155
$(C_5H_5)_2Ti(C_{10}H_8N_2)$	II	67
$(C_5H_5)_2TiNCO$	II	52
$(C_5H_5TiCN)_3$	II	52
$(C_5H_5TiSCN)_3$	II	52
$(\pi\text{-}C_3H_5)_4Zr$	II	12
$(C_5H_5)_2ZrCl_2$	II	171
$(C_5H_5)_2ZrClSiPh_3$	II	39
$(C_5H_5)_2Zr(H)BF_4$	II	93
$[(C_5H_5)_2ZrCl]_2O$	II	171
$(\pi\text{-}C_3H_5)_4Hf$	II	12
$(C_5H_5)_2HfClSiPh_3$	II	125b

TABLE II

COMPLEXES OF VANADIUM AND NIOBIUM

Compound	Section	References
$(C_5H_5)_2V$	III	*81, 148*
$C_5H_5VC_7H_7$	III	*117, 150*
$(C_6H_6)_2V$	III	*150*
$C_5H_5V(CO)_4$	III	*114, 199*
$(C_5H_5)_2V_2(CO)_5$	III	*77a*
$C_5H_5V(CO)_3P(NMe_2)_3$	III	*113*
$C_5H_5V(CO_2Me)_2$	III	*105*
$[C_5H_5VS_2C_2(CF_3)_2]_2$	III	*120*
$(C_5H_5)_2NbOCl$	III	*195*

TABLE III

COMPLEXES OF CHROMIUM, MOLYBDENUM, AND TUNGSTEN

Compound	Section	References
$(C_5H_5)_2Cr$	IV A 1	*81, 148*
$[C_5H_5Cr(CO)_3]_2$	IV A 2	*111*
$C_5H_4CPh_2Cr(CO)_3$	IV A 2	*47*
$[C_5H_5Cr(NO)_2]_2$	IV A 3	*114*
$[C_5H_5Cr(NO)SMe]_2$	IV A 3	*167*
$[C_5H_5Cr(NO)SPh]_2$	IV A 3	*167*
$[C_5H_5Cr(NO)OMe]_2$	IV A 3	*167*
$[C_5H_5Cr(NO)NMe_2]_2$	IV A 3	*167*
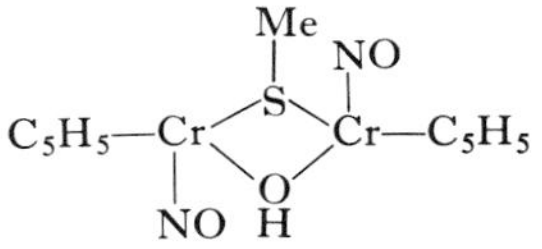	IV A 3	*167*
$C_5H_5CrC_6H_6$	IV A 4	*150*
$C_5H_5CrC_7H_7$	IV A 4	*76, 116, 150*
$(C_6H_6)_2Cr$	IV B	*54, 119, 165, 166*
$C_{12}H_{11}DCr$	IV B	*63*
$C_{12}H_{10}D_2Cr$	IV B	*63*
$C_{12}H_9D_3Cr$	IV B	*63*
$C_{12}H_8D_4Cr$	IV B	*63*
$C_6H_6Cr(CO)_3$	IV B	*118, 165, 166*

TABLE III—*continued*

Compound	Section	References
OMe	IV B	*27*
$Me_2CHOC_6H_5Cr(CO)_3$	IV B	*27*
m-$(CO_2H)(OMe)C_6H_4Cr(CO)_3$	IV B	*136*
m-$(CO_2Me)(OMe)C_6H_4Cr(CO)_3$	IV B	*136*
p-$(Me)(OCHMe_2)C_6H_4Cr(CO)_3$	IV B	*27*
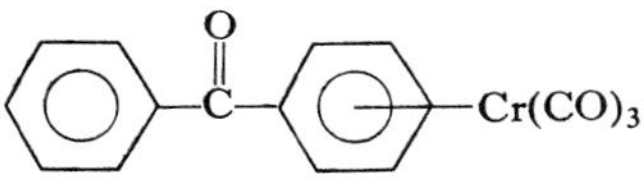	IV B	*136*
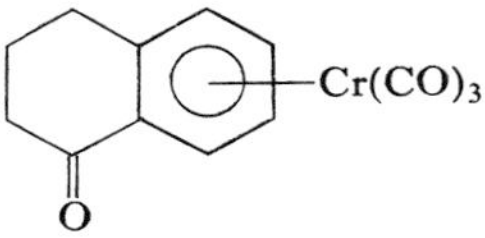	IV B	*136*
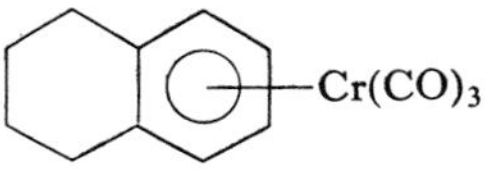	IV B	*136*
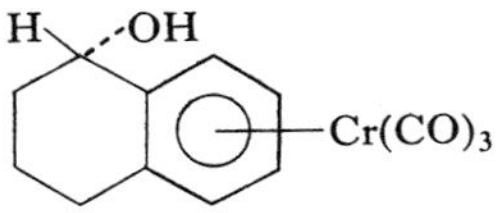	IV B	*58*
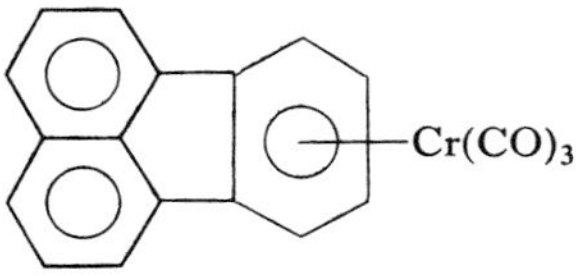	IV B	*55*
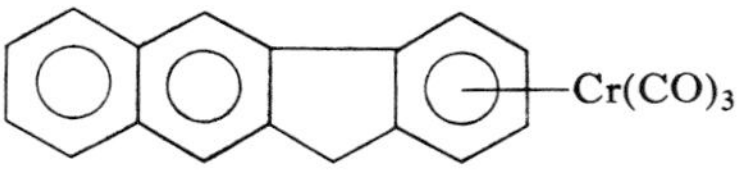	IV B	*55*
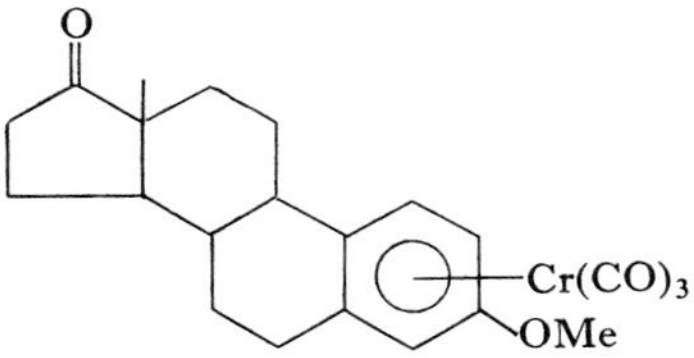	IV B	*27*

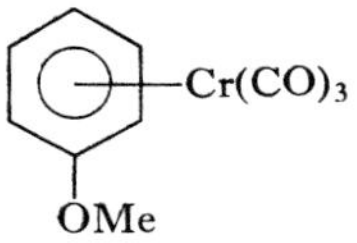

TABLE III—*continued*

Compound	Section	References
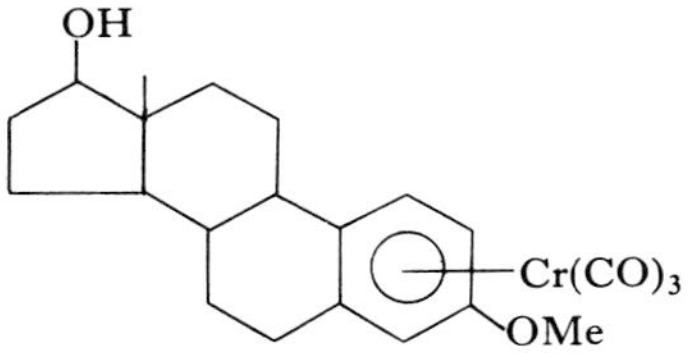	IV B	27
$[B_3N_3(CH_3)_6]Cr(CO)_3$	IV B	168
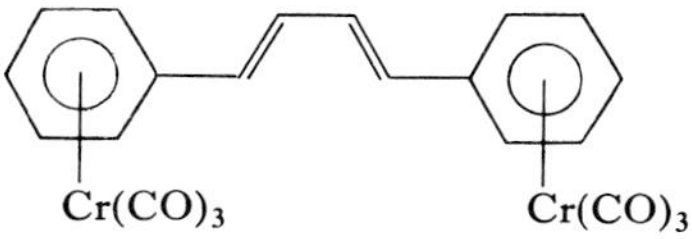	IV B	136
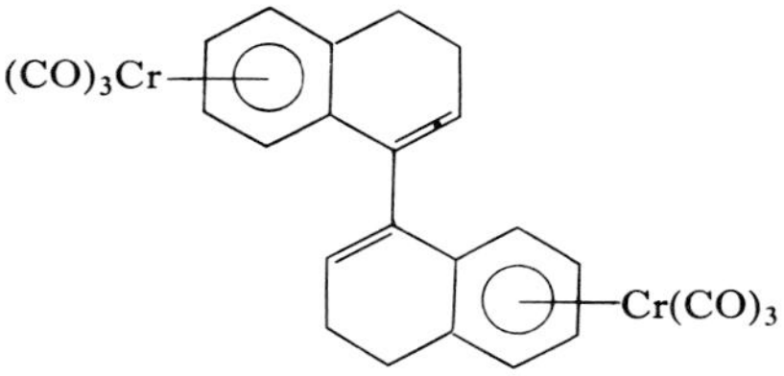	IV B	136
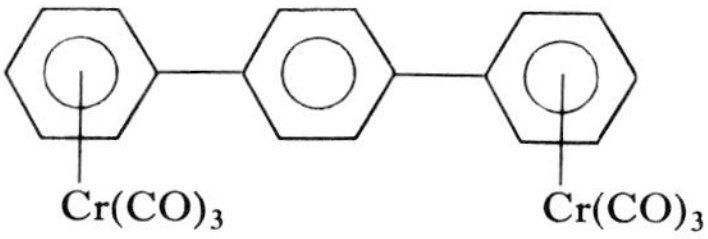	IV B	136
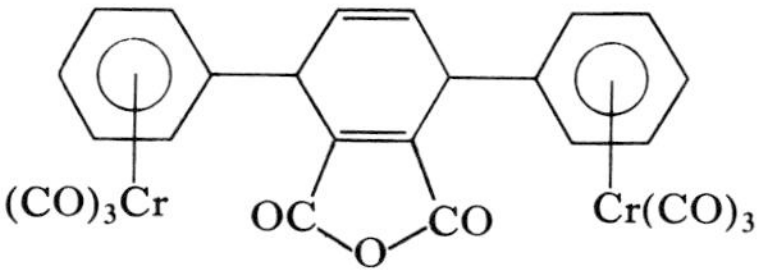	IV B	136
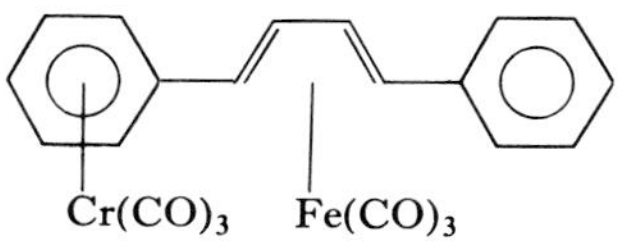	IV B	136
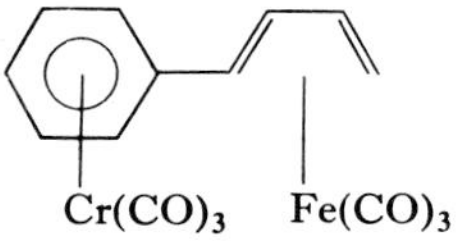	IV B	136

TABLE III—*continued*

Compound	Section	References
$(\pi-C_3H_5)_3Cr$	IV C	*12*

| 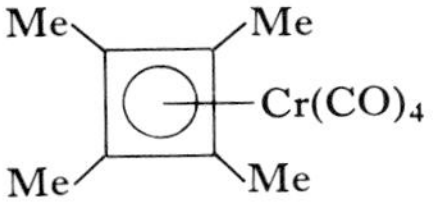 | IV C | *3* |

| 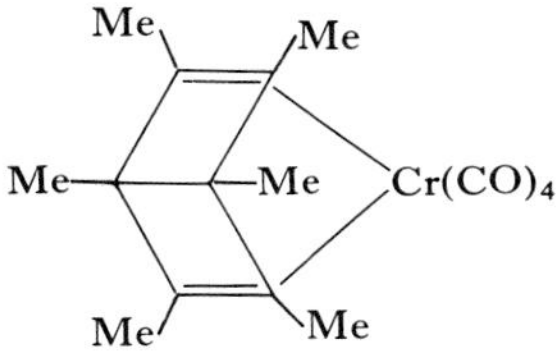 | IV C | *66, 72* |

$C_7H_8Cr(CO)_3$ (cycloheptatriene)	IV C	*119, 164a*
$C_7H_8Cr(CO)_4$	IV C	*27*
$C_8H_8Cr(CO)_3$	IV C	*109*
$C_7H_7CrC_7H_{10}$	IV C	*75*
$(CO)_5CrC(OMe)Me$	IV C	*6*
$(CO)_5CrC(OEt)Me$	IV C	*77*
$(CO)_5CrC(SPh)Me$	IV C	*128*
$(CO)_5CrC(NH_2)Me$	IV C	*128*
$(CO)_5CrC(OSiMe_3)Me$	IV C	*147*
$(CO)_5CrC(NHC_6H_{11})COMe$	IV C	*6*

| 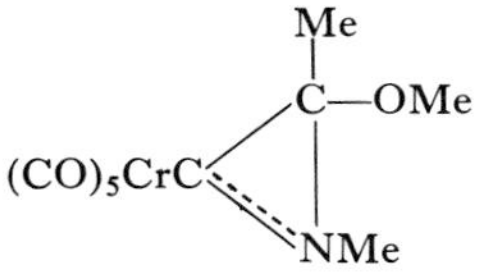 | IV C | *6* |

| 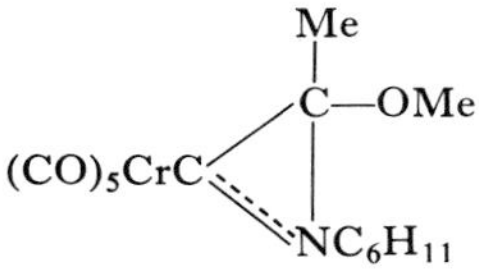 | IV C | *5, 6* |

| 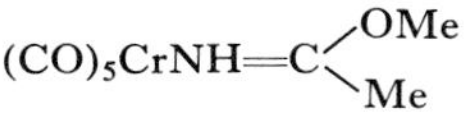 | IV C | *73* |

| 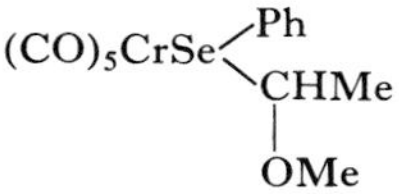 | IV C | *74* |

TABLE III—*continued*

Compound	Section	References
$C_5H_5Mo(CO)_3Br$	IV A 2	*182*
$C_5H_5Mo(CO)_3HgCl$	IV A 2	*139*
$C_5H_5Mo(CO)_3HgBr$	IV A 2	*139*
$C_5H_5Mo(CO)_3HgI$	IV A 2	*139*
$C_5H_5Mo(CO)_3HgSCN$	IV A 2	*139*
$C_5H_5Mo(CO)_3HgCo(CO)_4$	IV A 2	*139*
$C_5H_5Mo(CO)_3GeMe_3$	IV A 2	*41*
$C_5H_5Mo(CO)_3GeEt_3$	IV A 2	*40, 41*
$C_5H_5Mo(CO)_3Ge\text{-}n\text{-}Pr_3$	IV A 2	*41*
$C_5H_5Mo(CO)_3SiMe_3$	IV A 2	*39*
$C_5H_5Mo(CO)_3(CH_2)_3I$	IV A 2	*88*
$C_5H_5Mo(CO)_3CH_2SMe$	IV A 2	*115*
$C_5H_5Mo(CO)_3(CH_2)_4Br$	IV A 2	*114*
$C_5H_5Mo(CO)_3CH_2OCOMe$	IV A 2	*114*
$C_5H_5Mo(CO)_3CF_3$	IV A 2	*116*
$C_5H_5Mo(CO)_3C_3F_7$	IV A 2	*116*
$C_5H_5Mo(CO)_3COC_3F_7$ (and $-COCF_3$)	IV A 2	*116*
$[C_5H_5Mo(CO)_3]_2$	IV A 2	*111, 131, 182*
$[C_5H_5Mo(CO)_3]_2Hg$	IV A 2	*139*
$[C_5H_5Mo(CO)_3]_2(CF_2)_2$	IV A 2	*114*
$[(CH_3)_5C_5Mo(CO)_2]_2$	IV A 2	*121*
$C_5H_5Mo(CO)_2NO$	IV A 2	*200*
$C_5H_5Mo(CO)_2CH_2SMe$	IV A 2	*115*
$C_5H_5Mo(CO)_2C_7H_7$ (cycloheptatrienyl)	IV A 2	*114*
$C_5H_5Mo(CO)_2C_7H_7$ (benzyl)	IV A 2	*114*
$C_5H_5Mo(CO)_2(\pi\text{-}C_3H_5)$	IV A 2	*114*
$C_5H_5Mo(CO)_2(PEt_2H)GeEt_3$	IV A 2	*41*
$C_5H_5Mo(CO)_2(PPh_3)GeMe_3$	IV A 2	*41*
$C_5H_5Mo(CO)_2(PPh_3)GeEt_3$	IV A 2	*41*
$[C_5H_5Mo(NO)I_2]_2$	IV A 3	*107*
$[C_5H_5Mo(NO)S_2C_2(CF_3)_2]_2$	IV A 3	*120*
$[C_5H_5MoS_2C_2(CF_3)_2]_2$	IV A 4	*120*
$C_5H_5MoC_7H_7$	IV A 4	*76*
$C_9H_7Mo(CO)_3Me$	IV A 2	*118*
$C_9H_7Mo(CO)_3I$	IV A 2	*118*
$C_9H_7Mo(CO)_2(\pi\text{-}C_3H_5)$	IV A 2	*118*
$C_9H_7Mo(CO)_2(\pi\text{-}CH_2SMe)$	IV A 2	*118*
$[C_9H_9Mo(CO)_2]_2$	IV A 2	*108*
$C_6H_8Mo(CO)_2$	IV C	*119*
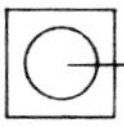 $Mo(CO)_4$	IV C	*3*

TABLE III—*continued*

Compound	Section	References
	IV C	*3*
	IV C	*72*
$C_7H_8Mo(CO)_3$(cycloheptatriene)	IV C	*164a*
$C_8H_{12}Mo(CO)_4$	IV C	*119*
$(MeCOCH=CH_2)_3Mo$	IV C	*109, 119*
$[(CF_3)_2C_2]_2MoNCMe$	IV C	*116*
$(C_5H_5)_2WH_2$	IV A 1	*148*
$C_5H_5W(CO)_3H$	IV A 2	*114*
$C_5H_5W(CO)_3HgCl$	IV A 2	*139*
$C_5H_5W(CO)_3HgBr$	IV A 2	*139*
$C_5H_5W(CO)_3HgI$	IV A 2	*139*
$C_5H_5W(CO)_3HgSCN$	IV A 2	*139*
$C_5H_5W(CO)_3HgCo(CO)_4$	IV A 2	*139*
$C_5H_5W(CO)_3SiMe_3$	IV A 2	*39*
$C_5H_5W(CO)_3GeMe_3$	IV A 2	*41*
$C_5H_5W(CO)_3GeEt_3$	IV A 2	*41*
$C_5H_5W(CO)_3GePh_3Pt(Ph_2PCH_2)_2$	IV A 2	*41*
$C_5H_5W(CO)_3SnMe_3$	IV A 2	*38*
$C_5H_5W(CO)_3COCH=CH_2$	IV A 2	*114*
$C_5H_5W(CO)_3CH_2C_5H_4N$	IV A 2	*114*
$C_5H_5W(CO)_3COCF_3$	IV A 2	*116*
$C_5H_5W(CO)_3COC_3F_7$	IV A 2	*116*
$[C_5H_5W(CO)_3]_2Hg$	IV A 2	*139*
$C_5H_5W(CO)_2C_7H_7$(cycloheptatrienyl)	IV A 2	*114, 123*
$C_5H_5W(CO)_2(PMe_3)GeMe_3$	IV A 2	*41*
$C_5H_5W(CO)_2(PEt_2H)GeMe_3$	IV A 2	*41*
$C_5H_5W(CO)_2(PPh_3)GeMe_3$	IV A 2	*41*
$[C_5H_5W(CO)S_2C_2(CF_3)_2]_2$	IV A 2	*120*
$C_5H_5W[S_2C_2(CF_3)_2]_2$	IV A 2	*120*
$C_6H_8W(CO)_2$	IV C	*119, 123*
$C_7H_8W(CO)_3$(cycloheptatriene)	IV C	*119, 164a*
$C_8H_{12}W(CO)_4$	IV C	*119*

TABLE III—*continued*

Compound	Section	References
$C_{10}H_{12}W(CO)_4$	IV C	*119*
$C_{10}H_{12}W(CO)_4$	IV C	*119, 123*
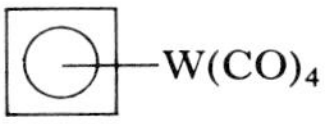 ⬚⊙—$W(CO)_4$	IV C	*3*
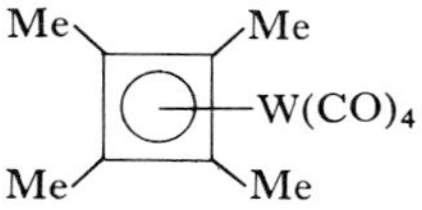 Me, Me / ⬚⊙—$W(CO)_4$ / Me, Me	IV C	*3*
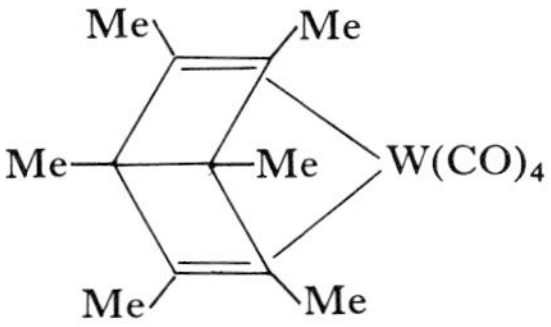 Me, Me / Me—⬡—Me $W(CO)_4$ / Me, Me	IV C	*72*
$(MeCOCH=CH_2)_3W$	IV C	*119, 122, 123*
$[(CF_3)_2C_2]_2WNCMe$	IV C	*116, 123*
$C_7H_8W(CO)_3(cycloheptatriene)$	IV C	*119*
$1,3,5\text{-}C_8H_{10}W(CO)_3$	IV C	*119*
$C_7H_8W(CO)_3(toluene)$	IV	*119*
$p\text{-}Me_2C_6H_4W(CO)_3$	IV	*119*
$1,3,5\text{-}Me_3C_6H_3W(CO)_3$	IV	*119*

TABLE IV

COMPLEXES OF MANGANESE, TECHNETIUM, AND RHENIUM

Compound	Section	References
$(C_5H_5)_2Mn$	V A 1	*81, 148*
$C_5H_5MnC_6H_6$	V A 1	*54, 150*
$C_5H_5Mn(CO)_3$	V A 3	*199*
$C_5H_4RMn(CO)_3$		
$\quad R=CHO$	V A 3	*37, 194*
$\quad R=COMe$	V A 3	*37*
$\quad R=CO_2H$	V A 3	*37, 136*
$\quad R=COCHN_2$	V A 3	*37*

TABLE IV—*continued*

Compound	Section	References
$R = CO(CH_2)_2CO_2H$	V A 3	*37*
$R = CO(CH_2)_3CO_2H$	V A 3	*37*
$R = COCH_2COCO_2Et$	V A 3	*37*
$R = CH:CHCO_2H$	V A 3	*37, 136*
$R = C(Me):CHCO_2H$	V A 3	*37*
$R = CH:CHCO_2Me$	V A 3	*37*
$R = C(Me):CHCO_2Et$	V A 3	*37*
$R = CH_2OH$	V A 3	*37*
$R = C(OH)(Me)CH_2CO_2Et$	V A 3	*37*
$R = (CH_2)_3C(OH)Ph_2$	V A 3	*37*
$R = (CH_2)_2CO_2H$	V A 3	*37, 136*
$R = CH(Me)CH_2CO_2H$	V A 3	*37*
$R = (CH_2)_4CO_2H$	V A 3	*37*
$R = CH(Me)CH_2CO_2Me$	V A 3	*37*
$(CO)_3Mn$—[indenyl structure]	V A 3	*133*
$(CO)_3Mn$—[indanone structure with O]	V A 3	*133*
$(CO)_3Mn$—[indanone structure with O]	V A 3	*133*
$(CO)_3Mn$—[hydroxy structure with OH]	V A 3	*133*
$(CO)_3Mn$—[dioxolane structure with O–O]	V A 3	*133*
$(CO)_3Mn$—[dibromo structure with Br, Br]	V A 3	*133*
$(CO)_3Mn$—[hydroxy tetrahydro structure with OH]	V A 3	*58*

TABLE IV—*continued*

Compound	Section	References
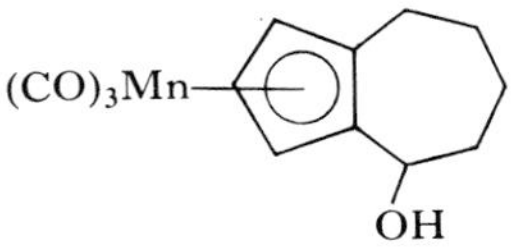	V A 3	133
$C_5H_5Mn(CO)_2CNH$	V A 3	78
$[C_5H_5Mn(NO)_2]_n$	V A 3	107
$C_5H_5Mn(NO)S_2C_2(CF_3)_2$	V A 3	120
$(C_5H_5)_3Mn_3(NO)_4$	V A 3	60a
$C_6(CH_3)_6MnC_6H_7$	V A 3	69
$[C_6(CH_3)_6MnC_6H_6]PF_6$	V A 3	69
$C_{10}H_8Mn_2(CO)_6$	V A 3	22
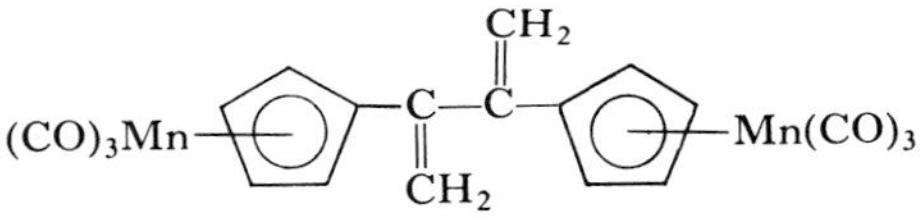	V A 3	136
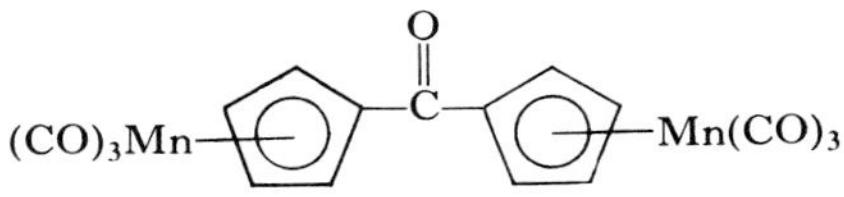	V A 3	136
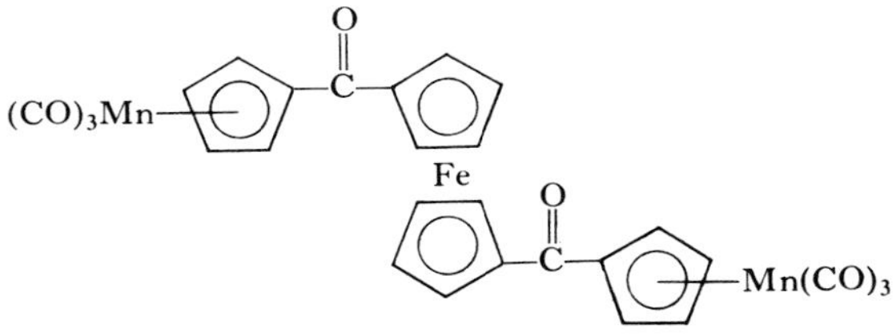	V A 3	136
$(C_5H_5)_2TcH$	V B	65, 148
$(C_5H_5)_2ReH$	V B	65, 81, 148
$C_5H_5ReC_6H_6$	V B	71
$C_5H_5Re(CO)_3$	V B	114
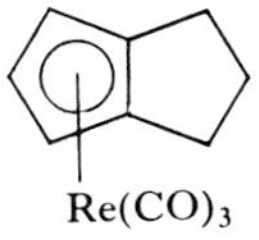	V B	102

TABLE V

FERROCENE DERIVATIVES[a]

Compound	Section	References
$(C_5H_5)_2Fe$	VI A 1	*141, 44, 54, 79, 81, 135, 148, 165, 166, 181, 189*
$[C_5Me_5]_2Fe$	VI A 1	*121*
Fe—Me	VI A 2	*170*
Fc—CN	VI A 2	*170*
Fc—CH$_2$OH	VI A 2	*58*
Fc—CD$_2$OH	VI A 2	*58*
Fc—(CH$_2$)$_2$OH	VI A 2	*58*
Fc—(CH$_2$)$_3$OH	VI A 2	*58*
Fc—(CH$_2$)$_4$OH	VI A 2	*58*
Fc—CH(OH)Me	VI A 2	*58, 170*
Fc—CD(OH)Me	VI A 2	*58*
Fc—CH(OH)Ph	VI A 2	*58*
Fc—CD(OH)Ph	VI A 2	*58*
Fc—CD(OH)CH$_2$D	VI A 2	*58*
Fc—C(OH)(Me)Ph	VI A 2	*58*
Fc—C(OH)(Me)CH$_2$Ph	VI A 2	*58*
Fc—C(OH)Ph$_2$	VI A 2	*186*
Fc—C(OH)CH$_2$CH=CH$_2$	VI A 2	*143*
Fc—CH$_2$NMe$_2$	VI A 2	*186*
Fc—CH$_2$Ph	VI A 2	*189*
Fc—C$_5$H$_5$	VI A 2	*189*
Fc—CH$_2$CO$_2$Me	VI A 2	*173*
Fc—(CH$_2$)$_2$CO$_2$Me	VI A 2	*173*
Fc—(CH$_2$)$_3$CO$_2$Me	VI A 2	*173*
Fc—(CH$_2$)$_4$CO$_2$Me	VI A 2	*173*
Fc—(CH$_2$)$_5$CO$_2$Me	VI A 2	*173*
Fc—CH$_2$CO$_2$Et	VI A 2	*173*
Fc—(CH$_2$)$_2$CO$_2$Et	VI A 2	*173*
Fc—(CH$_2$)$_3$CO$_2$Et	VI A 2	*173*
Fc—(CH$_2$)$_4$Et	VI A 2	*173*
Fc—(CH$_2$)$_5$Et	VI A 2	*173*
Fc—COMe	VI A 2	*135*
Fc—COPh	VI A 2	*135*
Fc—CO-p-MeOC$_6$H$_4$	VI A 2	*135*
Fc—CO(CH$_2$)$_2$CO$_2$Me	VI A 2	*170*
Fc—CONHMe	VI A 2	*135*
Fc—CO$_2$H	VI A 2	*44, 135*
Fc—CO$_2$D	VI A 2	*135*
Fc—CO$_2$Me	VI A 2	*135, 173*
Fc—CO$_2$Et	VI A 2	*173*

TABLE V—*continued*

Compound	Section	References
Fc—Ph	VI A 2	170
Fc—C$_6$H$_4$Cl	VI A 2	170
Fc—C$_6$H$_4$NO$_2$	VI A 2	170
Fc—o-C$_6$H$_4$CO$_2$H	VI A 2	134
Fc—o-C$_6$H$_4$CO$_2$Me	VI A 2	134, 172
Fc—m-C$_6$H$_4$CO$_2$Me	VI A 2	172
Fc—o-C$_6$H$_4$CO$_2$CD$_3$	VI A 2	172
Fc—o-C$_6$H$_4$CO$_2$Et	VI A 2	134, 172
Fc—m-C$_6$H$_4$CO$_2$Et	VI A 2	172
Fc—o-C$_6$H$_4$CO$_2$-n-Pr	VI A 2	134
Fc—o-C$_6$H$_4$CH$_2$CO$_2$Me	VI A 2	172
Fc—m-C$_6$H$_4$CH$_2$CO$_2$Me	VI A 2	172
Fc—p-C$_6$H$_4$CH$_2$CO$_2$Me	VI A 2	172
Fc—o-C$_6$H$_4$CH$_2$CO$_2$CD$_3$	VI A 2	172
Fc—o-C$_6$H$_4$COMe	VI A 2	172
Fc—m-C$_6$H$_4$COMe	VI A 2	172
Fc—p-C$_6$H$_4$COMe	VI A 2	172
Fc—o-C$_6$H$_4$COCD$_3$	VI A 2	172
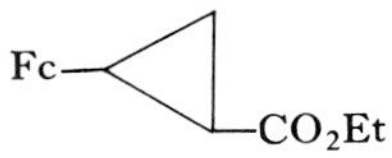 Fc—⟨cyclopropane⟩—CO$_2$Et	VI A 2	143
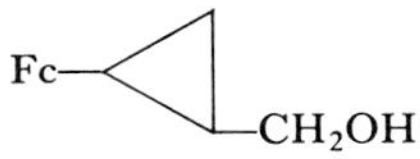 Fc—⟨cyclopropane⟩—CH$_2$OH	VI A 2	143
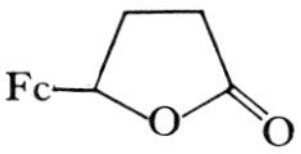 Fc—⟨lactone⟩	VI A 2	143
Fc—CH=CHCOMe	VI A 2	134
Fc—CH=CHCO$_2$H	VI A 2	134
Fc—CH=CHCO$_2$Me (*cis*)	VI A 2	134, 174
Fc—CH=CHCO$_2$Me (*trans*)	VI A 2	134, 174
Fc—CH=CH(CH$_2$)$_2$OCOMe	VI A 2	134
Fc—CH=CCl$_2$	VI A 2	134
Fc—CH=CBr$_2$	VI A 2	134
Fc—C≡C—CO$_2$H	VI A 2	134
Fc—(C≡C)$_4$Ph	VI A 2	179
Fc—NHCO$_2$Me	VI A 2	170
Fc—N=N—Ph ↓ O	VI A 2	154
Fc—SO$_3$H	VI A 2	44
Fc—SO$_2$NH$_2$	VI A 2	170

TABLE V—*continued*

Compound	Section	References
Fc—SiMe$_3$	VI A 2	*44*
Fc—SiPh$_3$	VI A 2	*44*
Fc—SiMe$_2$OEt	VI A 2	*44*
Fc—SiMe$_2$CH=CH$_2$	VI A 2	*44*
Fc—B(OH)$_2$	VI A 2	*44*

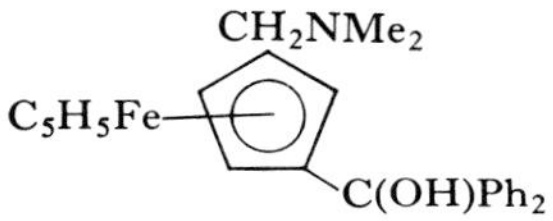

C$_5$H$_5$Fe, CH$_2$NMe$_2$, C(OH)Ph$_2$ — VI A 3a — *186*

C$_5$H$_5$Fe, Me, CH(OH)Me — VI A 3a — *58*

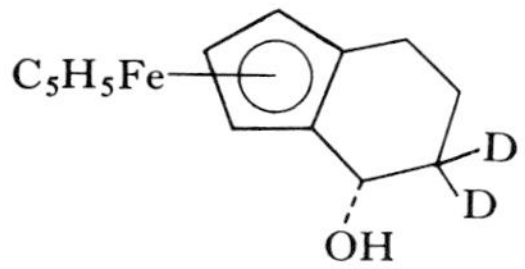

| | ψ-endo | VI A 3a | *58* |
| | ψ-exo | | |

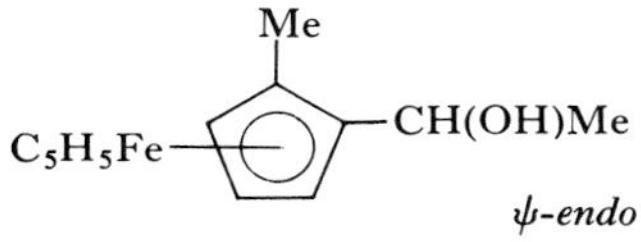

C$_5$H$_5$Fe, R, OH — endo — VI A 3a — *58*

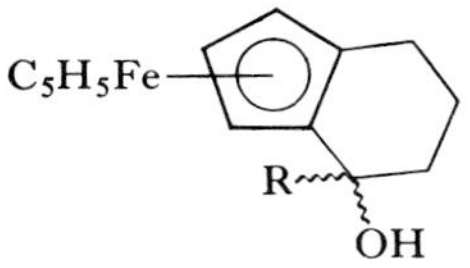

| | exo | VI A 3a | *58* |

(R = H, D)

C$_5$H$_5$Fe, D, D, OH — VI A 3a — *58*

C$_5$H$_5$Fe, R, Ph, OH — endo — VI A 3a — *58*

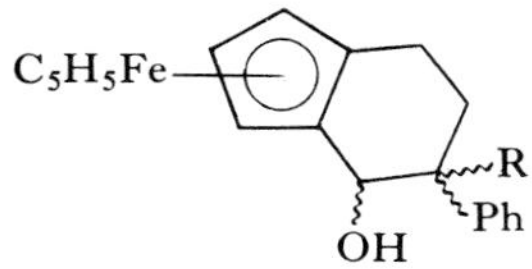

| | exo | VI A 3a | *58* |

(R = H, D)

C$_5$H$_5$Fe, Ph, OH — endo — VI A 3a — *58*

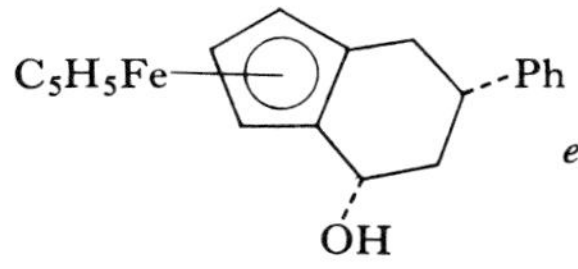

TABLE V—*continued*

Compound		Section	References
C_5H_5Fe—[structure]	*endo/endo*	VI A 3a	58
(R = H, D)	*exo/endo*	VI A 3a	58
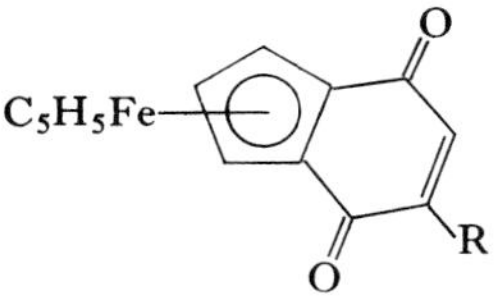 C_5H_5Fe—[structure]	(R = H, Ph)	VI A 3a	59
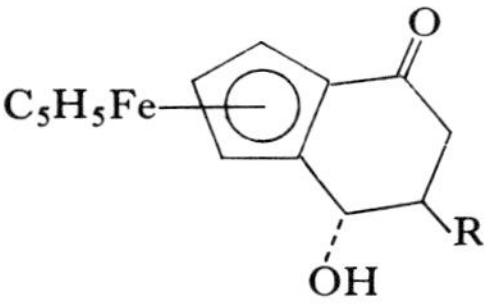 C_5H_5Fe—[structure]	(R = H, Ph, C_6H_{11})	VI A 3a	59
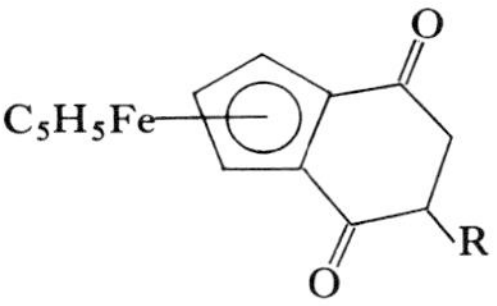 C_5H_5Fe—[structure]	(R = H, Ph)	VI A 3a	59
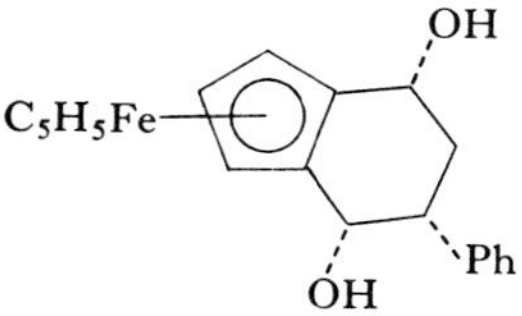 C_5H_5Fe—[structure]		VI A 3a	59
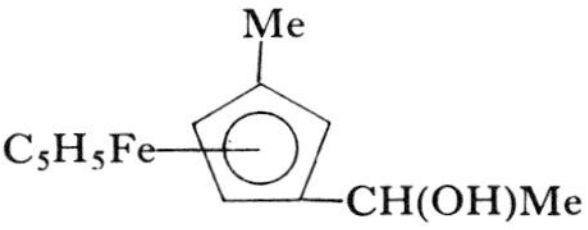 C_5H_5Fe—[structure]—CH(OH)Me		VI A 3a	58
$(C_5H_4Cl)_2Fe$		VI A 3a	44
$(C_5H_4Et)_2Fe$		VI A 3a	44, 189
$(C_5H_4Bu)_2Fe$		VI A 3a	44
$(C_5H_4SiMe_3)_2Fe$		VI A 3a	44
$[C_5H_4SiMe_2CH{=}CH_2]_2Fe$		VI A 3a	44

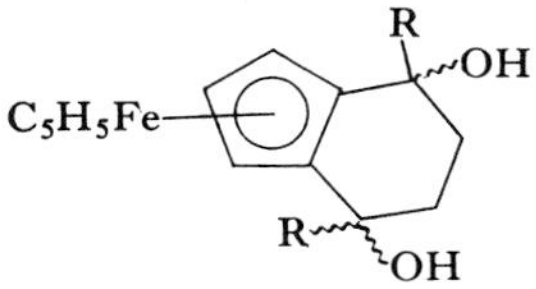

TABLE V—*continued*

Compound	Section	References
$[C_5H_4SiMe_2OEt]_2Fe$	VI A 3a	*44*
$[C_5H_4CH_2OH]_2Fe$	VI A 3a	*58*
$[C_5H_4CD_2OH]_2Fe$	VI A 3a	*58*
$[C_5H_4CH(OH)Me]_2Fe$	VI A 3a	*58*
$(C_5H_4CH_2C_6H_5)_2Fe$	VI A 3a	*189*
$(C_5H_4CH{=}CH_2)_2Fe$	VI A 3a	*174*
$[C_5H_4CD{=}CH_2)_2Fe$	VI A 3a	*174*
$[C_5H_4CH_2NMe_2]Fe[C_5H_4C(OH)Ph_2]$	VI A 3a	*186*

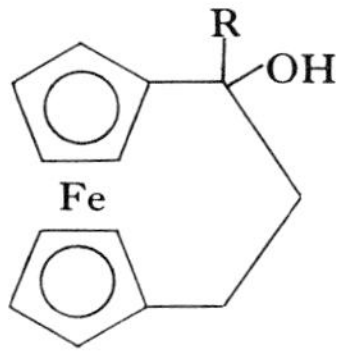

(R = H, D)

VI A 3a   *58*

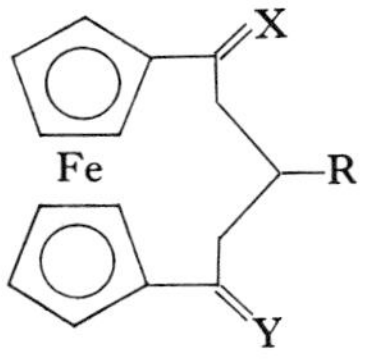

VI A 3a   *10*

R	X	Y
H	H_2	H_2
Ph	H_2	H_2
H	O	H_2
Ph	O	H_2
H	O	O

VI A 3a   *10*

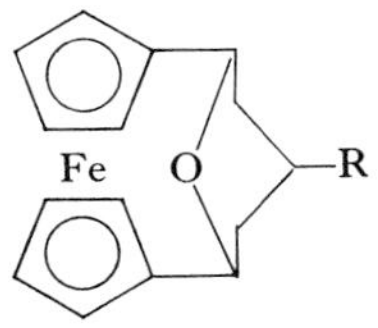

(R = H, Ph)

TABLE V—*continued*

Compound	Section	References
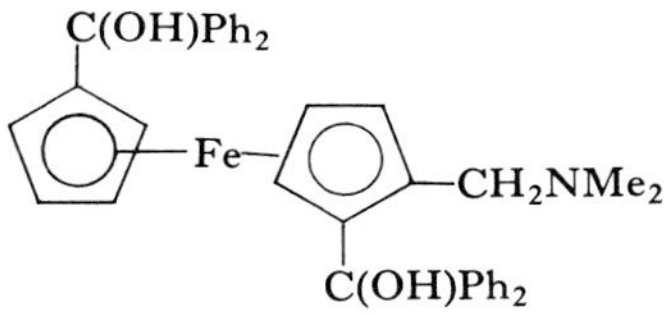 C(OH)Ph₂ Fe CH₂NMe₂ C(OH)Ph₂	VI A 3b	*186*
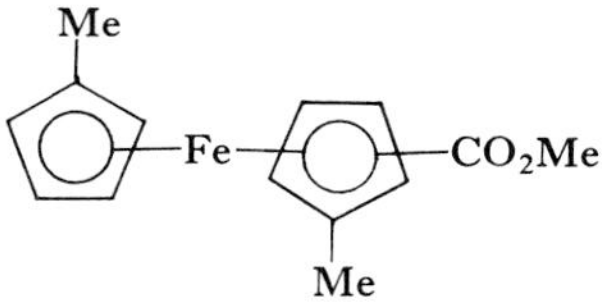 Me Fe CO₂Me Me	VI A 3b	*64*
[C₁₀H₇Et₃]Fe H OH Fe *endo/endo* *exo/endo*	VI A 3b	*189*
	VI A 3c	*58, 60*
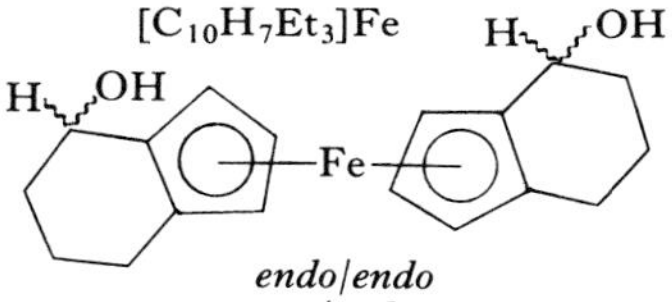 H OH Fe H OH *endo/endo* *exo/endo*	VI A 3c	*58, 60*
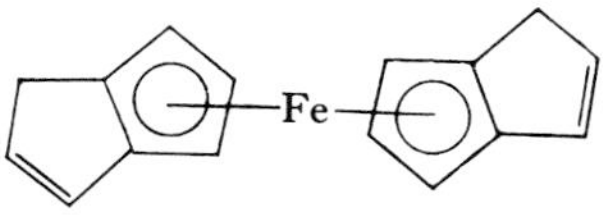 Fe	VI A 3c	*103*
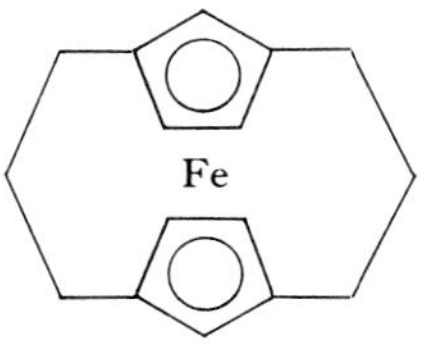 Fe	VI A 3c	*48*

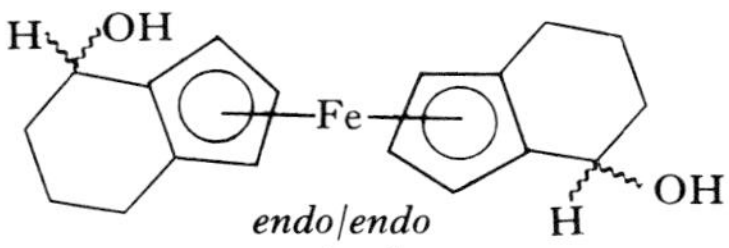

a Fc = C₅H₅FeC₅H₄.

TABLE VI

POLYFERROCENE COMPLEXES[a]

Compound	Section	References
Fc—Fc	VI A 4	*44, 189*
Ethylbiferrocenyl	VI A 4	*189*
Diethylbiferrocenyl	VI A 4	*189*
Triethylbiferrocenyl	VI A 4	*189*
Diferrocenylmercury	VI A 4	*169*
Benzylbiferrocenyl	VI A 4	*189*
Dibenzylferrocenyl	VI A 4	*189*
Fc—(C≡C)$_3$—Fc	VI A 4	*179*
Fc—(C≡C)$_4$—Fc	VI A 4	*179*

| 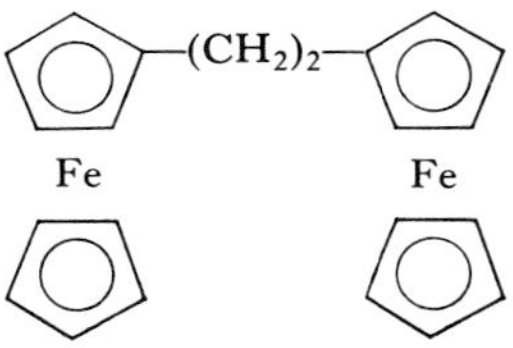 | VI A 4 | *48* |

| 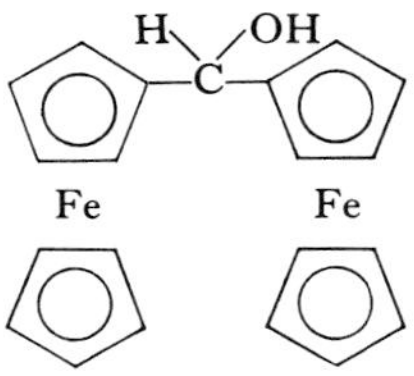 | VI A 4 | *58* |

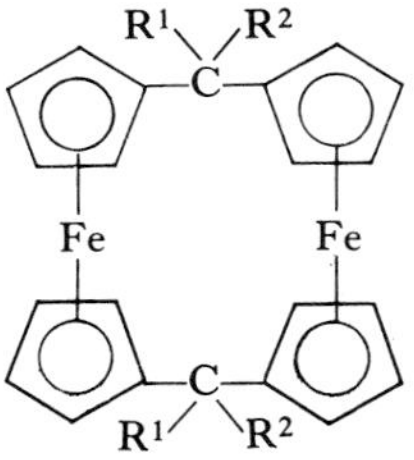

R^1	R^2		
H	H	VI A 4	*189*
H	Me	VI A 4	*27*
Me	Me	VI A 4	*197*
H	Ph	VI A 4	*27*
=O		VI A 4	*27*

TABLE VI—*continued*

Compound	Section	References
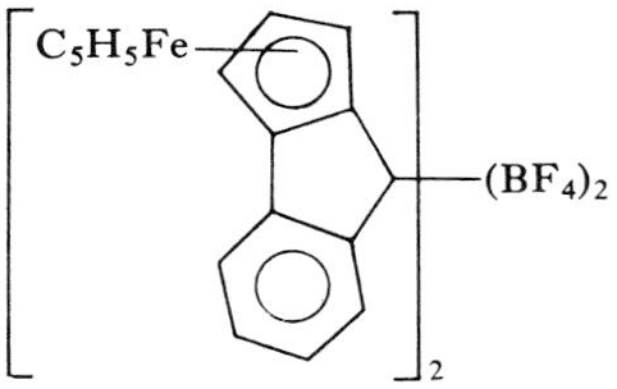 $\left[C_5H_5Fe \underset{}{\bigcirc} \right]_2$ (BF$_4$)$_2$	VI A 4	*35*
H$\left(\!-C_5H_4FeC_5H_4\!\right)_3$H	VI A 4	*189*
H$\left(\!-C_5H_4FeC_5H_4\!\right)_3CH_2$Ph	VI A 4	*189*
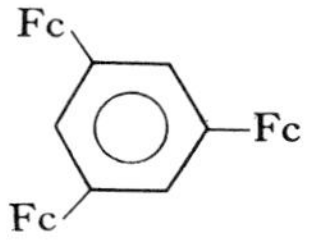 Fc, benzene ring with three Fc groups	VI A 4	*180*
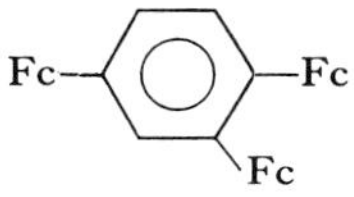 Fc—benzene ring with three Fc groups	VI A 4	*180*
Fc—CO—C=C—CO—Fc, Fc Fc	VI A 4	*175*
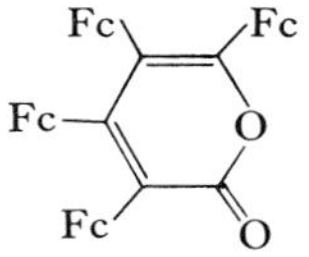 pyranone ring with four Fc groups	VI A 4	*175*
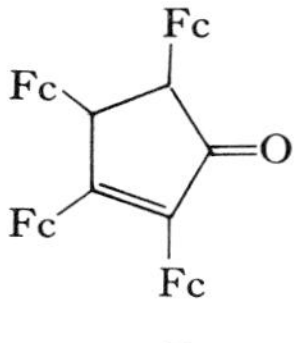 cyclopentenone ring with four Fc groups	VI A 4	*175*
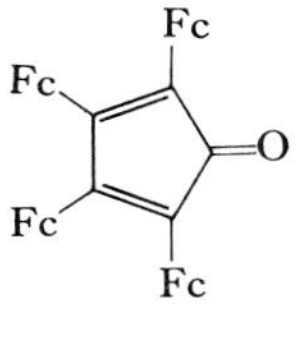 cyclopentadienone ring with four Fc groups	VI A 4	*175*
$\left(\!-Hg\!-C_5H_4FeC_5H_4\!\right)_x$	VI A 4	*169*

a Fc$=C_5H_5FeC_5H_4$.

TABLE VII

CYCLOPENTADIENYLIRON CARBONYL COMPLEXES

Compound	Section	References
$C_5H_5Fe(CO)_2Br$	VI A 5	*182*
$C_5H_5Fe(CO)_2HgCl$	VI A 5	*139*
$C_5H_5Fe(CO)_2HgBr$	VI A 5	*139*
$C_5H_5Fe(CO)_2HgI$	VI A 5	*139*
$C_5H_5Fe(CO)_2HgSCN$	VI A 5	*139*
$C_5H_5Fe(CO)_2HgCo(CO)_4$	VI A 5	*139*
$C_5H_5Fe(CO)_2SnPh_3$	VI A 5	*131*
$C_5H_5Fe(CO)_2COMe$	VI A 5	*114*
$C_5H_5FE(CO)_2COPh$	VI A 5	*114*
$C_5H_5Fe(CO)_2COC_3F_7$	VI A 5	*116*
$C_5H_5Fe(CO)_2CH_2C_5H_4N$	VI A 5	*114*
$C_5H_5Fe(CO)_2(CH_2)_2C_5H_{10}N$	VI A 5	*114*
$C_5H_5Fe(CO)_2SMe$	VI A 5	*114, 115*
$C_5H_5Fe(CO)_2SOCPh$	VI A 5	*114, 115*
$C_5H_5Fe(CO)_2C_6H_5$	VI A 5	*114*
$C_5H_5Fe(CO)_2C_6F_5$	VI A 5	*112, 116*
$C_5H_5Fe(CO)_2-3,4-H_2C_6F_3$	VI A 5	*112, 116*
$C_5H_5Fe(CO)_2-HC_6F_4$	VI A 5	*116*
$C_5H_5Fe(CO)_2-p-CF_3C_6H_4$	VI A 5	*112, 116*
$C_5H_5Fe(CO)_2CH_2C_6H_5$	VI A 5	*24, 26*
$C_5H_5Fe(CO)_2CH_2C_6F_5$	VI A 5	*24, 26*
$C_5H_5Fe(CO)_2CH_2C_6H_4Me$	VI A 5	*88*
$C_5H_5Fe(CO)_2COCH{=}CHPh$	VI A 5	*114*
$C_5H_5Fe(CO)_2CH_2OCOMe$	VI A 5	*114*
$C_5H_5Fe(CO)_2C{:}CPh$	VI A 5	*85*
$C_5H_5Fe(CO)_2CH{:}CHC_6F_4H$	VI A 5	*31*
$C_5H_5Fe(CO)_2CH{:}CHCF_3$	VI A 5	*31*
$C_5H_5Fe(CO)_2CF{:}CFCF{:}CF_2$	VI A 5	*84*
$C_5H_5Fe(CO)_2C_6F_9$	VI A 5	*114*
$C_5H_5Fe(CO)_2SPh$	VI A 5	*45*
$C_5H_5Fe(CO)_2SC_6F_5$	VI A 5	*45*
$C_5H_5Fe(CO)_2NC_5F_4$	VI A 5	*46*
$[C_5H_5Fe(CO)_2]_2$	VI A 5	*131, 182*
$[C_5H_5Fe(CO)_2]_2Hg$	VI A 5	*131, 139*
$[C_5H_5Fe(CO)_2]_2SnCl_2$	VI A 5	*131*
$[C_5H_5Fe(CO)SMe]_2$	VI A 5	*115, 167*
$[C_5H_5Fe(CO)SPh]_2$	VI A 5	*167*
$C_5H_5Fe(CO)_2CH_2C_6F_4Fe(CO)_2C_5H_5$	VI A 5	*26*
$\{C_5H_5Fe[P(OPh)_3]\}_2$	VI A 5	*156*
$[C_5H_5FeCO]_4$	VI A 5	*104*
$C_5H_5FeC_4H_4N$	VI A 5	*117*
$C_5H_5FeC_9H_7$	VI A 5	*118*
$C_5H_5FeC_9H_9$	VI A 5	*118*
$C_5H_5FeC_9H_{11}$	VI A 5	*118*
$(C_9H_7)_2Fe$	VI A 5	*118*

TABLE VIII

OLEFIN- AND ACETYLENE-IRON COMPLEXES

Compound	Section	References
Olefin-Fe(CO)$_4$ *Olefin*		
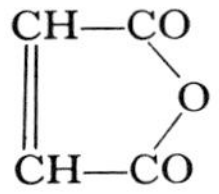 $\begin{array}{c} \text{CH---CO} \\ \parallel \qquad\quad \text{O} \\ \text{CH---CO} \end{array}$	VI A 6	*86*
$\begin{array}{c} \text{MeOOC---CH} \\ \parallel \\ \qquad\text{CH---COOMe} \end{array}$	VI A 6	*86*
$\begin{array}{c} \text{CH---COOMe} \\ \parallel \\ \text{CH---COOMe} \end{array}$	VI A 6	*86*
$\begin{array}{c} \text{CHCl} \\ \parallel \\ \text{CHCl} \end{array}$	VI A 6	*86*
$\begin{array}{c} \text{CHCl} \\ \parallel \\ \text{ClHC} \end{array}$	VI A 6	*86*
$\begin{array}{c} \text{CCl}_2 \\ \parallel \\ \text{CH}_2 \end{array}$	VI A 6	*86*
$\begin{array}{c} \text{CCl}_2 \\ \parallel \\ \text{CHCl} \end{array}$	VI A 6	*86*
$\begin{array}{c} \text{CHBr} \\ \parallel \\ \text{CHBr} \end{array}$	VI A 6	*86*
$\begin{array}{c} \text{CHMe} \\ \parallel \\ \text{CH}_2 \end{array}$	VI A 6	*86*
$\begin{array}{c} \text{CHOEt} \\ \parallel \\ \text{CH}_2 \end{array}$	VI A 6	*86*

TABLE VIII—*continued*

Compound	Section	References
CHPh ‖ CH_2	VI A 6	*86*
CHCN ‖ CH_2	VI A 6	*86*
CHCl ‖ CH_2	VI A 6	*86*
CHBr ‖ CH_2 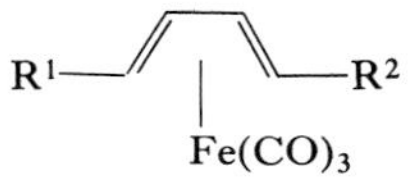	VI A 6	*86*

R^1—⟋═⟍—R^2
|
$Fe(CO)_3$

R^1	R^2		
Me	CO_2H	VI A 6	*136*
Me	CO_2D	VI A 6	*136*
Me	$CONH_2$	VI A 6	*136*
Me	$COND_2$	VI A 6	*136*
CO_2H	CO_2H	VI A 6	*136*
CO_2Me	CO_2Me	VI A 6	*136*

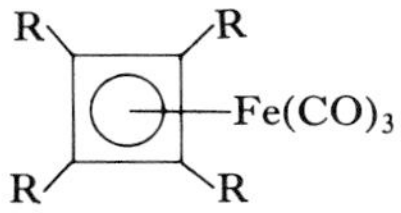

	Section	References
R=H	VI A 6	*3, 62, 145, 176*
R=Me 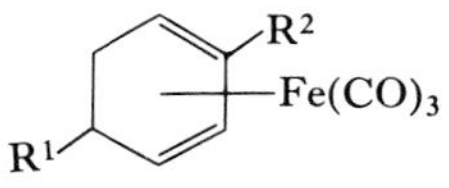	VI A 6	*3*

R^1	R^2		
H	H	VI A 6	*200*
OH	H	VI A 6	*87*
OH	OMe	VI A 6	*87*
$CHAc_2$	H	VI A 6	*87*
$CHAc_2$	OMe	VI A 6	*87*

TABLE VIII—*continued*

Compound	Section	References
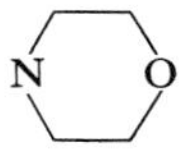 (N–O morpholine ring structure)	VI A 6	*87*
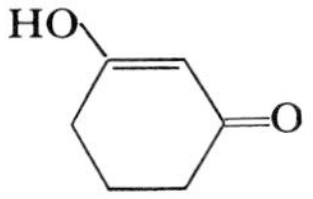 (HO-substituted cyclohexenone with O)	VI A 6	*87*
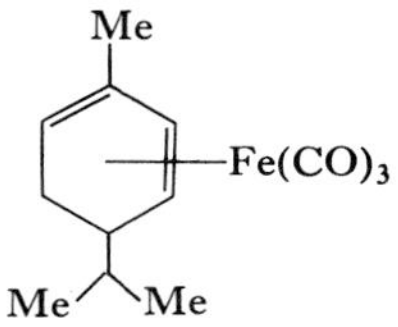 (Me-substituted cyclohexadiene–Fe(CO)₃ with Me–CH–Me)	VI A 6	*87*
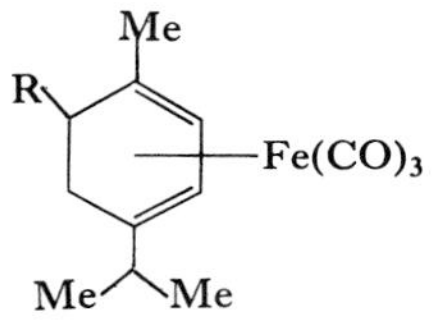 (R, Me-substituted cyclohexadiene–Fe(CO)₃ with Me–CH–Me, R = H, D)	VI A 6	*87*
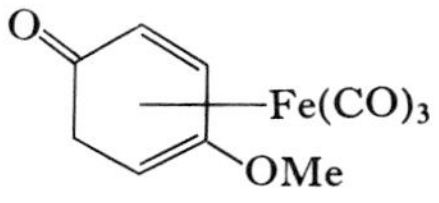 (O-substituted cyclohexadiene–Fe(CO)₃ with OMe)	VI A 6	*21*
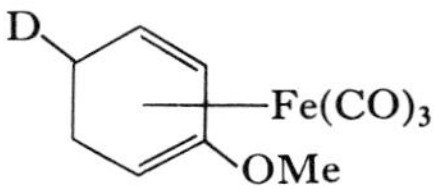 (D-substituted cyclohexadiene–Fe(CO)₃ with OMe)	VI A 6	*21*
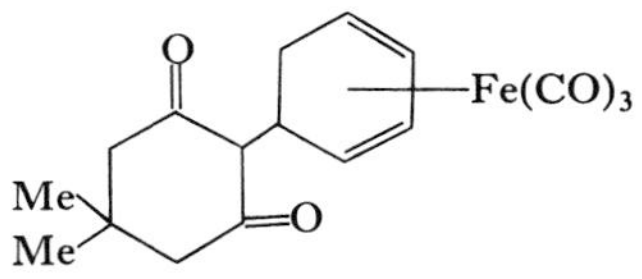 (dimethyl dioxocyclohexane linked to cyclohexadiene–Fe(CO)₃)	VI A 6	*87*
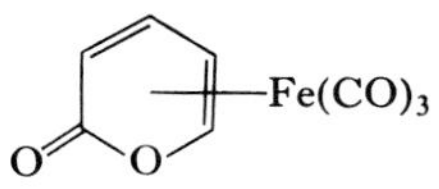 (pyranone ring–Fe(CO)₃)	VI A 6	*176*

TABLE VIII—*continued*

Compound	Section	References

| 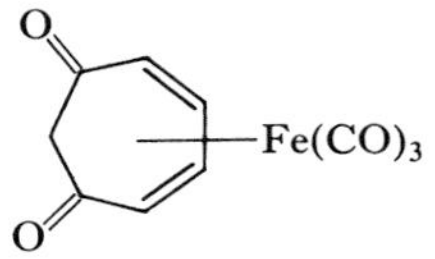 | VI A 6 | *21* |

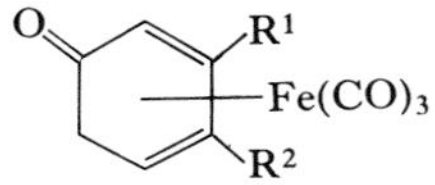

R^1	R^2		
H	H	VI A 6	*27*
H	OMe	VI A 6	*27*
Me	H	VI A 6	*27*

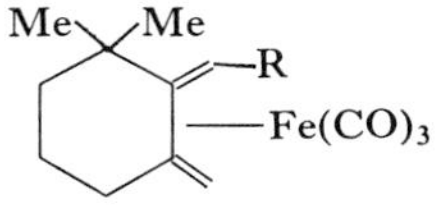		
$R = CH_2COMe$	VI A 6	*36*
$R = CH:C(Me)CH_2CO_2Et$	VI A 6	*36*

| 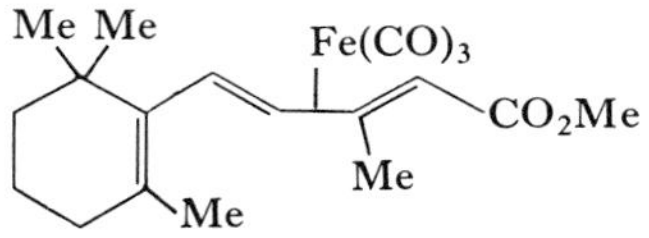 | VI A 6 | *36* |

Vitamin A acetate $Fe(CO)_3$	VI A 6	*20*
Acetylergosterol-$Fe(CO)_3$	VI A 6	*27*

| 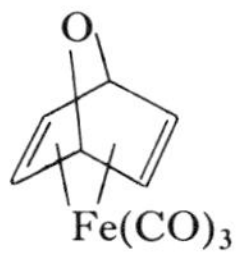 | VI A 6 | *130* |

Cyclooctatetraene-$Fe(CO)_3$	VI A 6	*119*

| 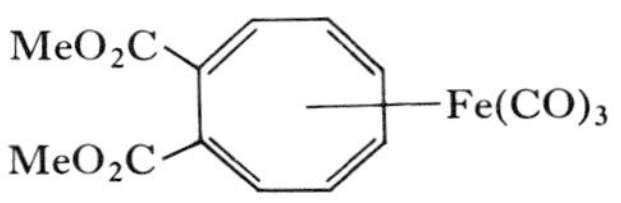 | VI A 6 | *83* |

TABLE VIII—*continued*

Compound	Section	References
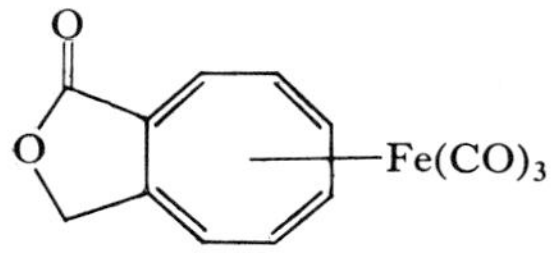 —Fe(CO)$_3$	VI A 6	*83*
CycloheptadieneFe$_2$(CO)$_4$(SMe)$_2$	VI A 6	*115*
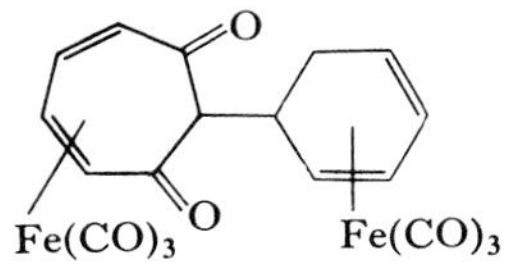	VI A 6	*21*
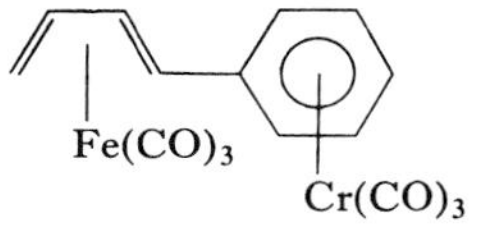	VI A 6	*136*
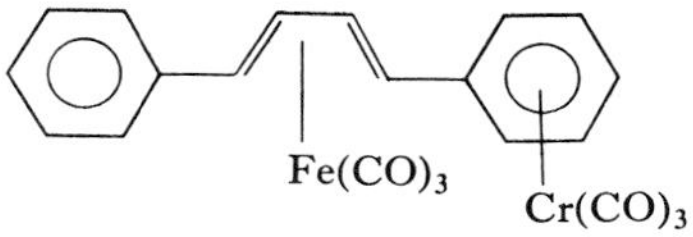	VI A 6	*136*
Azulene Fe$_2$(CO)$_5$	VI A 6	*111*
[Azulene Fe(CO)$_2$]	VI A 6	*111*
Acenaphthylene Fe$_2$(CO)$_5$	VI A 6	*111*
C$_{12}$H$_{16}$Fe(CO)$_2$ (allene dimer)	VI A 6	*111, 119*
(CH$_2$:C:C:CH$_2$)Fe$_2$(CO)$_6$	VI A 6	*111, 152*
(MeCH:C:C:CHMe)Fe$_2$(CO)$_6$	VI A 6	*158*
(CH$_2$:C:C:C:C:CH$_2$)Fe$_3$(CO)$_{7-8}$	VI A 6	*158*
(C$_7$H$_9$OH)Fe(CO)$_3$	VI A 6	*119*
C$_{14}$H$_{14}$F$_6$Fe(CO)$_3$	VI A 6	*116*
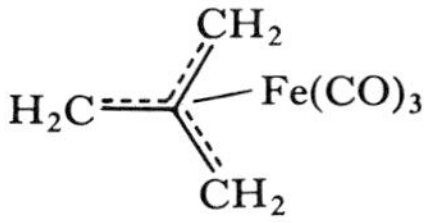	VI A 6	*62*
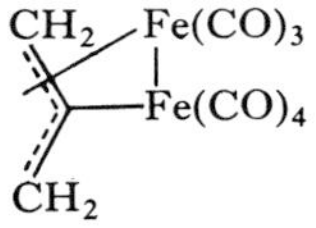	VI A 6	*13*
(Ph)$_2$C:C(FeCO$_4$)$_2$	VI A 6	*145*

TABLE IX

FLUOROCARBON-IRON COMPLEXES

Compound	Section	References
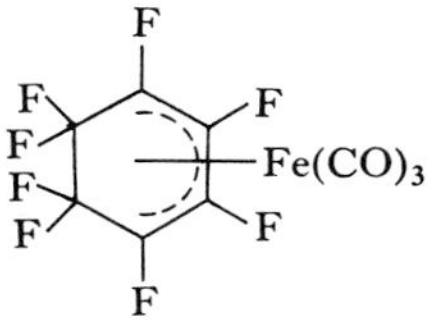	VI A 6	*91*
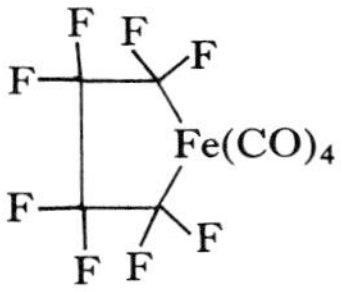	VI A 6	*91*

TABLE X

RUTHENIUM AND OSMIUM COMPLEXES

Compound	Section	References
$(C_5H_5)_2Ru$	VI B	*81, 148*
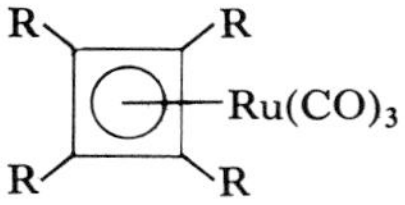		
R = Me	VI B	*3*
R = Ph	VI B	*183*
$C_8H_8Ru(CO)_3$	VI B	*29, 50*
$C_8H_8Ru_2(CO)_6$	VI B	*27*
$(C_8H_8)_2Ru_3(CO)_4$	VI B	*29*
$Me_3C_{10}H_5Ru_4(CO)_9$	VI B	*43*
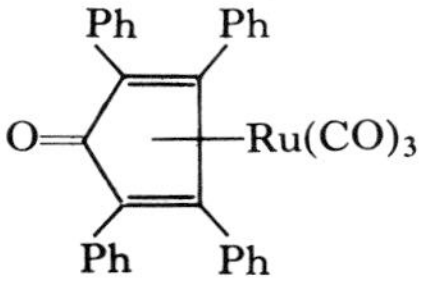	VI B	*30*

TABLE X—*continued*

Compound	Section	References
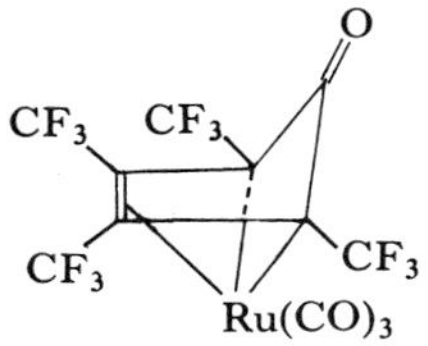	VI B	*30*
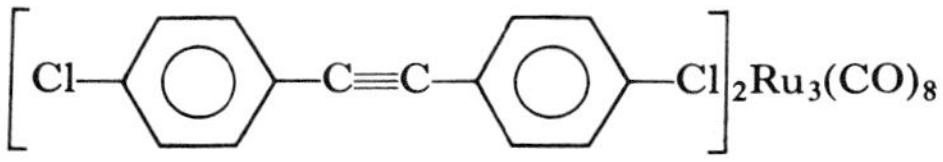	VI B	*183*
$Ru_6C(MeC_6H_5)(CO)_{14}$	VI B	*100*
$Ru_6C(Me_2C_6H_4)(CO)_{14}$	VI B	*100*
$Ru_6C(Me_3C_6H_3)(CO)_{14}$	VI B	*100*
$(C_5H_5)_2Os$	VI B	*148*

TABLE XI

COMPLEXES OF COBALT, RHODIUM, AND IRIDIUM

Compound	Section	References
$(C_5H_5)_2Co$	VI C	*81, 148, 165, 166*
$C_5H_5Co(CO)_2$	VI C	*165, 166, 199*
$(C_5H_5)_4Co_4(CO)_2$	VI C	*104*
$C_5H_5Co(CO)(GeCl_3)_2$	VI C	*129*
$C_5H_5CoCH_3NN:NNCH_3$	VI C	*153*
$C_5H_5CoC_6H_5NN:NNC_6H_5$	VI C	*153*
$C_5H_5CoS_2C_2(CF_3)_2$	VI C	*120*
$C_5H_5CoC_{14}H_{14}F_6$	VI C	*112, 116*
$(C_5H_5Co)_2(N\text{-}tert\text{-}Bu)_2CO$	VI C	*160*

TABLE XI—*continued*

Compound	Section	References
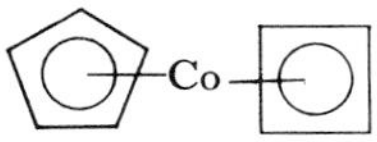	VI C	*4, 177*
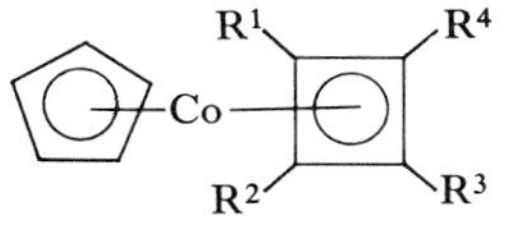		
$R^1 = R^2 = Ph$, $R^3 = R^4 = SiMe_3$	VI C	*90*
$R^1 = R^3 = Ph$, $R^2 = R^4 = SiMe_3$	VI C	*90*
$R^1 = R^3 = Ph$, $R^2 = R^4 = Me$	VI C	*90*
$R^1 = R^3 = Ph$, $R^2 = R^4 = OMe$	VI C	*90*
$R^1 = R^2 = Ph$, $R^3 = R^4 = COMe$	VI C	*90*
$R^1 = R^3 = Ph$, $R^4 = SnPh_3$	VI C	*90*
$R^1 = R^3 = Ph$, $R^2 = R^4 = CHO$	VI C	*90*
$R^1 = R^2 = Ph$, $R^3 = R^4 = CHO$	VI C	*90*
$R^1 = R^3 = Ph$, $R^2 = R^4 = CF_3$	VI C	*90*
$R^1 = R^3 = Ph$, $R^2 = R^4 = H$	VI C	*90*
$R^1 = R^2 = Ph$, $R^3 = R^4 = H$	VI C	*90*
$Co_2(CO)_4(CF_3C\equiv CH)_3$	VI C	*56*
$Co_2(CO)_6(CF_3C\equiv CH)$	VI C	*56*
$(\pi\text{-}C_3H_5)_3Rh$	VI D	*11*
$[(\pi\text{-}C_3H_5)_2RhCl]_2$	VI D	*132*
$[(\pi\text{-}MeC_3H_4)_2RhCl]_2$	VI D	*132*
$C_5H_5RhS_2C_2(CF_3)_2$	VI D	*120*
$C_5H_5Rh(C_2H_4)_2$	VI D	*117*
$C_5H_5RhC_8H_{12}$	VI D	*117*
$[C_5H_5Rh(CO)]_3$	VI D	*146*
$(C_5H_5)_4Rh_3H$	VI D	*68, 70*
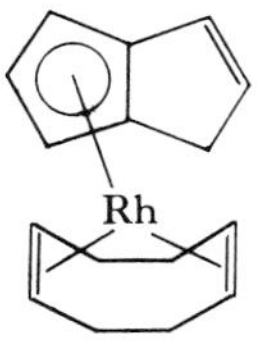	VI D	*103*
$(\pi\text{-}C_3H_5)_3Ir$	VI D	*42*
$C_5H_5IrS_2C_2(CF_3)_2$	VI D	*120*
$C_5H_5IrC_8H_{12}$	VI D	*117*

TABLE XII

COMPLEXES OF NICKEL, PALLADIUM, AND PLATINUM

Compound	Section	References
$(\pi\text{-}C_3H_5)_2Ni$	VI E 3	12, 49, 198
$(\pi\text{-}2\text{-}MeC_3H_4)_2Ni$	VI E 3	49
$(C_5H_5)_2Ni$	VI E 1	79, 81, 95, 148, 165, 166, 181
$C_5H_5Ni(CO)C_2F_5$	VI E 1	140
$C_5H_5NiC_4H_7$	VI E 3	140
$C_5H_5NiC_{13}H_{17}$	VI E 3	117
C_5H_5NiNO	VI E 1	79
$C_5H_5NiS_2C_2(CF_3)_2$	VI E 1	120
$[C_5H_5Ni(CO)]_2$	VI E 1	182
$[C_5H_5NiSMe]_2$	VI E 1	167
$[C_5H_5NiSPh]_2$	VI E 1	167
$(C_5H_5Ni)_3(CO)_2$	VI E 1	182
$(C_5H_5Ni)_3N\text{-}tert\text{-}Bu$	VI E 1	160
$[C_6H_4NiCOI_2]$	VI E 2	82
$(C_8H_{12})_2Ni$	VI E 2	147
$(B_9C_{12}H_{11})_2Ni$	VI E 2	89
$Me_8C_8NiFe(CO)_3$	VI E 2	32
$(CF_3C\vdots CCF_3)_2Ni_4(CO)_3$	VI E 2	125
$(\pi\text{-}C_3H_5)_2Pd$	VI E 3	12
$[\pi\text{-}C_3H_5PdCl]_2$	VI E 3	132

Compound	Section	References
$\left[\text{Me}\cdots\text{—Pd}\diagdown\text{Cl} \right]_2$ CH(OMe)(Me)	VI E 3	132
$\left[\text{Me—}\cdots\text{—Pd}\diagdown\text{Cl} \right]_2$ CH(OMe)(Me)	VI E 3	132
$[C_9H_{17}OPdCl]_2$	VI E 3	132
$[C_{17}H_{17}OPdCl]_2$	VI E 3	132
$C_6H_{10}PdCl_2$	VI E 2	132
$C_7H_8PdCl_2$	VI E 2	132
$Ph_4C_4PdCl_2$	VI E 2	15,16
$C_5H_5Pd(\pi\text{-}C_3H_5)$	VI E 3	117
$C_5H_5PdC_{10}H_{12}OMe$	VI E 3	117
$(\pi\text{-}C_3H_5)_2Pt$	VI E 3	12
$(\sigma\text{-}C_3H_5)_2Pt(PPh_3)_2$	VI E 3	12

TABLE XII—*continued*

Compound	Section	References
$(C_8H_{12})_2Pt$	VI E 2	*68*
$C_5H_5PtMe_3$	VI E 1	*117*
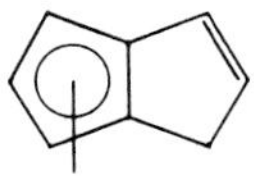 $PtMe_3$	VI E 1	*103*
$(C_5H_5)_2Pt$	VI E 1	*69a*

TABLE XIII

APPEARANCE POTENTIALS

Ion	m/e	A.P. (eV)	Parent molecule	References
$(C_5H_5)_2Mg^+$	154	7.76 ± 0.1	—	*81*
$C_5H_5Mg^+$	89	10.98 ± 0.1	$(C_5H_5)_2Mg$	*81*
Mg^+	24	14.36 ± 0.2	$(C_5H_5)_2Mg$	*81*
$C_5H_5V(CO)_4^+$	228	8.2 ± 0.3	—	*199*
$C_5H_5VC_7H_7^+$	207	7.24 ± 0.1	—	*150*
$(C_6H_6)_2V^+$	207	6.26 ± 0.1	—	*150*
$C_5H_5V(CO)_2^+$	172	9.7 ± 0.3	$C_5H_5V(CO)_4$	*199*
$C_5H_5VCO^+$	144	10.7 ± 0.3	$C_5H_5V(CO)_4$	*199*
$C_6H_6V^+$	129	11.0 ± 0.2	$C_5H_5VC_7H_7$	*150*
		10.5 ± 0.2	$(C_6H_6)_2V$	*150*
$C_5H_5V^+$	116	12.68 ± 0.1	$(C_5H_5)_2V$	*81*
		12.65 ± 0.1	$(C_5H_5)_2V$	*148*
		14.2 ± 0.2	$C_5H_5V(CO)_4$	*199*
		12.9 ± 0.2	$C_5H_5VC_7H_7$	*150*
$C_3H_3V^+$	90	18.9 ± 0.3	$C_5H_5V(CO)_4$	*199*
V^+	51	18.32 ± 0.2	$(C_5H_5)_2V$	*81*
		14.5 ± 0.5	$(C_5H_5)_2V$	*148*
		19.4 ± 0.4	$C_5H_5V(CO)_4$	*199*
		13.6 ± 0.3	$C_5H_5VC_7H_7$	*150*
		13.8 ± 0.3	$(C_6H_6)_2V$	*150*
$C_6H_6Cr(CO)_3^+$	214	7.39 ± 0.1	—	*165*
$C_5H_5CrC_7H_7^+$	208	5.96 ± 0.1	—	*150*

TABLE XIII—*continued*

Ion	m/e	A.P. (eV)	Parent molecule	References
$(C_6H_6)_2Cr^+$	208	5.70 ± 0.1	—	*165*
		5.91 ± 0.1	—	*150*
$C_5H_5CrC_6H_6^+$	195	6.13 ± 0.1	—	*150*
$(C_5H_5)_2Cr^+$	182	6.26 ± 0.1	—	*148*
		6.91 ± 0.2	—	*81*
$C_6H_6Cr^+$	130	8.8 ± 0.2	$(C_6H_6)_2Cr$	*150*
		9.2 ± 0.2	$(C_6H_6)_2Cr$	*165*
		10.8 ± 0.2	$C_6H_6Cr(CO)_3$	*165*
$C_5H_5Cr^+$	117	12.81 ± 0.1	$(C_5H_5)_2Cr$	*148*
		13.6 ± 0.1	$(C_5H_5)_2Cr$	*81*
		12.7 ± 0.2	$C_5H_5CrC_7H_7$	*150*
		9.3 ± 0.2	$C_5H_5CrC_6H_6$	*150*
Cr^+	52	14.6 ± 0.3	$(C_5H_5)_2Cr$	*148*
		16.15 ± 0.3	$(C_5H_5)_2Cr$	*81*
		12.2 ± 0.3	$C_5H_5CrC_7H_7$	*150*
		10.8 ± 0.3	$(C_6H_6)_2Cr$	*150*
		13.9 ± 0.3	$C_5H_5CrC_6H_6$	*150*
$C_5H_5Mn(CO)_3^+$	204	8.3 ± 0.4	—	*199*
$C_5H_5MnC_6H_6^+$	198	$6.8–7.1$	—	*54*
		6.92 ± 0.1	—	*150*
$(C_5H_5)_2Mn^+$	185	7.25 ± 0.1	—	*81*
		7.32 ± 0.1	—	*148*
$C_5H_5MnCO^+$	148	9.8 ± 0.3	$C_5H_5Mn(CO)_3$	*199*
$C_6H_6MnH^+$	134	12.1	$C_5H_5MnC_6H_6$	*54*
$C_5H_5Mn^+$	120	11.25 ± 0.2	$(C_5H_5)_2Mn$	*81*
		11.09 ± 0.1	$(C_5H_5)_2Mn$	*148*
		12.0 ± 0.3	$C_5H_5Mn(CO)_3$	*199*
		12.3	$C_5H_5MnC_6H_6$	*54*
		9.4 ± 0.2	$C_5H_5MnC_6H_6$	*150*
Mn^+	55	14.05 ± 0.2	$(C_5H_5)_2Mn$	*81*
		13.6 ± 0.3	$(C_5H_5)_2Mn$	*148*
		15.9 ± 0.3	$C_5H_5Mn(CO)_3$	*199*
		17.9	$C_5H_5MnC_6H_6$	*54*
		14.1 ± 0.3	$C_5H_5MnC_6H_6$	*150*
$C_6H_8Fe(CO)_3^+$	220	8.0 ± 0.2	—	*200*
$(C_5H_5)_2Fe^+$	186	7.05 ± 0.1	—	*81*
		7.15 ± 0.1	—	*148*
		6.99	—	*79*
$C_5H_5FeC_3H_3^+$	160	13.27 ± 0.1	$(C_5H_5)_2Fe$	*148*
$C_5H_5Fe^+$	121	14.38 ± 0.3	$(C_5H_5)_2Fe$	*81*
		13.78 ± 0.1	$(C_5H_5)_2Fe$	*148*
		12.8 ± 0.1	$(C_5H_5)_2Fe$	*165*

TABLE XIII—*continued*

Ion	m/e	A.P. (eV)	Parent molecule	References
$C_3H_3Fe^+$	95	18.9 ±0.1	$(C_5H_5)_2Fe$	*148*
Fe^+	56	17.1 ±0.9	$(C_5H_5)_2Fe$	*81*
		14.4 ±0.5	$(C_5H_5)_2Fe$	*148*
$(C_5H_5)_2Co^+$	189	6.2 ±0.3	—	*81*
		6.21±0.1	—	*148*
		5.95±0.1	—	*165*
$C_5H_5Co(CO)_2^+$	180	8.3 ±0.2	—	*199*
		7.78±0.1	—	*165*
$C_5H_5CoCO^+$	152	10.1 ±0.2	$C_5H_5Co(CO)_2$	*199*
$C_5H_5Co^+$	124	14.20±0.1	$(C_5H_5)_2Co$	*81*
		14.00±0.1	$(C_5H_5)_2Co$	*148*
		12.3 ±1	$(C_5H_5)_2Co$	*165*
		11.7 ±0.2	$C_5H_5Co(CO)_2$	*199*
		10.28±0.2	$C_5H_5Co(CO)_2$	*165*
$C_3H_3Co^+$	98	16.8 ±0.3	$C_5H_5Co(CO)_2$	*199*
		17.62±0.1	$(C_5H_5)_2Co$	*148*
Co^+	59	16.1 ±0.6	$(C_5H_5)_2Co$	*81*
		14.66±0.2	$(C_5H_5)_2Co$	*148*
		16.8 ±0.3	$C_5H_5Co(CO)_2$	*199*
$(C_5H_5)_2Ni^+$	188	7.06±0.1	$(C_5H_5)_2Ni$	*81*
		7.16±0.1	—	*148*
		6.75	—	*79*
$C_5H_5NiC_3H_3^+$	162	12.19±0.1	$(C_5H_5)_2Ni$	*148*
$C_5H_5NiNO^+$	153	8.50	—	*79*
$C_5H_5Ni^+$	123	12.67±0.1	$(C_5H_5)_2Ni$	*81*
		12.59±0.1	$(C_5H_5)_2Ni$	*148*
		11.9 ±1	$(C_5H_5)_2Ni$	*165*
$C_3H_3Ni^+$	97	17.16±0.2	$(C_5H_5)_2Ni$	*148*
Ni^+		14.32±0.2	$(C_5H_5)_2Ni$	*81*
		13.65±0.2	$(C_5H_5)_2Ni$	*148*
$C_5H_5Mo(CO)_2NO^+$	249	8.1 ±0.2	—	*200*
$(C_5H_5)_2TcH^+$	230	7.13±0.05	—	*148*
$(C_5H_5)_2Tc^+$	229	7.86±0.1	$(C_5H_5)_2TcH$	*148*
$(C_5H_5)_2Ru^+$	231	7.81±0.1	—	*148*
$C_5H_5RuC_3H_3^+$	205	13.7 ±0.1	$(C_5H_5)_2Ru$	*148*
$C_5H_5Ru^+$	166	14.3 ±0.2	$(C_5H_5)_2Ru$	*148*
Ru^+	101	16.1 ±0.5	$(C_5H_5)_2Ru$	*148*
$(C_5H_5)_2WH_2^+$	316	6.49±0.1	—	*148*
$(C_5H_5)_2ReH^+$	318	6.76±0.05	—	*148*
$(C_5H_5)_2Re$	317	7.86±0.1	$(C_5H_5)_2ReH$	*148*
$(C_5H_5)_2Os^+$	322	7.59±0.1	—	*148*

TABLE XIV

IONIZATION POTENTIALS OF LIGANDS

Ligand	I.P. (eV)
CO	14.1
NO	9.25
C_5H_5	8.72
C_6H_6	9.6
C_6H_8	8.4

TABLE XV

COMPARISON OF IONIZATION POTENTIALS OF METALLOCENES WITH
IONIZATION POTENTIALS OF THE FREE METAL ATOMS

Ion	I.P. (eV)	I.P.(M^+) (eV)	$\varDelta V$
$(C_5H_5)_2Mg^+$	7.76 ± 0.1	7.64	$+0.1$
$C_5H_5V(CO)_4^+$	8.2 ± 0.3	6.74	$+1.5$
$C_5H_5VC_7H_7^+$	7.24 ± 0.1	—	$+0.5$
$(C_6H_6)_2V^+$	6.26 ± 0.1	—	-0.5
$(C_5H_5)_2V^+$	7.33 ± 0.1	—	$+0.6$
	7.56 ± 0.1	—	$+0.8$
$C_6H_6Cr(CO)_3^+$	7.39 ± 0.1	6.76	$+0.6$
$C_5H_5CrC_7H_7^+$	5.96 ± 0.1	—	-0.8
$(C_6H_6)_2Cr^+$	5.70 ± 0.1	—	-1.1
$C_5H_5CrC_6H_6^+$	5.91 ± 0.1	—	-0.85
$(C_5H_5)_2Cr^+$	6.26 ± 0.2	—	-0.50
	6.91 ± 0.1	—	$+0.15$
$C_5H_5Mo(CO)_2NO^+$	8.1 ± 0.2	7.18	$+0.9$
$(C_5H_5)_2WH_2^+$	6.49 ± 0.1	7.98	-1.5
$C_5H_5Mn(CO)_3^+$	8.3 ± 0.4	7.43	$+0.9$
$C_5H_5MnC_6H_6^+$	$6.8-7.1$	—	-0.5
	6.92 ± 0.1	—	-0.5
$(C_5H_5)_2Mn^+$	7.25 ± 0.1	—	-0.15
	7.32 ± 0.1	—	-0.1
$(C_5H_5)_2TcH^+$	7.13 ± 0.05	7.28	-0.15
$(C_5H_5)_2ReH^+$	6.76 ± 0.05	7.87	-1.1
$C_6H_8Fe(CO)_3^+$	8.0 ± 0.2	7.90	$+0.1$

TABLE XV—*continued*

Ion	I.P. (eV)	I.P.(M$^+$) (eV)	ΔV
$(C_5H_5)_2Fe^+$	6.99	—	−0.9
	7.05 ± 0.1	—	−0.85
	7.15 ± 0.1	—	−0.75
$(C_5H_5)_2Ru^+$	7.81 ± 0.1	7.36	+0.45
$(C_5H_5)_2Os^+$	7.59 ± 0.1	8.7	−1.1
$(C_5H_5)_2Co^+$	5.95 ± 0.1	7.86	−1.9
	6.21 ± 0.1	—	−1.65
	6.2 ± 0.3	—	−1.7
$C_5H_5Co(CO)_2{}^+$	7.78 ± 0.1	—	−0.1
	8.3 ± 0.2	—	+0.4
$(C_5H_5)_2Ni^+$	6.75	7.63	−0.9
	7.06 ± 0.1	—	−0.6
	7.16 ± 0.1	—	−0.5
C_5H_5NiNO	8.50	—	+0.9

TABLE XVI

BOND DISSOCIATION ENERGIES

Process	eV	kcal/mole	References
$(C_5H_5)_2Mg^+ \rightarrow C_5H_5Mg^+ + C_5H_5 \cdot$	3.22	74	*81*
$C_5H_5Mg^+ \rightarrow Mg^+ + C_5H_5 \cdot$	3.38	78	*81*
$(C_5H_5)_2V^+ \rightarrow C_5H_5V^+ + C_5H_5 \cdot$	5.32	123	*148*
	5.12	118	*81*
$C_5H_5VC_7H_7{}^+ \rightarrow C_5H_5V^+ + C_7H_7 \cdot$	1.4	32	*148*
$(C_6H_6)_2V^+ \rightarrow C_6H_6V^+ + C_6H_6$	4.2	97	*150*
$C_5H_5V^+ \rightarrow V^+ + C_5H_5 \cdot$	1.4	32	*199*
	1.8	41	*148*
	5.64	130	*81*
$C_3H_3V^+ \rightarrow V^+ + C_3H_3 \cdot$	0.5	12	*199*
$(C_5H_5)_2Cr^+ \rightarrow C_5H_5Cr^+ + C_5H_5 \cdot$	6.55	151	*148*
	6.7	154	*81*
$C_5H_5CrC_7H_7{}^+ \rightarrow C_5H_5Cr^+ + C_7H_7 \cdot$	3.4	78	*150*
$(C_6H_6)_2Cr^+ \rightarrow C_6H_6Cr^+ + C_6H_6$	2.4	55	*150*
	3.5 ± 0.3	81 ± 8	*163*
$C_5H_5CrC_6H_6{}^+ \rightarrow C_5H_5Cr^+ + C_6H_6$	2.9	67	*150*

TABLE XVI—*continued*

Process	eV	kcal/mole	References
$C_5H_5Cr^+ \rightarrow Cr^+ + C_5H_5 \cdot$	1.8	41	*148*
	2.45	55	*81*
$(C_5H_5)_2Mn^+ \rightarrow C_5H_5Mn^+ + C_5H_5 \cdot$	3.77	87	*148*
	4.0	92	*81*
$C_5H_5MnC_6H_6 \rightarrow C_5H_5Mn^+ + C_6H_6$	5	122	*54*
$C_5H_5Mn^+ \rightarrow Mn^+ + C_5H_5 \cdot$	2.5	57	*148*
	2.8	64	*81*
	0.08	2	*199*
$(C_5H_5)_2Fe^+ \rightarrow C_5H_5Fe^+ + C_5H_5 \cdot$	6.63	153	*148*
	7.33	169	*81*
	5.8 ± 1.1	134 ± 25	*165*
$C_5H_5Fe^+ \rightarrow Fe^+ + C_5H_5 \cdot$	0.6	14	*148*
	2.7	62	*81*
$(C_5H_5)_2Co^+ \rightarrow C_5H_5Co^+ + C_5H_5 \cdot$	7.79	180	*148*
	8.0	184	*81*
	6.3 ± 1.1	146 ± 25	*165*
$C_5H_5Co^+ \rightarrow Co^+ + C_5H_5 \cdot$	0.66	15	*148*
	1.9	44	*81*
	1.2	29	*199*
$C_3H_3Co^+ \rightarrow Co^+ + C_3H_3 \cdot$		0	*199*
$(C_5H_5)_2Ni^+ \rightarrow C_5H_5Ni^+ + C_5H_5 \cdot$	5.43	125	*148*
	5.6	129	*81*
	5.2 ± 1.1	119 ± 25	*165*
$C_5H_5Ni^+ \rightarrow Ni^+ + C_5H_5 \cdot$	1.06	24	*148*
	1.65	38	*81*

TABLE XVII

Heats of Formation

Compound	$-\Delta H_f(g)$ from electron impact data (kcal/mole)	$-\Delta H_f°$ from thermochemical data (kcal/mole)
$(C_5H_5)_2V$	179	—
$C_5H_5VC_7H_7$	163	—
$(C_6H_6)_2V$	159	22.7 ± 4^a
$(C_5H_5)_2Cr$	179	—
$C_5H_5Cr_6C_6H_6$	163	102^b
$C_5H_5CrC_7H_7$	124	—

TABLE XVII—*continued*

Compound	$-\Delta H_f(g)$ from electron impact data (kcal/mole)	$-\Delta H_f°$ from thermochemical data (kcal/mole)
$(C_6H_6)_2Cr$	92	53.7 ± 15^a
$(C_5H_5)_2Mn$	143	—
$C_5H_5MnC_6H_6$	154	
$(C_5H_5)_2Fe$	150	51.3 ± 1.3^a
$(C_5H_5)_2Co$	156	—
$(C_5H_5)_2Ni$	138	80.8 ± 1.1^a
$C_5H_5V(CO)_4$	139	—
$C_5H_5Mn(CO)_3$	67	—
$C_5H_5Co(CO)_2$	15	—

[a] Skinner (185).
[b] A. Reckziegel, Dissertation, Universität München, 1962, quoted in Müller and D'Or (150).

TABLE XVIII

HEATS OF FORMATION OF SOME FRAGMENT
IONS[a]

Ion	ΔH_f (ion) (kcal/mole)
$C_5H_5V(CO)_2^+$	137
$C_5H_5VCO^+$	187
$C_5H_5V^+$	294
$C_3H_3V^+$	348
$C_5H_5MnCo^+$	216
$C_5H_5Mn^+$	289
$C_5H_5CoCO^+$	244
$C_5H_5Co^+$	308
$C_3H_3Co^+$	371
$CoCO^+$	254

[a] Winters and Kiser (199).

TABLE XIX

TRANSITION METAL COMPLEXES FOR WHICH MASS SPECTRAL DATA HAVE BEEN REPORTED
SINCE BRUCE'S REVIEW[a]

Compound	Information	References
	(a) Metal Carbonyls	
$V(CO)_6$	M$^+$, fragmentation and I.P's	18
$Cr(CO)_6$	M$^+$, fragmentation and I.P's	18, 166
$Mo(CO)_6$	M$^+$, fragmentation and I.P's	18, 166
$W(CO)_6$	M$^+$, fragmentation and I.P's	18, 166
$Mn_2(CO)_{10}$	M$^+$, fragmentation and I.P's	190
$MnRe(CO)_{10}$	M$^+$, fragmentation and I.P's	97, 190
$Re_2(CO)_{10}$	M$^+$, fragmentation and I.P's	187, 190
$Fe(CO)_5$	M$^+$, fragmentation and I.P's	18, 86, 166
$Fe_2(CO)_9$	M$^+$ and fragmentation	97
$Fe_2Ru_2(CO)_{12}$	M$^+$	201
$Fe_2Ru(CO)_{12}$	M$^+$ and fragmentation	201
$Os_3(CO)_{12}$	M$^+$ and fragmentation	97
$Co_2(CO)_8$	M$^+$ and fragmentation	97
$Co_4(CO)_{12}$	M$^+$ and fragmentation	97
$Rh_4(CO)_{12}$	M$^+$ and fragmentation	97
$Ni(CO)_4$	M$^+$ and fragmentation	18, 166, 178
	(b) Metal Carbonyl and Nitrosyl Derivatives	
$CH_3SCH_2Mo(CO)_3CH_2NCO$	Fragmentation	114
$(CO)_4MnCOCH_2CH_2SCH_3$	M$^+$ and fragmentation	115
$(CH_3)_3SnNCW(CO)_5$	M$^+$ and fragmentation	106
$Mn(CO)_5Cl$	M$^+$ and fragmentation	57
$Mn(CO)_5I$	M$^+$ and fragmentation	57
$CH_3Mn(CO)_5$	M$^+$ and fragmentation	138
$CD_3Mn(CO)_5$	Partial fragmentation	138
$CH_2CNMn(CO)_5$	Partial fragmentation	138
$SCNMn(CO)_5$	M$^+$ and fragmentation	138
$C_6H_5Mn(CO)_5$	M$^+$ and fragmentation	138
$Ph_3SnMn(CO)_5$	Partial fragmentation	138
$PhBr_2SnMn(CO)_5$	Partial fragmentation	138
$HMn(CO)_5$	M$^+$ and fragmentation	138
$Br_2GaMn(CO)_5$	M$^+$	92
$CH_3COMn(CO)_4NH_3$	M$^+$ and fragmentation	138
$CD_3COMn(CO)_4NH_3$	Partial fragmentation	138
$Mn(CO)_4NO$	M$^+$ and fragmentation	138
$Mn_2(CO)_8I_2$	M$^+$ and fragmentation	54
$(CH_3)_2Sn[Mn(CO)_5]_2$	M$^+$ and fragmentation	193
$ClSn[Mn(CO)_5]_3$	M$^+$ and fragmentation	193
$H_3Mn_3(CO)_{12}$	M$^+$ and fragmentation	78a, 99, 187
$H_7B_2Mn_3(CO)_{10}$	M$^+$ and fragmentation	187

TABLE XIX—*continued*

Compound	Information	References
$HMnRe_2(CO)_{14}$	M^+ and fragmentation	187
$Re(CO)_5Cl$	M^+ and fragmentation	57
$Re(CO)_5I$	M^+ and fragmentation	57
$Re_2(CO)_8Cl_2$	M^+ and fragmentation	57
$Re_2(CO)_8I_2$	M^+ and fragmentation	57
$HRe_3(CO)_{14}$	M^+ and fragmentation	187
$H_3Re_3(CO)_{12}$	M^+ and fragmentation	99, 187
$D_2HRe_3(CO)_{12}$	M^+ and fragmentation	187
$D_3Re_3(CO)_{12}$	M^+ and fragmentation	187
$Fe(CO)_4Cl_2$	M^+ and fragmentation	57
$Fe(CO)_4ClSnCl_3$	M^+	129a
trans-$Fe(CO)_4(GeCl_3)_2$	M^+	129a
trans-$Fe(CO)_4(GeI_3)_2$	M^+	129a
$[Fe(CO)_4GeCl_2]_2$	M^+	129a
$Fe(NO)_4Cl_2$	M^+ and fragmentation	96
$Fe(NO)_4Br_2$	M^+ and fragmentation	96
$HFeCo_3(CO)_{12}$	M^+ and fragmentation	137, 137a
$H_2FeRu_3(CO)_{13}$	M^+ and fragmentation	201
$Ru_2(CO)_6Cl_4$	M^+ and fragmentation	28
$Ru_2(CO)_6I_2$	M^+	28a
$Ru_6(CO)_{17}C$	M^+	43, 100
$Me_{10}Sn_4Ru_2(CO)_6$	$[M-2\,Me]^+$, fragmentation	51
cis-$[Me_3Sn]_2Ru(CO)_4$	M^+ and fragmentation	51
trans-$Me_3SiRu(CO)_4 \cdot Ru(CO)_4SiMe_3$	M^+	51
$HRuCo_3(CO)_{12}$	M^+ and fragmentation	137, 137a
$Os_4O_4(CO)_{12}$	M^+ and fragmentation	97
$HOs_3(CO)_{10}OH$	M^+ and fragmentation	101
$HOs_3(CO)_{10}OMe$	M^+ and fragmentation	101
$Os_3(CO)_{10}(OMe)_2$	M^+ and fragmentation	101
$Co(CO)_3NO$	M^+ and fragmentation	166
$Co_2(NO)_4Cl_2$	M^+ and fragmentation	96
$Co_2(NO)_4Br_2$	M^+ and fragmentation	96
$Co_2(NO)_4I_2$	M^+ and fragmentation	96
$Co_2(NO)_4BrCl$	M^+ and fragmentation	96
$(C_5H_7O_2)_2Sn[Co(CO)_4]_2$	$[M-CO]^+$	162
$(C_5H_7O_2)_2SnCo_2(CO)_7$	M^+	162
$(CH_3CO_2)_2Sn[Co(CO)_4]_2$	$[M-2\,CO]^+$	163
$XCCo_3(CO)_9$ (X = H, F, Cl, Br, Me, CF_3, Ph)	M^+ and fragmentation	137a
$HFeCo_3(CO)_{12}$	M^+ and fragmentation	137
$HRuCo_3(CO)_{12}$	M^+ and fragmentation	137
$FSn[Co(CO)_4]_3$	M^+ and fragmentation	163
$Sn[Co(CO)_4]_4$	$[M-CO]^+$ and fragmentation	163
$Rh_2(CO)_4Cl_2$	M^+ and fragmentation	57

TABLE XIX—*continued*

Compound	Information	References
	(c) Fluorocarbon Complexes	
$C_5F_6ClMn(CO)_5$	M^+ and fragmentation	25

	M^+ and fragmentation	17

Compound	Information	References
$CF_3Mn(CO)_5$	M^+ and fragmentation	138
$CHF_2CF_2CF:CFMn(CO)_5$	M^+	191
$CF_3CF:C(CHF_2)Mn(CO)_5$	M^+	191, 192
$CF_2:CFCH(CF_3)Mn(CO)_5$	M^+	191, 192
$CF_2:CFCF:CFMn(CO)_5$	M^+	84
$Me_3GeCF_2CF:CFMn(CO)_5$	M^+	84
$NC_5F_4Mn(CO)_5$	M^+	46
$N_3C_3F_2Mn(CO)_5$	M^+	46
o-$FC_6H_4Mn(CO)_5$	Base peak $FC_6H_4Mn^+$	138
m-$FC_6H_4Mn(CO)_5$	Base peak $FC_6H_4Mn^+$	138
p-$FC_6H_4Mn(CO)_5$	Base peak $FC_6H_4Mn^+$	138
p-$FC_6H_4CH_2Mn(CO)_5$	M^+ and fragmentation	88
$C_6F_5CH_2Mn(CO)_5$	M^+ and fragmentation	88
$C_5H_6ClRe(CO)_5$	M^+ and fragmentation	25
$CHF_2CF_2CF:CFRe(CO)_5$	M^+	192
$CF_3CH:CHRe(CO)_5$	M^+	31
$HC:CC_6F_4Re(CO)_5$	M^+	31

	M^+	84

	M^+	84

Compound	Information	References
$NC_5F_4Re(CO)_5$	M^+	46
$N_2C_4F_3Re(CO)_5$	M^+	46
$N_3C_3F_2Re(CO)_5$	M^+	46

TABLE XIX—*continued*

Compound	Information	References

(d) Nitrogen Complexes

$$Me-N\backslash Cr(CO)_5 / N-Me$$

Compound	Information	References
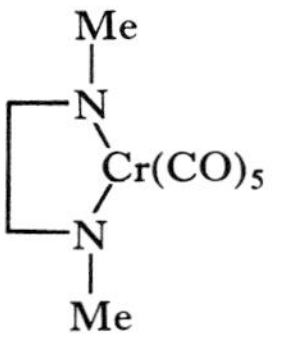 (structure above) Cr(CO)$_5$	M$^+$ and fragmentation	157a
C$_6$H$_5$NH$_2$Mo(CO)$_3$	M$^+$	197a
(C$_6$H$_5$NH$_2$)$_3$Mo(CO)$_3$	M$^+$	197a
(C$_6$H$_5$N)$_2$CHN$_2$Fe(CO)$_3$	M$^+$ and fragmentation	7
(CH$_3$)$_2$N$_4$Fe(CO)$_3$	M$^+$ and fragmentation	53
PhN=C(Me)—C(Me)=NPhFe(CO)$_3$	M$^+$	159
PhCH$_2$NC$_6$H$_5$Fe$_2$(CO)$_6$	M$^+$ and fragmentation	8
(C$_6$H$_5$N)$_2$Fe$_2$(CO)$_7$	M$^+$ and fragmentation	94
Fe$_2$(CO)$_6$(NH$_2$)$_2$	M$^+$ and fragmentation	80
Et$_2$NCOFe$_2$(CO)$_6$	M$^+$ and fragmentation	78b

(e) Phosphorus Complexes

Compound	Information	References
Cr(CO)$_5$P(NMe$_2$)$_3$	M$^+$ and fragmentation	23, 113
trans-Cr(CO)$_4$[P(NMe$_2$)$_3$]$_2$	M$^+$ and fragmentation	23, 113
trans-Cr(CO)$_4$[P(OMe)$_3$]$_2$	M$^+$ and fragmentation	23
(CO)$_5$CrPMe$_2$PMe$_2$Cr(CO)$_5$	M$^+$ and fragmentation	98
(CO)$_4$Cr(PMe$_2$)$_2$Cr(CO)$_4$	M$^+$ and fragmentation	98
Et$_3$P(CO)$_3$Cr(PMe$_2$)$_2$Cr(CO)$_3$PEt$_3$	M$^+$ and fragmentation	98
(CO)$_5$MoP(NMe$_2$)$_3$	M$^+$ and fragmentation	113
(NMe$_2$)$_3$PMo(CO)$_4$P(NMe$_2$)$_3$	M$^+$ and fragmentation	113
(CO)$_5$MoPMe$_2$·PMe$_2$Mo(CO)$_5$	M$^+$ and fragmentation	98
(CO)$_4$Mo(PMe$_2$)$_2$Mo(CO)$_4$	M$^+$ and fragmentation	98
PEt$_3$P(CO)$_3$Mo(PMe$_2$)$_2$Mo(CO)$_3$PEt$_3$	M$^+$ and fragmentation	98
(Me$_2$N)$_3$PW(CO)$_4$P(NMe$_2$)$_3$	M$^+$ and fragmentation	113
(CO)$_5$WPMe$_2$PMe$_2$W(CO)$_5$	M$^+$ and fragmentation	98
(CO)$_4$W(PMe$_2$)$_2$W(CO)$_4$	M$^+$ and fragmentation	98
(CO)$_4$Mn(PMe$_2$)$_2$Mn(CO)$_4$	M$^+$ and fragmentation	98
(CO)$_4$Mn(PPh$_2$)$_2$Mn(CO)$_4$	M$^+$ and fragmentation	98
(CO)$_4$FeP(NMe$_2$)$_3$	M$^+$ and fragmentation	23, 113
(CO)$_4$FeP(OMe)$_3$	M$^+$ and fragmentation	23
(CO)$_3$Fe[P(NMe$_2$)$_3$]$_2$	M$^+$ and fragmentation	113
(CO)$_3$Fe[P(OMe)$_3$]$_2$	M$^+$ and fragmentation	23
(CO)$_4$FePMe$_2$·PMe$_2$Fe(CO)$_4$	M$^+$ and fragmentation	98
(CO)$_3$Fe(PMe$_2$)$_2$Fe(CO)$_3$	M$^+$ and fragmentation	98
(CO)$_3$Fe(PPh$_2$)$_2$Fe(CO)$_3$	M$^+$ and fragmentation	98

TABLE XIX—*continued*

Compound	Information	References
$(CO)_3FeP(C_6F_5)_2Fe(CO)_3$	M^+ and fragmentation	*144*
$(CO)_3Fe(PMe_2)_2Fe(CO)_2PPh_3$	M^+ and fragmentation	*98*
$(CO)_3Fe(PMe_2)_2Fe(CO)_2P(C_6H_{11})_3$	M^+ and fragmentation	*98*
$Cl(CO)_3Fe(PMe_2)_2Fe(CO)_3Cl$	M^+ and fragmentation	*98*
$Br(CO)_3Fe(PMe_2)_2Fe(CO)_3Br$	M^+ and fragmentation	*98*
$I(CO)_3Fe(PMe_2)_2Fe(CO)_3I$	M^+ and fragmentation	*98*
$Fe(PF_3)_5$	M^+ and fragmentation	*128b*
$(PF_3)_3Co(PF_2)_2Co(PF_3)_3$	M^+ and fragmentation	*128a*
$Ni(PF_3)_4$	M^+ and fragmentation	*127*
$(PPh_3)_2Pd(CO)_3$	No metal-containing ions	*157*
$(Ph_2PCH_2)_2Pt(H)GePh_3$	M^+ and fragmentation	*40*
$(PPh_3)_2Pt(CO)_3$	No metal-containing ions	*157*
$PPh_3Pt(CO)_3$(solvent)	No metal-containing ions	*157*
[solvent$=CH_2Cl_2$; $(CH_3)_2CO$; C_6H_6]		
$[(PEt_3)_2PtGePh_3]_2$	Ions with Pt_2Ge_2 isotope patterns observed	*41*

(f) Sulfur and Selenium Complexes

Compound	Information	References
$W(CO)_4CH_3S(CH_2)_2SCH_3$	M^+ and fragmentation	*115*
$Mn(CO)_5SO_2CH_3$	M^+ and fragmentation	*138*
$Mn(CO)_4CH_2SCH_3$	M^+ and fragmentation	*115*
$[Mn(CO)_4SMe]_2$	M^+ and fragmentation	*2*
$[Mn(CO)_4SEt]_2$	M^+ and fragmentation	*57*
$[Mn(CO)_4S-n-Bu]_2$	M^+ and fragmentation	*57*
$[Mn(CO)_3SMe]_4$	M^+ and fragmentation	*2*
$[Mn(CO)_3SEt]_4$	M^+ and fragmentation	*57*
$[Mn(CO)_3S-n-Bu]_4$	M^+ and fragmentation	*57*
$[Mn(CO)_3SPh]_4$	M^+ and fragmentation	*2*
$(PhS)_3Br[Mn(CO)_3]_4$	M^+ and fragmentation	*2*
$[Mn(CO)_3SeMe]_4$	M^+	*1*
$[Mn(CO)_3SePh]_4$	M^+	*1*
$H_2C_2S_2Mn_2(CO)_6$	M^+ and fragmentation	*125a*
$[Re(CO)_4SPh]_2$	M^+ and fragmentation	*57*
$[Re(CO)_3SPh]_3$	M^+ and fragmentation	*57*
$[Re(CO)_3SPh]_4$	M^+ and fragmentation	*57*
$[Re(CO)_3SeMe]_4$	M^+	*1*
$[Re(CO)_3SePh]_4$	M^+	*1*
$[Fe(NO)_2SR]_2$	M^+ and fragmentation	*95*
(R $=$ Me, Et, *n*-Bu, Ph)		
$[Fe(CO)_3SMe]_2$	M^+ and fragmentation	*57, 115*
$[Fe(CO)_3SR]_2$	M^+ and fragmentation	*57*
(R $=$ Et, Bu)		
$[Fe(CO)_3SPh]_2$	M^+ and fragmentation	*45, 57*

TABLE XIX—*continued*

Compound	Information	References
$Fe_2(CO)_6S_2C_2H_4$	M^+ and fragmentation	*115*
$Fe_2(CO)_6S_2C_2F_4$	M^+ and fragmentation	*112, 115*
$Fe_2(CO)_6S_2C_4F_6$	M^+ and fragmentation	*112*
$Fe_2(CO)_6S_2C_7H_6$	M^+ and fragmentation	*115*
$Fe_3(CO)_9S_2$	M^+ and fragmentation	*188*
$[Co(NO)_2SR]_2$	M^+ and fragmentation	*96*
(R = Et, *n*-Bu, Ph)		
$[Rh(CO)_2SC_6F_5]_2$	M^+	*45*

a Bruce (*27*).

REFERENCES

1. Abel, E. W., Crosse, B. C., and Hutson, G. V., *J. Chem. Soc.*, *A* p. 2014 (1967).
2. Ahmad, M., Knox, G. R., Preston, F. J., and Reed, R. I., *Chem. Commun.* p. 138 (1967).
3. Amiet, R. G., Reeves, P. C., and Pettit, R., *Chem. Commun.* p. 1208 (1967).
4. Amiet, R. G., and Pettit, R., *J. Am. Chem. Soc.* **86**, 1059 (1968).
5. Aumann, R., and Fischer, E. O., *Angew. Chem. Intern. Ed. Engl.* **6**, 879 (1967).
6. Aumann, R., and Fischer, E. O., *Chem. Ber.* **101**, 954 (1968).
7. Bagga, M. M., Baikie, P. E., Mills, O. S., and Pauson, P. L., *Chem. Commun.* p. 1106 (1967).
8. Bagga, M. M., Flannigan, W. T., Knox, G. R., Pauson, P. L., Preston, F. J., and Reed, R. I., *J. Chem. Soc.*, *C* p. 36 (1968).
9. Barr, T. H., and Watts, W. E., *J. Organometal. Chem.* (*Amsterdam*) **9**, P3 (1967).
10. Barr, T. H., and Watts, W. E., *Tetrahedron* **24**, 3219 (1968).
11. Becconsall, J. K., and O'Brien, S., *Chem. Commun.* p. 720 (1966).
12. Becconsall, J. K., Job, B. E., and O'Brien, S., *J. Chem. Soc.*, *A* p. 423 (1967).
13. Ben-Shoshan, R., and Pettit, R. *Chem. Commun.* p. 247 (1968).
14. Beynon, J. H., "Mass Spectrometry and its Applications to Organic Chemistry." Elsevier, Amsterdam, 1960.
15. Beynon, J. H., Cookson, R. C., Hill, R. R., Jones, D. W., Saunders, R. A., and Williams, A. E., *J. Chem. Soc.* p. 7052 (1965).
16. Beynon, J. H., Curtis, R. F., and Williams, A. E., *Chem. Commun.* p. 237 (1966).
17. Bicher, R. E. J., Booth, M. R., and Clark, H. C., *Inorg. Nucl. Chem. Letters* **3**, 71 (1967).
18. Bidinosti, D. R., and McIntyre, N. S., *Can. J. Chem.* **45**, 641 (1967).
19. Biemann, K., "Mass Spectrometry: Applications to Organic Chemistry." McGraw-Hill, New York, 1962.
20. Birch, A. J., and Fitton, H., *J. Chem. Soc.*, *C*. p. 2060 (1966).
21. Birch, A. J., Cross, P. E., Lewis, J., White, D. A., and Wild, S. B., *J. Chem. Soc.*, *A* p. 332 (1968).

328   M. CAIS and M. S. LUPIN

22. Bird, P. H., and Churchill, M. R., *Chem. Commun.* p. 145 (1968).
23. Braterman, P. S., *J. Organometal. Chem.* (*Amsterdam*) **11**, 198 (1968).
24. Bruce, M. I., *Inorg. Nucl. Chem. Letters* **3**, 157 (1967).
25. Bruce, M. I., *J. Organometal. Chem.* (*Amsterdam*) **10**, 95 (1967).
26. Bruce, M. I., *J. Organometal. Chem.* (*Amsterdam*) **10**, 495 (1967).
27. Bruce, M. I., *Advan. Organometal. Chem.* **6**, 273 (1968).
28. Bruce, M. I., and Stone, F. G. A., *J. Chem. Soc., A* p. 1238 (1967).
28a. Bruce, M. I., Cooke, M., and Green, M., *J. Organometal. Chem.* (*Amsterdam*) **13**, 227 (1968).
29. Bruce, M. I., Cooke, M., Green, M., and Stone, F. G. A., *Chem. Commun.* p. 523 (1967).
30. Bruce, M. I., and Knight, J. R., *J. Organometal. Chem.* (*Amsterdam*) **12**, 411 (1968).
31. Bruce, M. I., Harbourne, D. A., Waugh, F., and Stone, F. G. A., *J. Chem. Soc., A* p. 895 (1968).
32. Bruce, R., Moseley, K., and Maitlis, P. M., *Can. J. Chem.* **45**, 2011 (1967).
33. Budzikiewicz, H., Djerassi, C., and Williams, D. H., "Interpretation of Mass Spectra of Organic Compounds." Holden-Day, San Francisco, California, 1964.
34. Burton, R., Pratt, L., and Wilkinson, G., *J. Chem. Soc.* p. 4290 (1960).
35. Cais, M., Modiano, A., and Raveh, A., *J. Am. Chem. Soc.* **87**, 5607 (1965).
36. Cais, M., and Maoz, N., *J. Organometal. Chem.* (*Amsterdam*) **5**, 370 (1966).
37. Cais, M., Lupin, M. S., Maoz, N., and Sharvit, J., *J. Chem. Soc., A* p. 3086 (1968).
38. Cardin, D. J., and Lappert, M. F., *Chem. Commun.* p. 506 (1966).
39. Cardin, D. J., Keppie, S. A., Kingston, B. M., and Lappert, M. F., *Chem. Commun.* p. 1035 (1967).
40. Carrick, A., and Glockling, F., *J. Chem. Soc., A* p. 40 (1967).
41. Carrick, A., and Glockling, F., *J. Chem. Soc., A* p. 913 (1968).
42. Chini, P., and Martinengo, S., *Inorg. Chem.* **6**, 837 (1967).
43. Churchill, M. R., and Bird, P. H., *J. Am. Chem. Soc.* **90**, 800 (1968).
44. Clancy, D. J., and Spilners, I. J., *Anal. Chem.* **34**, 1839 (1962).
45. Cooke, J., Green, M., and Stone, F. G. A., *J. Chem. Soc., A* p. 170 (1968).
46. Cooke, J., Green, M., and Stone, F. G. A., *J. Chem. Soc., A* p. 173 (1968).
47. Cooper, R. K., Fischer, E. O., and Semmlinger, W., *J. Organometal. Chem.* (*Amsterdam*) **9**, 333 (1967).
48. Cordes, C., and Rinehart, K. L., *Abstr. 150th Meeting Am. Chem. Soc., Atlantic City* p. 37-S (1965).
49. Corey, E. J., Hegedus, L. S., and Semmelhack, M. F., *J. Am. Chem. Soc.* **90**, 2417 (1968).
50. Cotton, F. A., Davison, A., and Musco, A., *J. Am. Chem. Soc.* **89**, 6796 (1967).
51. Cotton, J. D., Knox, S. A. R., and Stone, F. G. A., *Chem. Commun.* p. 965 (1967).
52. Coutts, R. S. P., and Wailes, P. C., *Inorg. Nucl. Chem. Letters* **3**, 1 (1967).
53. Dekker, M., and Knox, G. R., *Chem. Commun.* p. 1243 (1967).
54. Denning, R. G., and Wentworth, R. A. D., *J. Am. Chem. Soc.* **88**, 4619 (1966).
55. Deubzer, B., Fischer, E. O., Fritz, H. P., Kreiter, C. G., Kriebitzsch, N., Simmons, H. D., and Willeford, B. R., *Chem. Ber.* **100**, 3084 (1967).
56. Dickson, R. S., and Yawney, D. B. W., *Australian J. Chem.* **20**, 77 (1967).
57. Edgar, K., Johnson, B. F. G., Lewis, J., Williams, I. G., and Wilson, J. M., *J. Chem. Soc., A* p. 379 (1967).
58. Egger, H., *Monatsh. Chem.* **97**, 602 (1966).
59. Egger, H., and Falk, H., *Monatsh. Chem.* **97**, 1590 (1966).

60. Egger, H., and Falk, H., *Tetrahedron Letters* p. 437 (1966).
60a. Elder, R. C., Cotton, F. A., and Schunn, R. A., *J. Am. Chem. Soc.* **89**, 3645 (1967).
61. Elschenbroich, C., *J. Organometal. Chem.* (*Amsterdam*) **14**, 157 (1968).
62. Emerson, G. F., Watts, L., and Pettit, R., *J. Am. Chem. Soc.* **87**, 131 (1965).
63. Emerson, G. F., Ehrlich, K., Giering, W. P., and Lauterbur, P. C., *Am. J.Chem. Soc.* **88**, 3172 (1966).
64. Falk, H., Haller, G., and Schlögl, K., *Monatsh. Chem.* **98**, 592 (1967).
65. Fischer, E. O., and Schmidt, M. W., *Angew. Chem. Intern. Ed. Engl.* **6**, 93 (1967).
66. Fischer, E. O., Kreiter, C. G., and Berngruber, W., *Angew. Chem. Intern. Ed. Engl.* **6**, 634 (1967).
67. Fischer, E. O., and Amtmann, R., *J. Organometal. Chem.* (*Amsterdam*) **9**, P15 (1967).
68. Fischer, E. O., Mills, O. S., Paulus, E. F., and Wawersik, H., *Chem. Commun.* p. 643 (1967).
69. Fischer, E. O., and Schmidt, M. W., *Chem. Ber.* **100**, 3782 (1967).
69a. Fischer, E. O., and Schuster-Woldan, H., *Chem. Ber.* **100**, 705 (1967).
70. Fischer, E. O., and Wawersik, H., *Chem. Ber.* **101**, 150 (1968).
71. Fischer, E. O., and Wehner, H. W., *Chem. Ber.* **101**, 454 (1968).
72. Fischer, E. O., Berngruber, W., and Kreiter, C. G., *Chem. Ber.* **101**, 824 (1968).
73. Fischer, E. O., and Aumann, R., *Angew. Chem. Intern. Ed. Engl.* **6**, 181 (1967); *Chem. Ber.* **101**, 963 (1968).
74. Fischer, E. O., and Kiener, V., *Angew. Chem. Intern. Ed. Engl.* **6**, 961 (1967).
75. Fischer, E. O., Reckziegel, A., Müller, J., and Göser, P., *J. Organometal. Chem.* (*Amsterdam*) **11**, P13 (1968).
76. Fischer, E. O., and Wehner, H. W., *J. Organometal. Chem.* (*Amsterdam*) **11**, P29 (1968).
77. Fischer, E. O., and Maasböl, A., *J. Organometal. Chem.* (*Amsterdam*) **12**, P15 (1968).
77a. Fischer, E. O., and Schneider, R. J. J., *Angew. Chem. Intern. Ed. Engl.* **6**, 569 (1967).
78. Fischer, E. O., and Schneider, R. J. J., *J. Organometal. Chem.* (*Amsterdam*) **12**, P27 (1968).
78a. Fischer, E. O., and Aumann, R., *J. Organometal. Chem.* (*Amsterdam*) **8**, P1 (1967).
78b. Flannigan, W. T., Knox, G. R., and Pauson, P. L., *Chem. & Ind.* (*London*) p. 1094 (1967).
79. Foffani, A., Pignataro, S., Distefano, G., and Innorta, G., *J. Organometal. Chem.* (*Amsterdam*) **7**, 473 (1967).
80. Frey, V., Hieber, W., and Mills, O. S., *Z. Naturforsch.* **23b**, 105 (1968).
81. Friedman, L., Irsa, I. P., and Wilkinson, G., *J. Am. Chem. Soc.* **77**, 3689 (1955).
82. Gowling, E. W., Kettle, S. F. A., and Sharples, G. M., *Chem. Commun.* p. 21 (1968).
83. Grubbs, R., Breslow, R., Herber, R., and Lippard, S. J., *J. Am. Chem. Soc.* **89**, 6864 (1967).
84. Green, M., Mayne, N., and Stone, F. G. A., *J. Chem. Soc., A* p. 902 (1968).
85. Green, M. L. H., and Mole, T., *J. Organometal. Chem.* (*Amsterdam*) **12**, 404 (1968).
86. Gustorf, E. K. von, Henry, M. C., and McAdoo, D. J., *Ann. Chem.* **707**, 190 (1967).
87. Haas, M. A., and Wilson, J. M., *J. Chem. Soc., B* p. 104 (1968).
88. Hawthorne, J. D., Mays, M. J., and Simpson, R. N. F., *J. Organometal. Chem.* (*Amsterdam*) **12**, 407 (1968).
89. Hawthorne, M. F., Young, D. C., Andrews, T. D., Howe, D. V., Pilling, R. L., Pitts, A. D., Reintjes, M., Warren, L. F., and Wegner, P. A., *J. Am. Chem. Soc.* **90**, 879 (1968).
90. Helling, J. F., Rennison, S. C., and Merijan, A., *J. Am. Chem. Soc.* **89**, 7141 (1967).

91. Hoehn, H. H., Pratt, L., Watterson, K. F., and Wilkinson, G., *J. Chem. Soc.* p. 2738 (1961).
92. Hoyano, J., Patmore, D. J., and Graham, W. A. G., *Inorg. Nucl. Chem. Letters* **4**, 201 (1968).
93. James, B. D., Nanda, R. K., and Wallbridge, M. G. H., *Chem. Commun.* p. 849 (1966).
94. Jarvis, J. A. J., Job, B. E., Kilbourn, B. T., Mais, R. H. B., Owston, P. G., and Todd, R. F., *Chem. Commun.* p. 1149 (1967).
95. Jiru, P., and Kuchynka, K., *Z. Physik. Chem.* (*Frankfurt*) [N.S.] **25**, 286 (1960).
96. Johnson, B. F. G., Lewis, J., Williams, I. G., and Wilson, J. M., *J. Chem. Soc., A* p. 338 (1967).
97. Johnson, B. F. G., Lewis, J., Williams, I. G., and Wilson, J. M., *J. Chem. Soc., A* p. 341 (1967).
98. Johnson, B. F. G., Lewis, J., Wilson, J. M., and Thompson, D. T., *J. Chem. Soc., A* p. 1445 (1967).
99. Johnson, B. F. G., Johnston, R. D., Lewis, J., and Robinson, B. H., *J. Organometal. Chem.* (*Amsterdam*) **10**, 105 (1967).
100. Johnson, B. F. G., Johnston, R. D., and Lewis, J., *Chem. Commun.* p. 1057 (1967).
101. Johnson, B. F. G., Lewis, J., and Kilty, P. A., *Chem. Commun.* p. 180 (1968).
102. Joshi, K. K., Mais, R. H. B., Nyman, F., Owston, P. G., and Wood, A. M., *J. Chem. Soc., A* p. 318 (1968).
103. Katz, T. J., and Mrowca, J. J., *J. Am. Chem. Soc.* **89**, 1105 (1967).
104. King, R. B., *Inorg. Chem.* **5**, 2227 (1966).
105. King, R. B., *Inorg. Chem.* **5**, 2231 (1966).
106. King, R. B., *Inorg. Chem.* **6**, 25 (1967).
107. King, R. B., *Inorg. Chem.* **6**, 30 (1967).
108. King, R. B., *Chem. Commun.* p. 986 (1967).
109. King, R. B., *J. Organometal. Chem.* (*Amsterdam*) **8**, 139 (1967).
110. King, R. B., *J. Am. Chem. Soc.* **85**, 1587 (1963).
111. King, R. B., *J. Am. Chem. Soc.* **88**, 2075 (1966).
112. King, R. B., *J. Am. Chem. Soc.* **89**, 6368 (1967).
113. King, R. B., *J. Am. Chem. Soc.* **90**, 1412 (1968).
114. King, R. B., *J. Am. Chem. Soc.* **90**, 1417 (1968).
115. King, R. B., *J. Am. Chem. Soc.* **90**, 1429 (1968).
116. King, R. B., *Appl. Spect*, **23**, 137 (1969).
117. King, R. B., *Appl. Spect.* **23**, 148 (1969).
118. King, R. B., *Canad. J. Chem.* **47**, 559 (1969).
119. King, R. B., *Appl. Spect.* (1969) (in press).
120. King, R. B., and Bisnette, M. B., *Inorg. Chem.* **6**, 469 (1967).
121. King, R. B., and Bisnette, M. B., *J. Organometal. Chem.* (*Amsterdam*) **8**, 287 (1967).
122. King, R. B., and Fronzaglia, A., *Chem. Commun.* p. 274 (1966).
123. King, R. B., and Fronzaglia, A., *Inorg. Chem.* **5**, 1837 (1966).
124. King, R. B., and Stone, F. G. A., *J. Am. Chem. Soc.* **82**, 4557 (1960).
125. King, R. B., Bruce, M. I., Phillips, J. R., and Stone, F. G. A., *Inorg. Chem.* **5**, 684 (1966).
125a. King, R. B., and Eggers, C. A., *Inorg. Chem.* **7**, 1214 (1968).
125b. Kingston, B. M., and Lappert, M. F., *Inorg. Nucl. Chem. Letters* **4**, 371 (1968).
126. Kiser, R. W., "Introduction to Mass Spectrometry and its Applications." Prentice-Hall, Englewood Cliffs, New Jersey, 1965.
127. Kiser, R. W., Krassoi, M. A., and Clark, R. J., *J. Am. Chem. Soc.* **89**, 3653 (1967).

128. Klabunde, U., and Fischer, E. O., *J. Am. Chem. Soc.* **89**, 7141 (1967).
128a. Kruck, T., and Lang, W., *Angew. Chem. Intern. Ed. Engl.* **6**, 454 (1967).
128b. Kruck, T., and Prasch, A., *Z. Anorg. Allgem. Chem.* **356**, 118 (1968).
129. Kummer, R., and Graham, W. A. G., *Inorg. Chem.* **7**, 523 (1968).
129a. Kummer, R., and Graham, W. A. G., *Inorg. Chem.* **7**, 1208 (1968).
130. Landesberg, J. M., and Sieczkowski, J., *J. Am. Chem. Soc.* **90**, 1655 (1968).
131. Lewis, J., Manning, A. R., Miller, J. R., and Wilson, J. M., *J. Chem. Soc. A* p. 1663 (1966).
132. Lupin, M. S., and Cais, M., *J. Chem. Soc., A* p. 3095 (1968).
133. Lupin, M. S., and Cais, M., unpublished results (1968).
134. Lupin, M. S., Sharvit, J., and Cais, M., *Israel J. Chem.* **7**, 73 (1969).
135. Mandelbaum, A., and Cais, M., *Tetrahedron Letters* p. 3847 (1964).
136. Maoz, N., Mandelbaum, A., and Cais, M., *Tetrahedron Letters* p. 2087 (1965).
137. Mays, M. J., and Simpson, R. N. F., *Chem. Commun.* p. 1024 (1967).
137a. Mays, M. J., and Simpson, R. N. F., *J. Chem. Soc., A* p. 1444 (1968).
138. Mays, M. J., and Simpson, R. N. F., *J. Chem. Soc., A* p. 1936 (1967).
139. Mays, M. J., and Robb, J. D., *J. Chem. Soc., A* p. 329 (1968).
140. McBride, D. W., Dudek, E., and Stone, F. G. A., *J. Chem. Soc.* p. 1752 (1964).
141. McLafferty, F. W., *Anal. Chem.* **28**, 306 (1956).
142. McLafferty, F. W., ed., "Mass Spectrometry of Organic Ions." Academic Press, New York, 1963.
143. Mechtler, H., and Schlögl, K., *Monatsh. Chem.* **97**, 574 (1967).
144. Miller, J. M., *J. Chem. Soc., A* p. 828 (1967).
145. Mills, O. S., and Redhouse, A. D., *Chem. Commun.* p. 444 (1966).
146. Mills, O. S., and Paulus, E. F., *J. Organometal. Chem. (Amsterdam)* **10**, 331 (1967).
147. Moser, E., and Fischer, E. O., *J. Organometal. Chem. (Amsterdam)* **12**, 1 (1968).
148. Müller, J., and D'Or, L., *J. Organometal. Chem. (Amsterdam)* **10**, 313 (1967).
149. Müller, J., and Göser, P., *Angew. Chem. Intern. Ed. Engl.* **6**, 364 (1967).
150. Müller, J., and Göser, P., *J. Organometal. Chem. (Amsterdam)* **12**, 163 (1968).
151. Nakamura, A., Kim, P-J., and Hagihara, N., *J. Organometal. Chem. (Amsterdam)* **3**, 7 (1965).
152. Nakamura, A., Kim, P-J., and Hagihara, N., *J. Organometal. Chem. (Amsterdam)* **6**, 420 (1966).
153. Nakamura, A., and Otsuka, S., private communication (1968).
154. Nesmeyanov, A. N., Nikitina, T. V., and Perevalova, E. G., *Izv. Akad. Nauk SSSR, Ser. Khim.* p. 197 (1964); *Bull. Acad. Sci. USSR, Ser. Chem. (English Transl.)* p. 184 (1964).
155. Nesmayanov, A. N., Dubovitskü, V. A., Nogina, O. V., and Bochkarev, V. N., *Dokl. Akad. Nauk SSSR* **165**, 125 (1965); *Proc. Acad. Sci. USSR (English Transl.)* **165**, 1083 (1965).
156. Nesmeyanov, A. N., Chapovsky, Yu. A., and Ustynyuk, Yu. A., *J. Organometal. Chem. (Amsterdam)* **9**, 345 (1967).
157. Nyman, C. J., Wymore, C. E., and Wilkinson, G., *J. Chem. Soc., A* p. 561 (1968).
157a. Oféle, K., *J. Organometal. Chem. (Amsterdam)* **12**, P42 (1968).
158. Otsuka, S., Nakamura, A., and Yoshida, T., *Bull. Chem. Soc. Japan* **40**, 1266 (1967).
159. Otsuka, S., Yoshida, T., and Nakamura, A., *Inorg. Chem.* **6**, 20 (1967).
160. Otsuka, S., Nakamura, A., and Yoshida, T., *Inorg. Chem.* **7**, 261 (1968).
161. Patmore, D. J., and Graham, W. A. G., *Chem. Commun.* p. 7 (1967).
162. Patmore, D. J., and Graham, W. A. G., *Inorg. Chem.* **6**, 1879 (1967).

163. Patmore, D. J., and Graham, W. A. G., *Inorg. Chem.* **7**, 771 (1968).
164. Pauson, P. L., Cotton, F. A., and Wilkinson, G., *J. Am. Chem. Soc.* **76**, 1970 (1954).
164a. Pidcock, A., and Taylor, B. W., *J. Chem. Soc.*, *A* p. 877 (1967).
165. Pignataro, S., and Lossing, F. P., *J. Organometal. Chem. (Amsterdam)* **10**, 531 (1967).
166. Pignataro, S., and Lossing, F. P., *J. Organometal. Chem. (Amsterdam)* **11**, 571 (1968).
167. Preston, F. J., and Reed, R. I., *Chem. Commun.* p. 51 (1966).
168. Prinz, R., and Werner, H., *Angew. Chem. Intern. Ed. Engl.* **6**, 91 (1967).
169. Rausch, M. D., *J. Org. Chem.* **28**, 3337 (1963).
170. Reed, R. I., and Tabrizi, F. M., *Appl. Spectry.* **17**, 124 (1963).
171. Reid, A. F., Shannon, J. S., Swan, J. M., and Wailes, P. C., *Australian J. Chem.* **18**, 173 (1965).
172. Roberts, D. T., Little, W. F., and Bursey, M. M., *J. Am. Chem. Soc.* **89**, 4917 (1967).
173. Roberts, D. T., Little, W. F., and Bursey, M. M., *J. Am. Chem. Soc.* **89**, 6156 (1967).
174. Roberts, D. T., Little, W. F., and Bursey, M. M., *J. Am. Chem. Soc.* **90**, 973 (1968).
175. Rosenblum, M., Brawn, N., and King, B., *Tetrahedron Letters* p. 4421 (1967).
176. Rosenblum, M., and Gatsonis, C., *J. Am. Chem. Soc.* **89**, 5074 (1967).
177. Rosenblum, M., and North, B., *J. Am. Chem. Soc.* **90**, 1060 (1968).
178. Schildcrout, S. M., Pressley, G. A., and Stafford, F. E., *J. Am. Chem. Soc.* **89**, 1617 (1967).
179. Schlögl, K., and Steyrer, W., *J. Organometal. Chem. (Amsterdam)* **6**, 399 (1966).
180. Schlögl, K., and Soukup, H., *Tetrahedron Letters* p. 118 (1967).
181. Schumacher, E., and Taubenest, R., *Helv. Chim. Acta* **47**, 1525 (1964).
182. Schumacher, E., and Taubenest, R., *Helv. Chim. Acta* **49**, 1447 (1966).
183. Sears, C. T., and Stone, F. G. A., *J. Organometal. Chem. (Amsterdam)* **11**, 644 (1968).
184. Shannon, J. S., and Swan, J. M., *Chem. Commun.* p. 33 (1965).
185. Skinner, H. A., *Advan. Organometal. Chem.* **2**, 49 (1964).
186. Slocum, D. W., Lewis, R., and Mains, G. J., *Chem. & Ind. (London)* p. 2095 (1966).
187. Smith, J. M., Mehner, K., and Kaesz, H. D., *J. Am. Chem. Soc.* **89**, 1759 (1967).
188. Smith, S. R., Krauze, R. A., and Dudek, G. O., *J. Inorg. & Nucl. Chem.* **29**, 1533 (1967).
189. Spilners, I. J., and Pellegrini, J. R., *J. Org. Chem.* **30**, 3800 (1965).
190. Svec, H. J., and Junk, G. A., *J. Am. Chem. Soc.* **89**, 2836 (1967).
191. Tattershall, B. W., Rest, A. J., Green, M., and Stone, F. G. A. *Angew. Chem. Intern. Ed. Engl.* **6**, 878 (1967).
192. Tattershall, B. W., Rest, A. J., Green, M., and Stone, F. G. A., *J. Chem. Soc.*, *A* p. 899 (1968).
193. Thompson, J. A. J., and Graham, W. A. G., *Inorg. Chem.* **6**, 1365 (1967).
194. Tirosh, N., Modiano, A., and Cais, M., *J. Organometal. Chem. (Amsterdam)* **5**, 357 (1966).
195. Treichel, P. M., and Werber, G. P., *J. Am. Chem. Soc.* **90**, 1753 (1968).
196. Tyerman, W. J. R., Kato, M., Kebarle, P., Masamune, S., Strausz, O. P., and Gunning, H. E., *Chem. Commun.* p. 497 (1967).
197. Watts, W. E., *J. Am. Chem. Soc.* **88**, 855 (1966).
197a. Werner, H., and Prinz, R., *Chem. Ber.* **100**, 265 (1967).
198. Wilke, G., and Bogdanovič, B., *Angew. Chem.* **73**, 756 (1961).
199. Winters, R. E., and Kiser, R. W., *J. Organometal. Chem. (Amsterdam)* **4**, 190 (1965).
200. Winters, R. E., and Kiser, R. W., *J. Phys. Chem.* **69**, 3198 (1965).
201. Yawney, D. B. W., and Stone, F. G. A., *Chem. Commun.* p. 619 (1968).

SUPPLEMENTARY REFERENCES

Since completion of the article, the following references have appeared in which the mass spectra of organometallic compounds have been reported in more or less detail (M W indicates molecular weight determination only).

Bruce, M. I., *J. Chem. Soc., A* p. 1459 (1968). $C_5H_5Fe(CO)_2X$ ($X = C_6H_5$, p-C_6F_4H).

Bruce, M. I., Harbourne, D. A., Waugh, F., and Stone, F. G. A., *J. Chem. Soc., A* p. 356 (1968). $C_5H_5Fe(CO)_2R$ ($R = C_2 . CF_3$, $C_2C_6H_5$, $C_2C_6F_5$), $Re(CO)_5R$ ($R = C_2C_6H_5$, $C_2(C_6F_5)$).

Brunner, H., *J. Organometal. Chem.* (*Amsterdam*) **12**, 517 (1968). (C_5H_5Co $NO)_2$.

Cardin, D. J., Keppie, S. A., and Lappert, M. F., *Inorg. Nucl. Chem. Letters* **4**, 365 (1968). $C_5H_5M(CO)_3$ $GeMe_3$ ($M = Cr$, Mo, W). $C_5H_5M(CO)_3$ $SnMe_3$ ($M = Cr$, Mo). (MW)

Cooke, M., Green, M., and Kirkpatrick, D., *J. Chem. Soc., A* p. 1507 (1968). $C_5H_5Fe(CO)_2X$ ($X = P(C_6F_5)_2$, $As(C_6H_5)_2$: $C_5H_5Mo(CO)_2R$ ($R = P(C_6F_5)_2$, $As(C_6F_5)_2$).

Coutts, R. S. P., and Wailes, P. C., *Australian J. Chem.* **21**, 1181 (1968). ($C_5H_5Ti)_2$ ($CO)_3$.

DePuy, C. H., Kobal, V. M., and Gibson, D. H., *J. Organometal. Chem.* (*Amsterdam*) **13**, 266 (1968). Iron carbonyl complexes of spiro [2.4] hepta-4,6-diene.

Goodfellow, R. J., Green, M., Mayne, N., Rest, A. J., and Stone, F. G. A., *J. Chem. Soc., A* p. 177 (1968). $(CO)_5Re$ $C(CF_3) : C(CF_3) . CF_2 . CH : C(CF_3)$ $Re(CO)_5$ (M W).

Hawthorne, M. F., and George, T. A., *J. Am. Chem. Soc.* **89**, 7114 (1967); **90**, 1661 (1968). C_5H_5 Co $B_7C_2H_9$ (M W).

Katz, T. J., Balogh, V., and Shulman, J., *J. Am. Chem. Soc.* **90**, 734 (1968). bis(*as*-indacenyliron).

Lewis, J., and Parkins, A. W., *J. Chem. Soc., A* p. 1150 (1967). Substituted cycloocta-1,5-diene complexes of cyclopentadienyl cobalt and cyclopentadienylrhodium (M W).

Mills, O. S., and Nice, J. P., *J. Organometal. Chem.* (*Amsterdam*) **10**, 337 (1967). $(C_5H_5)_2$ $Rh_2(CO)_3$.

O'Brien, S., *Chem. Commun.* p. 757 (1968). $(CO)_2$ (π-C_3H_5)Rh.

Preston, F. J., and Reed, R. I., *Org. Mass Spectrom.* **1**, 71 (1968). Mercapto, alkoxo and amido bridged cyclopentadienyl nitrosyls and carbonyls of Cr, Fe and Ni.

Roth, W. R., and Meier, J. D., *Tetrahedron Letters* p. 2053 (1967). *o*-Quinodimethan $Fe(CO)_3$ and 2,2 dimethylisoindene $Fe(CO)_3$.

Author Index

Numbers in parentheses are reference numbers and indicate that an author's work is referred to although his name is not cited in the text. Numbers in italics show the page on which the complete reference is listed.

A

Abeck, W., 120(137), *162*

Abel, E. W., 15(67, 69), *27*, 118(1, 3, 4, 6), 129(1), 130(2), 139(3, 4, 5, 6), 140(2), 141(5, 6, 7), 142(2, 6), *159*, 239(1), 326(1), *327*

Abraham, M. H., 173(1), 174(2), *205*

Adams, D. M., 144(8), *159*

Adams, R. M., 88(1), *113*

Addison, C. C., 120(9), *159*

Ahmad, M., 326(2), *327*

Allen, G., 174(3), *205*

Allred, A. L., 176(87), 186, *207*

Amiet, R. G., 231(3), 266(3), 275, 291(3), 292(3), 293(3), 294(3), 307(3), 311(3), 313(4), *327*

Amma, E. L., 179(145), *209*

Amtmann, R., 214(67), 287(67), *329*

Anders, U., 5(5), *24*

Anderson, J. S., 21(98), 22(100), *28*, 118(70), 158(70), *161*

Andrews, T. D., 88(18), 90(18), 91(18), 92(18), 93(18), 94(11, 18), 96(16), 97(12, 18), 98(12, 18), 99(18), 100(18), 105(18), 107(18), 109(18), *114*, 314(89), *329*

Angelici, R. J., 140(10), 142(10), *159*

Angolletta, M., 154(193), 156(193), 157(11, 193), 158(11), *159*, *164*

Appel, H., 14(61), *26*

Ariyaratne, J. K. P., 151(12, 13, 14), *159*

Ashby, E. C., 168(4), 170(5), *205*

Atwood, J. L., 203(6), *205*

Aumann, R., 8(34), *25*, 234(5, 6, 73), 291(5, 6, 73), 322(78a), *327*, *329*

B

Bacciarelli, S., 129(32), *160*

Bader, G., 13(57), 19(57, 87), *26*, *27*

Badin, E. J., *205*

Bagga, M. M., 325(7, 8), *327*

Baikie, P. E., 325(7), *327*

Baker, E. B., 169(88, 146), *207*, *209*

Barnett, K. W., 119(234), 136(231), 151(234), *165*

Barney, A. L., 82(116), *86*

Barr, T. H., 255(9, 10), 301(10), *327*

Barraclough, C. G., 21(99), *28*

Bartenstein, C., 10(83), 17(83), 18(83), 19(83), *27*

Basolo, F., 13(58), *26*, 119(15), 145(16), *159*

Bau, R., 8(36), 9(36), *26*

Bauld, N. L., 31(31), 45, 47(57), *84*

Baur, K., 13(56), *26*

Bayer, R., 5(6), *24*, 143(69), *161*

Beachley, O. T., 193(8), *205*

Becconsall, J. K., 215(12), 233(12), 277(11), 281(12), 282(12), 283(12), 287(12), 291(12), 313(11), 314(12), *327*

Beck, W., 4(VII), 5(VII), 6(VII, VIII), 7(VII), 8(VII, 26), 9(VII, 43), 10(VII, 79), 11(VII, 43, 48), 17(VII, 79), 18(VII), 19(VII, 90), 20(VII), 22(105), 23(VII, 90, 105, 110), *24*, *25*, *26*, *27*, *28*, 120(122), 139(17), 141(122), *159*, *162*

Becker, E., *24*

Beckert, O., 154(77), *161*

Beckert, W. F., 37(27), 47(60), *84*

Behrens, H., 4(10), 5, 6(18), 7(10), 11, 15(71), 16(75), 18, *24*, *25*, *27*, 119(22), 132(19, 20), 133(18, 20), 135(21), 138(22), 140(22), 142(22), *159*, *160*

Bell, N. A., 171(9, 10), *205*

Bennett, M. A., 130(2), 140(2), 142(2), *159*

Ben-Shoshan, R., 270(13), 310(13), *327*

Benson, S. W., 59(89), *85*

Berngruber, W., 231(66, 72), 291(66, 72), 293(72), 294(72), *329*

Berry, T. E., 105(30), *114*

Bertelli, D. J., 126(60), 147(60), *160*

335

Subject Index

Cumulative List of Contributors

Cumulative List of Titles